Springer Series in Reliability Engineering

Series Editor

Hoang Pham, Department of Industrial and Systems Engineering, Rutgers University, Piscataway, NJ, USA

Today's modern systems have become increasingly complex to design and build, while the demand for reliability and cost effective development continues. Reliability is one of the most important attributes in all these systems, including aerospace applications, real-time control, medical applications, defense systems, human decision-making, and home-security products. Growing international competition has increased the need for all designers, managers, practitioners, scientists and engineers to ensure a level of reliability of their product before release at the lowest cost. The interest in reliability has been growing in recent years and this trend will continue during the next decade and beyond.

The Springer Series in Reliability Engineering publishes books, monographs and edited volumes in important areas of current theoretical research development in reliability and in areas that attempt to bridge the gap between theory and application in areas of interest to practitioners in industry, laboratories, business, and government.

Indexed in Scopus and EI Compendex

Interested authors should contact the series editor, Hoang Pham, Department of Industrial and Systems Engineering, Rutgers University, Piscataway, NJ 08854, USA. Email: hopham@rci.rutgers.edu, or Anthony Doyle, Executive Editor, Springer, London. Email: anthony.doyle@springer.com.

More information about this series at http://www.springer.com/series/6917

Kim Phuc Tran
Editor

Control Charts and Machine Learning for Anomaly Detection in Manufacturing

Editor
Kim Phuc Tran
ENSAIT& GEMTEX
University of Lille
Lille, France

ISSN 1614-7839 ISSN 2196-999X (electronic)
Springer Series in Reliability Engineering
ISBN 978-3-030-83818-8 ISBN 978-3-030-83819-5 (eBook)
https://doi.org/10.1007/978-3-030-83819-5

This Springer imprint is published by the registered company Springer Nature Switzerland AG
The registered company address is: Gewerbestrasse 11, 6330 Cham, Switzerland

Contents

Introduction to Control Charts and Machine Learning for Anomaly Detection in Manufacturing

Kim Phuc Tran

Abstract In this chapter, we provide an introduction to Anomaly Detection and potential applications in manufacturing using Control Charts and Machine Learning techniques. We elaborate on the peculiarities of process monitoring and Anomaly Detection with Control Charts and Machine Learning in the manufacturing process and especially in the smart manufacturing contexts. We present the main research directions in this area and summarize the structure and contribution of the book.

1 Scope of the Research Domain

Anomaly Detection is a set of major techniques with an aim to detect rare events or observations that deviate from normal behavior. Process monitoring and Anomaly Detection are becoming increasingly important to enhance reliability and productivity in manufacturing by detecting abnormalities early. For example, a vibration level in an electric motor exceeding the permissible threshold can be considered as an anomaly, it might not be considered as a fault. However, if the vibration level continues to rise and leads to motor destruction, it can be considered as faulty. Therefore, Anomaly Detection can provide advantages to manufacturing companies by reducing their downtime due to machine breakdowns by detecting a failure before this results in a catastrophic event that may cause degradation of the process and product (Lindemann et al. [1]). There have been various data-driven and model-based approaches to detect anomalies occurring in manufacturing systems. The most common approach to Anomaly Detection includes Control Charts and Machine Learning methods. In manufacturing, Control Charts are effective tools of Statistical Process Control (SPC) for continuously monitoring a process as well as detecting process abnormalities to improve and optimize the process. There are many different types of Control Charts that have been developed for this purpose. In addition, Machine Learning methods have been used a lot in detecting anomalies with different applications in manufacturing such as, detecting network attacks, detecting abnormal states in machines.

K. P. Tran (✉)
University of Lille, ENSAIT, GEMTEX, 59000 Lille, France
e-mail: kim-phuc.tran@ensait.fr

K. P. Tran (ed.), *Control Charts and Machine Learning for Anomaly Detection in Manufacturing*, Springer Series in Reliability Engineering,
https://doi.org/10.1007/978-3-030-83819-5_1

Finally, the interference of these two techniques is also found in literature such as the design of Control Charts, anomaly signal interpretation, and pattern recognition in Control Charts using Machine Learning techniques.

In recent years, the rapid development and wide application of advanced technologies have profoundly impacted industrial manufacturing. The recent development of information and communication technologies such as smart sensor networks and the Internet of Things (IoT) has engendered the concept of Smart Manufacturing (SM) that adds intelligence into the manufacturing process to drive continuous improvement, knowledge transfer, and data-based decision-making. In this context, the increasing volume and quality of data from production facilitate the extraction of meaningful information, predicting future states of the manufacturing system that would be impossible to obtain even by human experts. Due to recent advances in the field of SPC, there are a lot of advanced Control Charts that have been developed, thus SPC can become a powerful tool for handling many Big Data applications that are beyond the production line monitoring in the context of SM Qiu [2]. Also, there are many studies on Artificial Intelligence applications in SM that exploit the valuable information in data to facilitate process monitoring, defect prediction, and predictive maintenance Wang et al. [3]. Using multiple sensors to collect data during manufacturing enhances real-time monitoring and decision-making, but data quality should also be ensured before using it. In this case, we can use Anomaly Detection algorithms to remove outliers in the dataset. This is the first application of Anomaly Detection algorithms in smart manufacturing, in addition, it is also used a lot in different aspects of manufacturing operations such as Anomaly Detection in machine operations, detection of attacks in industrial systems, detection of mechanical anomalies before they affect product quality, ...Therefore, Anomaly Detection plays a really important part in smart manufacturing. Lopez et al. [4] categorized anomalies in machines, controllers, and networks along with their detection mechanisms, and unify them under a common framework to allows the identification of gaps in Anomaly Detection in SM systems that should be addressed in future studies solutions.

In summary, the existing knowledge on Anomaly Detection with the applications in manufacturing is classified as Machine Learning and statistical approach. The statistical Anomaly Detection approach like Control Charts can be developed with little computational effort. However, their effectiveness has been proven during a long period of industrial application. Therefore, an effort should be made to develop advanced Control Charts for application in modern industrial contexts, see Tsung et al. [5], Qiu [2], and Zwetsloot et al. [6] for some examples. However, in SM contexts where assumptions about data distribution and independence are violated, Anomaly Detection methods will come into play, although they require considerable computational effort and resources. For example, Nguyen et al. [7] have developed a novel deep hybrid model for Anomaly Detection for multivariate time series without using any assumptions for the distribution of prediction errors. The autoencoder LSTM (Long Short-Term Memory networks) is used as a feature extractor to extract important representations of the multivariate time series input and then these features are input to OCSVM (One Class Support Vector Machine) for detecting anomalies.

This model results in better performance compared to the performance from several previous studies. Therefore, efforts are needed to develop Machine Learning based Anomaly Detection methods that are suitable for applications in SM. Finally, there are studies that combine both techniques to develop hybrid methods to combine the strengths of both techniques Lee et al. [8], Qiu and Xie [9].

2 Main Features of This Book

The key features of this book are given as follows:

1. Machine Learning has many applications in the development, pattern recognition, and interpreting of Control Charts. Especially applying Machine Learning to design Control Charts to monitor and detect anomalies in non-Gaussian, auto-correlated processes, or Non-stationary processes are important topics.
2. Advanced Control Charts are designed for monitoring Time-Between-Events-and-Amplitude Data, First-order Binomial Autoregressive Process. All of these studies aim to address the monitoring of manufacturing processes where the assumption of independence is violated.
3. To monitor the processes correlated data, Machine Learning-based Control Charts for monitoring categorical event series and monitoring serially correlated data are introduced in this book. In contrast to other methods, these new methods are more efficient in monitoring the correlated process data.
4. To detect anomalies in processes with Machine Learning, Tree-based approaches, autoencoder approaches, L1 SVDD (Support Vector Data Description), and L2 SVDD approaches are given in detail.
5. As an industrial application of Anomaly Detection, a comprehensive review and new advances of feature engineering techniques and health indicator construction methods for fault detection and diagnostic of engineering systems are introduced.
6. The case studies presented and analyzed in each chapter will help researchers, students, and practitioners understand and know how to apply these advanced methods in practice. The design parameters, as well as some source code of the tests and algorithms, are also shared with readers.

3 Structure of the Book

This book uncovers fundamental principles and recent developments in the advanced Control Charts and new Machine Learning approaches for Anomaly Detection in the manufacturing process and especially in the smart manufacturing contexts. The purpose of this book is to comprehensively present recent developments of Anomaly Detection techniques in manufacturing and to systemize these developments in new taxonomies and methodological principles with the application in SM to shape this new research domain. By approaching Anomaly Detection by both statistics and Machine Learning, this book also promotes cooperation between the research communities on SPC and Machine Learning to jointly develop new Anomaly Detection approaches that are more suitable for the 4.0 industrial revolution. This book addresses the needs of both researchers and practitioners to uncover the challenges and opportunities of Anomaly Detection techniques with the applications to manufacturing. The book will also provide ready-to-use algorithms and parameter sheets so readers and practitioners can design advanced Control Charts and Machine Learning-based approaches for Anomaly Detection in manufacturing. Case studies will also be introduced in each chapter to help readers and practitioners easily apply these tools to real-world manufacturing processes. The book contains 10 chapters.

In the Introductory chapter "Introduction to Control Charts and Machine Learning for Anomaly Detection in Manufacturing," the book editor Kim Phuc Tran elaborates on the peculiarities of Anomaly Detection problems using Control Charts and Machine Learning. He determines recent research streams and summarizes the structure and contribution of the book.

Phuong Hanh Tran, Adel Ahmadi Nadi, Thi Hien Nguyen, Kim Duc Tran, and Kim Phuc Tran investigate in their chapter, "Application of Machine Learning in Statistical Process Control Charts: a survey and perspective," a survey and perspective about the development of Machine Learning-based Control Charts, Control Chart Pattern Recognition method using Machine Learning, and interpreting of out-of-control signals. This chapter fills the gap in the literature by identifying and analyzing research on the application of Machine Learning in statistical process Control Charts. The authors review and discuss open research issues that are important for this research stream.

Philippe Castagliola, Giovanni Celano, Dorra Rahali, and Shu Wu develop in their chapter, "Control Charts for Monitoring Time-Between-Events-and-Amplitude Data," a study to investigate several Time-Between-Events-and-Amplitude Data Control Charts and to open new research directions.

Maria Anastasopoulou and Athanasios C. Rakitzis develop in their chapter, "Monitoring a First-order Binomial Autoregressive Process with EWMA and DEWMA Control Charts," one-sided and two-sided EWMA (Exponentially Weighted Moving Average) and Double EWMA Control Charts for monitoring an integer-valued autocorrelated process with bounded support.

Christian H. Weiß develops in his chapter, "On Approaches for Monitoring Categorical Event Series" a survey of approaches for monitoring categorical event series. Also, rule-based procedures from Machine Learning are used for the monitoring of categorical event series, where the generated rules are used to predict the occurrence of critical events.

Xiulin Xie and Peihua Qiu develop in their chapter, "Machine Learning Control Charts for Monitoring Serially Correlated Data," an approach of using certain existing Machine Learning Control Charts together with a recursive data de-correlation procedure.

Tommaso Barbariol, Filippo Dalla Chiara, Davide Marcato and Gian Antonio Susto develop in their chapter, “A review of Tree-based approaches for Anomaly Detection,” a review of several relevant aspects of the methods, like computational costs and interpretability traits.

Anne-Sophie Collin and Christophe De Vleeschouwer develop in their chapter, “Joint use of skip connections and synthetic corruption for Anomaly Detection with autoencoders” a detection of abnormal structure in images based on the reconstruction of a clean version of this query image.

Edgard M. Maboudou-Tchao and Charles W. Harrison develop in their chapter, “A comparative study of L1 and L2 norms in Support Vector Data Descriptions” a comparative study of L1 and L2 norms in Support Vector Data Descriptions. They apply the L1 SVDD and L2 SVDD to a real-world dataset that involves monitoring machine failures in a manufacturing process.

Khanh T. P. Nguyen develops in her chapter, “Feature engineering and health indicator construction for fault detection and diagnostic” a comprehensive review and new advances of feature engineering techniques and health indicator construction methods for fault detection and diagnostic of engineering systems.

4 Conclusion

This book, consisting of 10 chapters, aims to address both research and practical aspects in Control Charts and Machine Learning with an emphasis on the applications. Each chapter is written by active researchers and experienced practitioners in the field aiming to connect the gap between theory and practice and to trigger new research challenges in Anomaly Detection with the applications in manufacturing. The strong digital transformation that has been taking place in manufacturing creates a lot of data with different structures from the process, Anomaly Detection with its applications becomes more important. This book is an important reference focused on advanced Machine Learning algorithms and Control Charts to help managers extract anomalies from process data, which can aid decision-making, early warning of failures, and help improve the quality and productivity in manufacturing.

References

1. Lindemann B, Fesenmayr F, Jazdi N, Weyrich M (2019) Anomaly detection in discrete manufacturing using self-learning approaches. Procedia CIRP 79:313–318
2. Qiu P (2020) Big data? Statistical process control can help! Am Stat 74(4):329–344
3. Wang J, Ma Y, Zhang L, Gao RX, Wu D (2018) Deep learning for smart manufacturing: methods and applications. J Manuf Syst 48:144–156
4. Lopez F, Saez M, Shao Y, Balta EC, Moyne J, Mao ZM, Barton K, Tilbury D (2017) Categorization of anomalies in smart manufacturing systems to support the selection of detection mechanisms. IEEE Rob Autom Lett 2(4):1885–1892

5. Tsung F, Zhang K, Cheng L, Song Z (2018) Statistical transfer learning: a review and some extensions to statistical process control. Qual Eng 30(1):115–128
6. Zwetsloot IM, Mahmood T, Woodall WH (2020) Multivariate time-between-events monitoring: an overview and some overlooked underlying complexities. Qual Eng 33(1):1–13
7. Nguyen HD, Tran KP, Thomassey S, Hamad M (2021) Forecasting and anomaly detection approaches using LSTM and LSTM autoencoder techniques with the applications in supply chain management. Int J Inf Manag 57:102282
8. Lee S, Kwak M, Tsui KL, Kim SB (2019) Process monitoring using variational autoencoder for high-dimensional nonlinear processes. Eng Appl Artif Intell 83:13–27
9. Qiu P, Xie X (2021) Transparent sequential learning for statistical process control of serially correlated data. Technometrics, (just-accepted), 1–29

Application of Machine Learning in Statistical Process Control Charts: A Survey and Perspective

Phuong Hanh Tran, Adel Ahmadi Nadi, Thi Hien Nguyen, Kim Duc Tran, and Kim Phuc Tran

Abstract Over the past decades, control charts, one of the essential tools in Statistical Process Control (SPC), have been widely implemented in manufacturing industries as an effective approach for Anomaly Detection (AD). Thanks to the development of technologies like the Internet of Things (IoT) and Artificial Intelligence (AI), Smart Manufacturing (SM) has become an important concept for expressing the end goal of digitization in manufacturing. However, SM requires a more automatic procedure with capabilities to deal with huge data from the continuous and simultaneous process. Hence, traditional control charts of SPC now find difficulties in reality activities including designing, pattern recognition, and interpreting stages. Machine Learning (ML) algorithms have emerged as powerful analytic tools and great assistance that can be integrating to control charts of SPC to solve these issues. Therefore, the purpose of this chapter is first to presents a survey on the applications of ML techniques in the stages of designing, pattern recognition, and interpreting of control charts respectively in SPC especially in the context of SM for AD. Second, difficulties and challenges in these areas are discussed. Third, perspectives of ML techniques-based control charts for AD in SM are proposed. Finally, a case study of an ML-based control chart for bearing failure AD is also provided in this chapter.

P. H. Tran · T. H. Nguyen · K. D. Tran
International Research Institute for Artificial Intelligence and Data Science, Dong A University, Danang, Vietnam

A. Ahmadi Nadi
Department of Statistics, Ferdowsi University of Mashhad, P. O. Box 1159, 91775 Mashhad, Iran

T. H. Nguyen
Laboratoire AGM, UMR CNRS 8088, CY Cergy Paris Université, 95000 Cergy, France

A. Ahmadi Nadi · K. P. Tran (✉)
University of Lille, ENSAIT, GEMTEX, 59000 Lille, France
e-mail: kim-phuc.tran@ensait.fr

K. P. Tran (ed.), *Control Charts and Machine Learning for Anomaly Detection in Manufacturing*, Springer Series in Reliability Engineering,
https://doi.org/10.1007/978-3-030-83819-5_2

1 Introduction

Together with the blooming flourish rapidly of data, including velocity, volume, and variety, anomaly detection (AD) has become a hot topic in recent years. The important role of AD has demonstrated throughout various studies in numerous different disciplines such as emergency hospital systems [1], traffic measurement [2], credit card fraud detection [3], and manufacturing industry [4, 5]. According to Chandola et al. [6], AD has seen as a term concern to find the instances that do not well conform to a defined notion of normal behavior. These instances are called anomalies or outliers or interchangeably. The beginning of the 19th century is considered as the milestones of the AD issue that has been dealt with by the statistical science community [7]. The requirement for early detection of anomalies in the process is necessary to ensure system performance and save time as well as cost for an organization.

It is worth mentioning that statistical process control (SPC) is an essential approach for AD that is widely applied in industry. The aim of this approach is to monitor and reduce variation in the process as soon as possible to guarantee high product quality at a minimal cost. In particular, the control chart, one of the fundamental tools of SPC first introduced by Shewhart [8] has been an effective tool to detect changes and anomalies of characteristics in the procedure. The contribution of the control chart is based on the idea to gives the producers a simple graphical tool for controlling production, i.e. having correction activities in a timely manner. This allows them to keep production centered on its target and to maintain its dispersion within the specified tolerance interval. However, numerous studies show that the implementation of traditional control charts meets some disadvantages in particular situations including designing [9–13], trend recognition [14–18], and interpreting [19–21] of control chart. A more specific discussion is presented as follows.

It is important to note that a disadvantage of traditional control charts have been discussed in the designing stage. One of the principles in designing a control chart by statistical traditional methods is that it has to under an assumption in which samples of the observed process are normally, independently, and identically distributed (i.i.d. assumption). For example: in the case of univariate process, this implies that the observed in-control process has a steady-state and is characterized by two fixed parameters as mean μ and standard deviation σ. They also lie on an assumption that the main parameters are known or estimated from the historical data. However, this approach faces difficulties in some real activities situations of industry process when considering in the new context as dynamic behavior environment or sampling regularly. Firstly, the normal population distribution assumption is unreal in many cases. Secondly, a variety of researches [9–13] showed the developed control charts using the assumption of independent observations have been enormous influenced by the presence of autocorrelation. Finally, the complex industry procedure could be dominated by various variables and it is impossible to know the covariance relationships before. This leads to false alarms appear many times. Therefore, efforts to develop advanced control charts using Machine Learning (ML) in the mentioned cases are necessary.

Besides, control chart pattern recognition (CCPR) is an important problem in SPC. A control chart is used for detecting whether a process is in-control or out-of-control. But one out-of-control state is found and is eliminated, it is necessary to have an observation, i.e., abnormal pattern recognition to well monitor the behavior of the process in the future. Numerous studies focus on CCPR issue from the middle of 1980s [22, 23]. The aim of the CCPR task is to diagnose nine common abnormal patterns, i.e. unnatural patterns in the process including upward trend, downward trend, upward shift, downward shift, cycles, runs, stratification, freak patterns, and freak points [24]. This activity in order to find out and prevent potential causes as soon as possible. CCPR can be performed by quality engineers in small production systems. However, along with the development of manufacturing systems especially SM, sensors are deployed everywhere with huge data sources to be collected and monitored, the application of ML to automating this task is an irreversible trend. Miao and Yang [17] revealed that the analysis of the statistical characteristics and shape features of the control chart pattern contribute to recognizing unreal patterns of the process through the relevant algorithm was classified. However, the application of Deep Learning (DL) methods to automatically extract features from the control chart has proven superior in the ability to recognize patterns, see Zan et al. [18] for more details. Since then, efforts in applying DL in this field are a very important research direction.

Finally, a very important issue that needs attention in SPC is the interpretation of out-of-control signals. Traditional univariate control charts have played a significant role in the literature to monitor the characteristic processes for ensuring the quality of the system. However, in real activities of industry, the truth is that the process was dominated by various characteristics in some cases. This issue was often solved by the way of using different univariate control charts. But this would lead to false alarms when these characteristics have a high correlation or sampling carry out in a short duration. Therefore, it is necessary to collect and monitor multivariate variables simultaneously, i.e. using multivariate statistical process control (MSPC). Hotelling's T^2 chart [25], Multivariate Cumulative Sum (MCUSUM) chart [26], and Multivariate Exponentially Weighted Moving Average (MEWMA) chart [27] are common multivariate control charts of MSPC used to solve the quality control problems. However, a challenge of these traditional multivariate control charts is that they are just only able to detect a shift in the process mean vector, i.e., out-of-control signals of the process. It is impossible to indicate which variable(s) or a group of variables is responsible for out-of-control signals of the process. Moreover, the MSPC requires more rapid identification in comparison with a univariate process that is beyond the capacity of traditional multivariate control chats. The interpretation of out-of-control signals can be considered a classification problem in ML. Therefore, the application of ML to develop methods to automatically interpret the out-of-control signals in the advanced multivariate control charts e.g. ML techniques-based control charts, has attracted a lot of efforts from researchers [28].

In short, thanks to the appearance of ML methods, these difficulties are solved. The application of ML in control charts is a new approach that is overcome these previous disadvantages or issues. Swift [23] and Shewhart [24] have seen as the pioneer researchers published ideas combining ML in a control chart. Recently, many pieces of research showed that CCPR based new ML algorithms have performance better than one based traditional statistical methods as well as conduct to estimate pattern parameters [14, 29, 30]. Besides, numerous authors also showed that ML methods are useful techniques applied to control charts to tackle the issues in the interpreting stage [19–21, 31]. Due to the various advantages of integrating ML techniques to control charts in SPC, we would like to encourage more studies to consider this approach. This can be seen as the alternative one to overcome the above limitations of traditional control charts. However, this is a lack of researches that focuses to give a general picture of these issues in literature. Therefore, the main objective of our chapter is to fill this gap. The remainder of this chapter is organized as follows. Section 2 briefly reviews the design of control chart-based ML methods. Section 3 makes a literature review relevant to CCPR. Section 4 presents the recent studies about the issue of the interpreting-based ML of control charts. Difficulties and challenges in these areas are discussed in Sect. 5. Section 6 proposed perspectives for ML techniques-based control charts for AD in SM. An experiment for a case study is proposed in Sect. 7. Finally, concluding remarks of the study are outlined in Sect. 8.

2 Machine Learning Techniques Based Control Charts for Process Monitoring

Control charts have been developed and applied a lot in many fields (see Fig. 1, taken from Web of Science), major publications in the fields of the engineering industry. Control charts provide a simple method that can be used to indicate whether the process is stable or not (in-control or out-of-control). In detail, it is a chronological graph whose dots represent the tracking of a characteristic of the process. A horizontal line represents the central value (the average). The lower control limit (LCL) and the upper control limit (UCL) are represented by two horizontal lines on either side of the mean. The values of a measured characteristic must be within these limits; otherwise, the process is out-of-control and must be examined. The main benefits of control charts are: (1) they increase productivity by the proportion of "good product" and decrease costs because there is less waste; (2) they give an estimate of the central tendency of the characteristic of interest, its variability, and the limits within which it varies; (3) control charts assist in the evaluation of the performance of a measurement system. One of the major advantages of the control chart is its ease of construction and use, an operator or engineer familiar with the technique of control charts can, in general, diagnose the cause of a problem. However, in order for the control chart to be a reliable and effective indicator of the status of the process, the production using the control chart should pay special attention to the type of chart used.

ML is a domain of Artificial Intelligence (AI), which consists of programming algorithms to learn automatically from data and experiences or by interaction with the environment. What makes ML really useful is the fact that the algorithm can "learn" and adapt its results based on new data without any a priori programming. There are

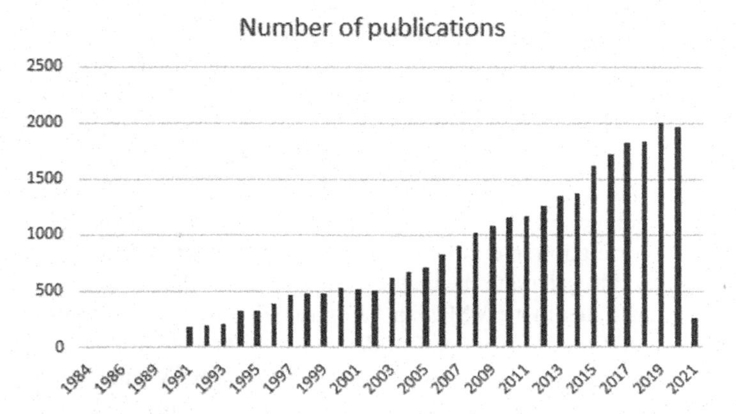

Fig. 1 Number of publications on control charts from 1984

three main branches: supervised learning, unsupervised learning, and reinforcement learning. The algorithm of supervised learning is to find correlations between input data (explanatory variables) and output data (predictable variables), for then infer the knowledge extracted on inputs with unknown outputs. Different from supervised learning, the technique of unsupervised learning must discover by itself the structure according to the data, which has only one dataset collected as input. This technique is used to divide data into groups of homogeneous items/datapoint. Finally, reinforcement learning is an area of ML concerned with how to make a sequence of decisions. In literature and practice, many researchers have combined techniques of ML and control charts. As mentioned above, by the ease of use of controls charts and the wide application of ML, this combination is increasingly researched and applied. This is because many types of problems that are arising during the implementation of control charts in nowadays complex processes can be effectively solved with the help of ML approaches (see for example Kang et al. [32] and Qiu and Xie [33]). One of the main contributions of applying ML techniques in designing control charts is that the modern (production, insurance, healthcare, and etc.) processes generate huge data sets with a large degree of diversity by means of modern measurement systems like sensors. In such situations, the traditional statistical monitoring methods fail to handle the monitoring procedure of such processes while ML techniques are able to provide impressive results [34]. This section will summarize the most common techniques for designing control charts with ML methods for process monitoring.

2.1 *Kernel-Based Learning Methods*

Kernel-based learning methods such as the Support Vector Machine (SVM) algorithm are extensively used and play major roles in the SPC activities, both in developing control charts and recognition of abnormal patterns, due to their remarkable solutions for existing problems. In brief, kernels have been applying in the ML area because, when it is difficult to do a task in the original problem space, the kernel method enables the practitioner to transform the problem space into another in which they can work easier. Recently, Apsemidis et al. [35] provided a comprehensive review on about 90 articles after 2002 that include the combination of kernel-based approaches with other ML techniques. Mashuri et al. [36] proposed a $Tr(R2)$ control chart based on the squared correlation matrix with the trace operator and used the kernel density estimation method to calculate the better control limit for the proposed chart. Chinnam [37] demonstrates that SVMs can be extremely effective in minimizing both Type-I errors and Type-II errors and in detecting shifts in both the non-correlated processes ou autocorrelated processes. A comparison of SVM and Neural Network (NN) for drug/nondrug classification has been done by Byvatov et al. [38] and it was demonstrated that the SVMs classifier yielded slightly higher prediction accuracy than NN. By the efficiency of SVMs, many researchers used this technique based on control charts. For example, Li and Jia [39] proposed a SVMs based model for fault identification in MEWMA control charts, they examined the effects of SVM parameters on classification performance and provided a SVM parameter optimization method.

Although the kernel-based ML algorithms are mainly applied as classifiers for dividing data into two or more classes, in most SPC implementations training data from one class (normal state) are only available and there is no information about the other class (abnormal state). This situation may arise from several reasons such as the general difficulties (lack of resources or time or cost) or even impossibility of collecting enough observations for the abnormal class to learn the ML algorithm [40]. To handle such situations, one-class classifiers are introduced. One-class classifier just learns from the normal training data and labeled the newly encountered data as in-class or out-of-class observations. Several one-class classifiers have been developed by researchers, while support vector data description (SVDD), the k nearest neighbor data description (KNNDD), and K-means data description (KMDD) one-class classifiers were only used to develop control charts. One of the first studies in this domain was conducted by Sun and Tsung [41] who designed a kernel distance-based chart (K chart) using SVDD algorithm, as a modified version of the original SVM for solving one-class classification problems, and concluded that the K chart outperforms conventional charts when the data distribution departs from normality. This work improved by Ning and Tsung [42] for non–normal process data. Sukchotrat et al. [43] developed a K chart that integrates a traditional control chart technique with a KNNDD algorithm, one of the one-class classification algorithms. Later, to examine the feasibility of using one-class classification-based control charts to handle multivariate and autocorrelated processes, Kim et al. [44] developed a K chart that uses original observations instead of residuals to monitor autocorrelated multivariate processes. Throughout a simulation study, they showed that the K charts outperform the T^2 control charts, and the performance of K charts is not significantly affected by the degrees of autocorrelation. Gani and Limam [45] examined the performance of the K chart and KNNDD chart through a real industrial application. They investigated the effectiveness of both charts in detecting out-of-control observations using the average run length (ARL) criterion. The results of this study show that the K chart is sensitive to small shifts in the mean vector, whereas the KNNDD chart is sensitive to moderate shifts in the mean vector. In addition, Gani and Limam [46] introduced a new chart using the KMDD algorithm and reported that their chart has a better performance in detecting small shifts of mean vector based on ARL than the K chart and KNNDD chart. To improve the performance of K-charts, Maboudou-Tchao et al. [47] used a one-class SVM technique based on the SVDD method for monitoring the mean vector based on Mahalanobis kernel. They used the Mahalanobis kernel as an alternative for Gaussian kernel and showed that the proposed method is more sensitive than SVDD using Gaussian kernel for detecting shifts in the mean vectors of three different multivariate distributions. They also demonstrated that the proposed method outperforms Hotelling's T^2 chart in multivariate normal cases.

Zhang et al. [48] developed a general monitoring framework for detecting location shifts in complex processes using the SVM model and MEWMA chart. Later, Wang et al. [49] developed SVM-based one-sided control charts to monitor a process with multivariate quality characteristics. They used the differential evolution (DE) algorithm to obtain the optimal parameters of the SVM model by minimizing mean absolute error. In this study, the performance of the control charts is investigated using a multivariate normal distribution and two non-normal distributions by considering different process shift scenarios. In addition, through an ARL analysis using the Monte Carlo simulations, they showed that the proposed chart has better performance than the distance-based control charts based on SVM studied by He et al. [50]. Recently, Maboudou-Tchao [51] introduced a least-squares one-class SVM (LS-OCSVM) for monitoring the mean vector of processes. They counted several advantages of their proposed monitoring approach over the existing SVDD chart provided by Sun and Tsung [41] and Hotelling's T^2 chart in terms of simplicity in computation and design, flexibility in implementation, and superiority in performance. For example, they claimed that the LS-OCSVM method can be easily extended to online monitoring. This feature is very beneficial when we are facing a large-scale training dataset that updates over time. The SVDD method uses a batch learning phase in which we learn on the entire training set and generate the best model at once. If new additional training data arrive, SVDD must be retrained from scratch. Using SVM techniques based on control charts to have a better performance can be found at many works, see for example, He et al. [50], Salehi et al. [52], Hu and Zhao [53], Gani et al. [54], Kakde et al. [43, 55], Jang et al. [56].

Regression analysis is a technique of supervised ML. It is based on the basic principles of physics that help predict the future from current data. It also helps to find the correlation between two variables to define the cause and effect relationship. However, there are different forms of regression, ranging from linear regression and complex regression. One of the regression variants which yields very good results is the support vector regression (SVR) method. This technique has been applied a lot in the construction of control charts, especially when the process variables are highly auto-correlated. For example, Issam and Mohamed [57] apply the SVR method for the construction of a residuals MCUSUM control chart to monitoring changes in the process mean vector. This charts are designed to detect small shifts in the process parameters and it performed better than the time series based control chart because it can handle non-linear relation between the controlled variables and do not use any restrictive assumption. In 2013, Du et al. [58] proposed one new Minimal Euclidean Distance (MED) based control chart for recognizing the mean shifts of auto-correlated processes. They also used SVR to predict the values of a variable in time series. The numerical results showed that the MED chart outperformed those of some statistics-based charts and the NN based control scheme for the small process mean shifts. Another example of a combination of SVR technique and Control charts, Gani et al. [54] designed a SVR-chart which using SVR to construct robust control charts for residuals. By comparing the behavior of ARL, the authors showed that the efficiency of this chart is better than ordinary least squares (OLS), and the partial least squares method.

Besides the above-mentioned supervised learning methods, unsupervised learning algorithms are another type of ML algorithms that is applied to analyze and cluster unlabelled datasets. Clustering is one of the most important unsupervised ML techniques, in which similar traits are used to make a prediction. The algorithm measures the proximity between each element based on defined criteria. K-Means is the most popular method of grouping input data, which allows us to set the value of K and order the data according to that value. The aim of the study of Silva et al. [59] is to apply the u-chart to find out the number of clusters in the K-means method on Automatic Clustering Differential Evolution (ACDE) in order to identify the behavior patterns and relations between the different attributes. These results in this work showed that the use of an u-chart increases the performance of ACDE. Another example of application clustering technique based on control charts in medicine, Thirumalai et al. [60] gave a prediction of diabetes disease for people of various age groups and genders by using cost optimization and control chart.

2.2 *Dimensionality Reduction*

For a given data, the higher the number of variables, the more complex the results will be, which will make it difficult to consolidate the data. Dimensionality reduction is considered a method of ML to overcome this difficulty. Instead of studying the data involved in a grand dimension, the technique of dimensionality reduction is to replace it with data in a smaller dimension. Roughly speaking, principal components analysis (PCA) is one of the most important methods of dimensionality reduction that transforms a large dataset of (possibly) correlated observations into a smaller data set of uncorrelated observations by minimizing information loss. Developing control charts based on the PCA method has been widely investigated in the literature. For example, Stefatos and Hamza [61] introduced a robust multivariate statistical control chart using the Kernel PCA (KPCA) method. They reported that the new chart is robust to outliers detection and performs better than some existing multivariate monitoring and control charts. Phaladiganon et al. [62] presented non-parametric PCA technique, kernel density estimation, and bootstrapping to establish the control limits of control charts. The proposed non-parametric PCA control charts performed better than the parametric PCA control charts in non-normal situations through the behavior of ARL. The PCA's technique is also used in Kullaa [63], the author showed that the sensitivity of the control chart to damage was substantially increased by further dimensionality reduction applying the PCA technique. Applying this technique, Lee et al. [64] developed a new KPCA-based non-linear process monitoring technique for tackling the nonlinear problem. Base on T^2 and squared prediction error (SPE) charts in the feature space, KPCA was applied to fault detection in two example systems: a simple multivariate process and the simulation benchmark of the biological waste-water treatment process. These examples demonstrated that the proposed approach can effectively capture nonlinear relationships in process variables and that, when used for process monitoring, it shows better performance than linear PCA. Using Hotelling's T^2 statistic, Ahsan et al. [65] implemented the KPCA method for simultaneously monitoring mixed (continuous and categorical) quality characteristics. In this study, it is demonstrated that the KPCA-based control charts have a great performance in terms of successful detection of the out-of-control observations in comparison with the conventional PCA mix charts discussed in Ahsan et al. [66]. Another study in the area of monitoring procedures of mixed quality characteristics based on the KPCA technique has been presented by Mashuri et al. [67]. Recently, Lee et al. [68] presented new multivariate control charts by Hotelling's T^2 statistics and Q statistic based on KPCA approach for rapidly detecting a worn cutting tool and thus avoiding catastrophic tool failures products with unacceptable surface finish, and defective product. Their proposed method converts raw multi-sensor data into principal component space, and then, the KPCA-modified data are used to calculate T^2 and Q values to develop control charts.

2.3 *Neural Network and Deep Learning*

Unlike linear models, the NN is based on a complex, divisional data model. It includes multiple layers to provide you with unique and precise output. However, the model is still based on linear regression but uses multiple hidden layers; this is why it is called a NN. In the paper of Arkat et al. [69], they designed a NN based model to forecast and construct residuals CUSUM chart for multivariate Auto-Regressive of order one, AR(1), processes. The comparison of the performance of the proposed method with the time series-based residuals chart and the auto-correlated MCUSUM chart was made. DL is a subset of ML, which is essentially a NN with multi layeres. Recently, Lee et al. [70] proposed a variational autoencoder (VAE) approach to monitor high-dimensional processes in the presence of non-linearity and non-normality assumptions. They demonstrated the effectiveness and applicability of the proposed VAE-based control charts in comparison with the existing latent variable-based charts through a simulation study and also via real data from a TFT-LCD manufacturing process. Chen and Yu [71] suggested a novel recurrent neural network (RNN) residual chart with a DL technique to recognize mean shifts in autocorrelated processes.

A comparison study with some typical methods such as special causes control chart and backpropagation network residual chart demonstrate that the RNN-based chart provides the best performance for monitoring mean shifts in autocorrelated manufacturing processes. The readers can find more reference about this technique based on control charts, for example, see Niaki and Abbasi [72], Chen et al. [73], and Diren et al. [28].

3 Machine Learning Techniques Based Control Chart Pattern Recognition

Entering the 21st century, the world has changed dramatically with the development of information technology, this is the beginning of the era of big data. This comes with a marked increase in the general interest in ML. The interpretation of control charts is mainly based on rough rules (i.e. heuristics) which depend greatly on the experience and judgment of the operator. It is therefore very important to make sure that they are well trained. Consequently, expert systems were born and developed in the industry. An expert system is software that is linked to at least two data sources: a database that contains a set of rules and a data flow that comes from the process to be controlled. The rules are based on the knowledge of experts in the field and are encoded as logical conditions. Everything is connected to a motor inference that applies the rules. The latter produces a result that is then communicated to users through a graphical interface and is used as a decision support tool. More precisely, an expert system is a software capable of answering questions, by reasoning from known facts and rules. However, the period of popularity of expert systems is relatively short, from the end of the 1980s, NNs are beginning to be used to automate the reading and interpretation of control charts (see Pugh [74]). Since that time, pattern recognition, in general, is dominated by ML, is widely developed. There are several motivations for using ML algorithms for CCPR purposes. The first and probably the main motivation is that several researchers demonstrated that the ML-based CCPR model outperforms their alternative models in many practical situations. For example, Li et al. [75] proposed a SVM-based CCPR framework and demonstrated that this model can accurately classify the source(s) of out-of-control signal and even outperforms the conventional multivariate control scheme. There also other motivations for applying ML-based CCPR models. For example, Guh [76] stated that the NN models are capable of learning and self-organizing and hence are useful in pattern recognition and can recall patterns learned from noisy or incomplete representations which are practically impossible to detect by operators, even with the assistance of an expert system. This makes the ML-based approaches suitable for CCPR because CCPs are generally contaminated by common cause variations. In addition, Diren et al. [28] reported that traditional CCPR models are not able to predict unexpected new situations while ML techniques that can effectively predict the unexpected new situations by learning from the historical data. This section reviews some important references about the most popular ML algorithms used in recognition of patterns on control charts including classification and regression tree (CART), decision trees (DTs), SVMs, NNs, and DL.

3.1 Regression Tree and Decision Tree Based CCPR

A DT is a decision support tool representing a set of choices in the graphic form of a tree. Geometrically, constructing a decision tree is to partition the space of data attributes in areas where each region represents a class. During prediction, when data is in this region then the decision tree assigns it the corresponding class. In literature, there are different methods to construct one or more decision trees from a learning data set. The common goal of each method is to determine the optimal test sequence to partition the space of attributes into homogeneous regions. Very recently, Zaman and Hassan [77] demonstrate the development of fuzzy heuristics and the CART technique for CCPR and compare their classification performance. The results show the heuristics Mamdani fuzzy classifier performed well in classification accuracy (95.76%) but slightly lower compared to the CART classifier (98.58%). This study opens opportunities for deeper investigation and provides a useful revisit to promote more studies into explainable AI.

3.2 Neural Network and Deep Learning Based CCPR

In the paper of Hachicha and Ghorbel [78], a survey of CCPR literature, the majority of the reviewed articles use the NN approach. It is reported that for the period 1988 to 2000, 9% of the revised publications use NNs and that for the period 2001 to 2010, that number climbed to 25%. This trend then accelerates for the period from 2010 to 2021 (see Fig. 2 and 3). This observation is supported by the number of articles published on NN which shows an average annual increase of 10–15% for this period (source from Web of Science). Pugh [74] was the first author to experiment with NN and control charts. He concludes that NN is as effective as traditional control charts for detecting changes in average values following a surge (by comparing the ARL) and NN was found to perform reasonably well under most conditions. This study constitutes the proof of concept of NN in CCPR. Pham and Oztemel [79] were the firsts described the structures of pattern recognition systems which made up of independent multi-layer perception. They found that these composite pattern recognition systems have better classification capabilities than their individual modules. Cheng [80] also concluded that hybrid networks are more efficient than networks singular. Addeh et al. [81] proposed a CCPR procedure based on optimized radial basis function neural network (RBFNN). The proposed method consists of four main modules: feature extraction, feature selection, classification and learning algorithm. In addition, traditional patterns that have considered in literature including the normal, cyclic, increasing trend, decreasing trend, upward shift and downward shift, they investigated the stratification and systematic patterns as well. They tested RBFNN-based CCPR model based on a dataset containing 1600 samples (200 samples from each pattern) and the results showed that the proposed method has a very good performance. Yu et al. [82] developed an effective and reliable DL method known as stacked denoising autoencoder (SDAE) for CCPR in manufacturing processes. Recently, Xu et al. [83] proposed an efficient one-dimensional Convolutional Neural Network (1D-CNN) to applied for CCPR purposes. They showed that their method achieves 98.96% average recognition accuracy after 30 repeated tests as well as has better generalization ability when there is an error between the estimated value and true value of mean or standard deviation, which are satisfactory results. Yang and Zhou [84] developed online CCPR systems using NN0 ensemble also neglecting how the correlation coefficient is biased when abnormal patterns occur, thus training one CCPR system for each of the studied autocorrelation levels. Fuqua and Razzaghi [85] proposed a cost-sensitive classification scheme within a deep convolutional neural network (CSCNN) to fill the literature gap of developing computationally-efficient methods of CCPR classification for large time-series datasets in the presence of imbalance. To show the benefits of the method, they conducted an extensive experimental study using both simulated and real-world datasets based on simple and complex abnormal patterns. For more information, see for examples some publications as Pham and Wani [86], Yang and Zhou [84].

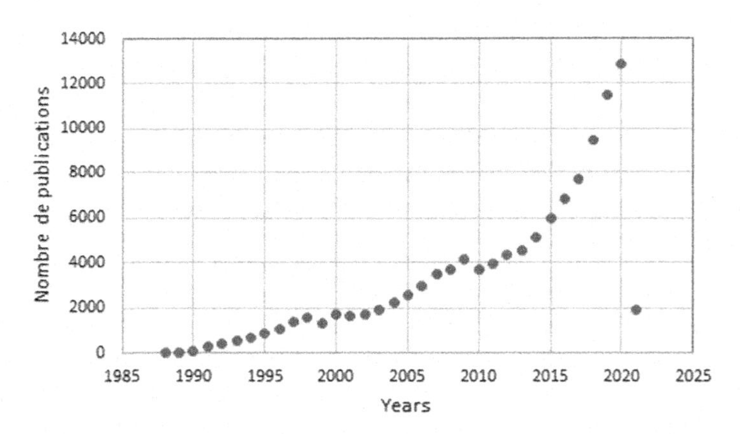

Fig. 2 Number of publications on NN from 1988

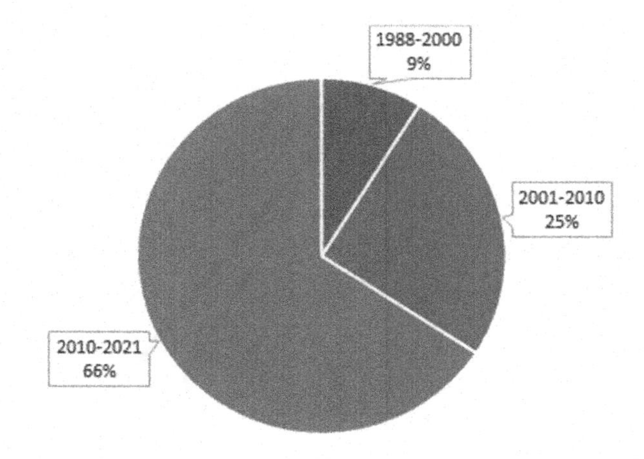

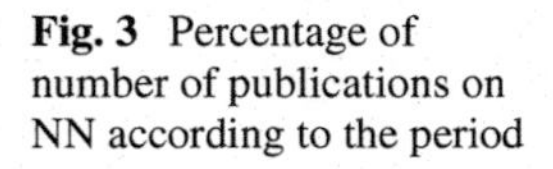
Fig. 3 Percentage of number of publications on NN according to the period

3.3 Support Vector Machines Based CCPR

SVM are new statistical learning techniques proposed by V. Vapnik in 1995. They help to address diverse issues as classification, regression, fusion, etc. The essential idea of SVM consists in projecting the data of the input space (belonging to two different classes) non-linearly separable in a space of greater dimension called space of characteristics in such a way that the data becomes linearly separable. In this space, the technique construction of the optimal hyperplane is used to calculate the function of classification separating the two classes (see Fig. 4). In other words, the algorithm creates a line or a hyperplane which separates the data into classes.

In this subsection, we will summarize some recent applications and extensions of SVM for the CCPR case. Ranaee et al. [87] study a novel hybrid intelligent system that includes three main modules, in which two modules, SVM technique is used to searching for the best value of the parameters that tune its discriminant function (kernel parameter selection) and upstream by looking for the best subset of features that feed the classifier. Simulation results show that the proposed algorithm has very high recognition accuracy. A hybrid independent component analysis (ICA) and SVM is proposed for CCPR by [88], the results showed that is able to effectively recognize mixture control chart patterns and outperform the single SVM models, which did not use an ICA as a preprocessor. Lin et al. [89] presented a SVM-based CCPR model for the online real-time recognition of seven typical types of abnormal patterns, assuming that the process observations come from an AR(1) model. Through an extensive simulation study, they showed that the proposed SVM-based CCPR model can effectively on-line recognize unnatural patterns in both independent and autocorrelated processes. In addition, they indicated that the new model has a better recognition accuracy and ARL performance than the existing learning vector quantization network CCPR model provided by Guh [76]. Du et al. [58] integrated wavelet transform and improved particle swarm optimization-based support vector machine (P-SVM) for online recognition of concurrent CCPR. In other research, original SVM demon-

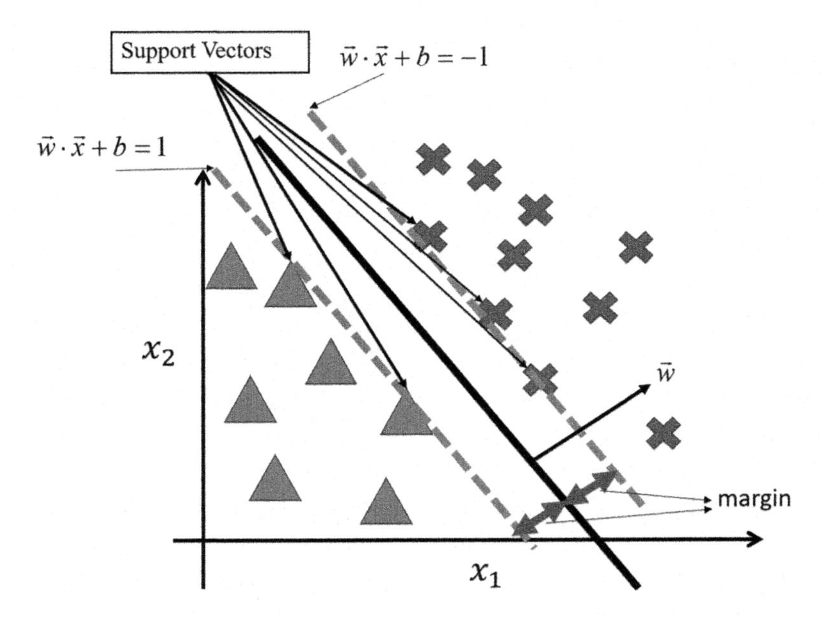

Fig. 4 Principle of SVM techniques

strates poor performance when applied directly to these problems. Xanthopoulos and Razzaghi [90] improve SVM by using weighted support vector machines (WSVM) for automated process monitoring and early fault diagnosis. They show the benefits of WSVM over traditional SVM, compare them under various fault scenarios. Readers can refer to many other references, see Wang [91], Ranaee and Ebrahimzadeh [92], Lin et al. [89], Zhou et al. [93], la Torre Gutierrez and Pham [94].

4 Interpreting Out-of-Control Signals Using Machine Learning

When the manufacturing process has more than two characteristics for monitoring, it should be often solve with different univariate control charts. However, when these characteristics have a high correlation or sampling in a short duration, the false alarms may be appeared. Therefore, it is necessary to use multivariable control charts for monitoring quality problems. Hotelling's T^2 chart [25], MCUSUM chart [26], and MEWMA chart [27] are common multivariate charts were used in MSPC. However, a challenge of these traditional multivariate control charts is that they are just only able to provide the general mean shifts in vector, i.e., out-of-control signals of the process. It is impossible for these charts to indicate which variable(s) or a group of variables is responsible for out-of-control signals of the process. Numerous researchers have paid attention to the topic which to find a variable or a number of variables or a set of variables responsible for the signals when a multivariate process is in the out-of-control state. From the past decades, the idea of integrating ML to multivariate control charts as an effectively approach. Recently, this approach seems more reasonable when the system of manufacturing has become more automatic. Thus, this section will give a look at the literature about ML methods for interpreting control charts in the multivariate process.

The first encouragement integrating ML methods to interpret signals of multivariate control charts in the quality control process has been discussed from the beginning of the 2000s with the publication of Wang and Chen [19]. Particularly, they used a neural-fuzzy model (a four-layer fully connected feed-forward network with a back-propagation training rule) for both detecting and classifying phases. An experiment for a bivariate process was conducted demonstrated that the proposed method reaches higher performance than the previous multivariate T^2 control chart. Lower out-of-control ARLs and more classification accuracy results of the proposed method have been recorded. Then, Low et al. [20] continued highlight NNs as the method contribute to detect more anomaly patterns and more sensitive than traditional charts through out-of-control ARL and the numerous abnormal instants detected. Chen and Wang [95] suggested using a model of NNs, a three-layer fully connected feed-forward network with a back-propagation training rule, based multivariate χ^2 control chart to investigate cause variable(s) of signals of bivariate process. The significant advantages are showed that the model can indicate both responsible variables (s) and the magnitude of the shifts in case the multivariate χ^2 control chart has sudden shifts in the mean vector. Niaki and Abbasi [21] suggested multilayer perceptron (MLP) network, a type of NNs, to classify patterns to explore variables or the set of variables that caused the fault of the process. The authors also make a comparison between MLP based Hotelling's T^2 multivariate Shewhart (MSCH) and based multivariate Shewhart (MS) chart, respectively. The results showed that the proposed MLP MSCH has a stronger performance. Cheng and Cheng [96] suggested to use 3-layer fully connected feed-forward network with a back-propagation training rule as an algorithm of NN for classifying out-of-control signals. The authors also recommend using SVMs which are considered as the method that has the same performance although it has more advantages than NN. On the contrary, Guh and Shiue [97] suggested using DT techniques instead of NNs based model to interpret which variable or group of variables has caused the out-of-control signals. They also demonstrated that the implementation of the DT approach gained results faster than 25 times than the NN one. According to Yu et al. [98], a selective NN ensemble approach named Discrete Particle Swarm Optimization (DPSOEN) algorithm has a significant performance to provide the source(s) of out-of-control. Alfaro et al. [99] proposed to use a multi-class exponential loss function (SAMME) algorithms, an extension of AdaBoost for classifying which variables have to be responsible for the out-of-control signals. They showed that the proposed method has more significant performance than ones in the study of Niaki and Abbasi [21]. Verron et al. [100] presented a Bayesian network-based control chart approach to detect and isolate fault variable(s) of a multivariate process. A DT learning-based model for bivariate process is recommended in a study of He et al. [101] to identify the cause of faults. Cheng and Lee [31] suggested using a SVM-based ensemble classification model for interpreting the out-of-control signals of a multivariate process by indicating the caused variable(s). An experiment comparison showed the significant performance of the proposed method in comparison with the single Support Vector Classification (SVC) model, bagging and, AdaBoost. Moreover, Carletti et al. [102] proposed Depth-based Isolation Forest Feature Importance (DIFFI) approach based Isolation Forest (IF) algorithm, the one from the idea as the DT to interpret the cause of faults in the process. The authors also make a comparison with the Permutation-based Importance (PIMP) approach. Recently, Song et al. [103] recommend using a NN) method like instance-based Naive Bayes (INB) algorithm to classify which variables are the cause of out-of-control signals. This is well implemented for both small and large number variables. This also overcomes two disadvantages of previous studies as independence assumption and ignorance of the features of a test instance. Furthermore, very recently, Diren et al. [28] conduct a study with a variety of ML techniques including Naive Bayes-kernel (NB-k), K-Nearest Neighbor (KNN), DT, NN, Multi-Layer Perceptron (MLP), and DL to find the variables responsible for the out-of-control signals based on types of faults. Performance comparison of these techniques is also explored. Recently, Salehi et al. [104] and Yu et al. [159] suggested to use hybrid learning methods for interpreting out-of-control signals in multivariate manufacturing processes.

5 Difficulties and Challenges for Application of Machine Learning in Statistical Process Control Charts

It is evident that firms and corporations are rapidly getting smarter by adding intelligence into their process to drive continuous improvement, knowledge transfer, and intelligent decision-making procedures. This increases the demand for advanced AI and SPC tools and also effectual effective integrated techniques in various production stages to decrease the cost of production, improve overall productivity, and process quality, reduce downtime, and etc. One of the most successful integrations is using ML algorithms, as an important subset of AI, in development, pattern recognition, and interpreting of control charts, as the main goals of SPC. To meet this need, several ML-based approaches have been developed by researchers and scientists that some of them are reviewed in the previous two sections. However, most of these tools have been introduced in laboratory environments and many difficulties and challenges still exist in their applications in practical environments. Implementation of an efficient ML algorithm that performs well in an industrial environment as well as produces reliable results is not very easy. Accordingly, it can be said that although ML is an efficient and widely-used technique for solving nowadays complex problems, like any other technique, it should be implemented as a solver due to its difficulties and challenges. Although data analysts may face a variety of challenges during the designing and implementation of ML algorithms in development, pattern recognition, and interpreting of control charts that we can not address them all here, however, in what follows, we will list some of them that are most appeared in daily operation problems.

5.1 Non-stationary Processes

Although several studies have been done for developing ML-based control charts, CCPR frameworks in the presence of autocorrelated observations (see for example Lin et al. [89], Chen and Yu [71], Kim et al. [44], and Yang and Zhou [84]), most of these studies are based on the assumption that the process is stationary. However, most processes in the manufacturing industries are non-stationary in particular for complex industrial processes which in general show non-stationary process characteristics, revealing a time-varying mean and/or variance or even time-varying autocovariance [105]. This phenomenon makes monitoring a complex task no matter the quality characteristic to be monitored is univariate or multivariate. Non-stationarity in processes' behavior frequently occurs due to several factors such as seasonal changes, processes that involve emptying and filling cycles, throughput changes, the presence of unmeasured disturbances, and also the nature of the process itself [106]. In these cases, interpreting out-of-control points is a challenge as studies on the topic almost always make assumptions about the distribution. Ketelaere et al. [107] presented examples of non-stationary processes from the industrial machinery monitoring context and agriculture industry. Another examples of non-stationary processes in industrial environments are discussed in Chen et al. [106] and Liu and Chen [108]. Monitoring non-stationary processes have its challenges and difficulties and it has to be done carefully since there are many hidden problems. For example, it is difficult to detect the abnormal patterns of non-stationary observations because they may be hidden by the normal non-stationary variations [105]. In addition, Lazariv and Schmid [109] showed that for some processes and change-point models, the ARL does not exist. This is a very important issue since the ARL is the most popular measure for the performance of control charts. In such situations, the traditional SPC techniques fail at monitoring such processes and it is important to have tools that can correctly detect changes in non-stationary processes [110].

5.2 Big Data Analysis

The term big data refers not only to the size or volume of data but also to the variety of data and the velocity of data. These features impose some challenging issues to the data analyst facing various big data monitoring problems. One of the main challenges for monitoring big data based on ML techniques is the training (Phase I) dataset that is expected to contain both in-control and out-of-control process observations [111].

It is known that completing Phase I is critical to successful Phase II monitoring and has a strong influence on the performance and suitability of the ML algorithm to get accurate results and to avoid false predictions. However, in SPC applications, we usually have an in-control dataset only and there is no information about out-of-control situations in the training data. We know that it is very important to provide a training data set that entirely represents the structure of the problem. To tackle this deficiency, the idea of artificial contrasts and one-class classification methods have been suggested by authors such as Tuv and Runger [112] and Sun and Tsung [41]. Another challenge in monitoring high dimensional data sets is the fact that not all of the monitored variables are likely to shift at the same time, thus, some method is necessary to identify the process variables that have changed. In high dimensional data sets, the decomposition methods used with multivariate control charts can become very computationally expensive Reis and Gins [113]. To serve the purpose, many scientists proposed feature selection techniques to monitor subsets of potentially faulty variables instead of monitoring a sequence of whole variables to improve detection performance (see for example Capizzi and Masarotto [114]. However, in such cases, the key questions that have not to be answered yet are (1) what kind(s) of features are appropriate to use for a specific big data monitoring problem, (2) how many features should be extracted for process monitoring, and (3) whether the original goals of process monitoring have been substantially compromised by using the selected features Qiu [111].

5.3 Monitoring Image Data

Thanks to the rapid developments of digital devices like sensors and computers and using them increasingly in industrial and medical applications, intelligent decision-making tools such as machine vision systems (MVS) has gradually taken the place of human-based inspections in many factories due to their ability to provide not only dimensional information but also information on product geometry, surface defects, surface finish, and other product and process characteristics Megahed et al. [115]. A MVS is a computer-based system for analyzing and processing image data that is provided by image-capturing devices (e.g., cameras, X-ray devices, or vision sensors). New studies show that implementing MVSs in industrial environments could be fully utilized to improve the quality of the product Zuo et al. [116]. In this regard, researchers developed a new interdisciplinary field of research by integrating MVS approaches and SPC principles. This new field applied SPC tools for monitoring the process quality using images. There are several applications in industrial and medical areas that image monitoring can be used to check the stability of the process state. For instance, monitoring the brightness of the cover in the printing process of a journal or monitoring the changes of tumors and vascular. Through an extensive review of image-based control charting methodologies, Megahed et al. [115] emphasized that using MVS-based monitoring procedure is superior to visual inspection with respect to, (1) monitoring processes with high production rates; (2) performing multiple simultaneous tasks with different objects; (3) their ability to cover all the ranges of the electromagnetic spectrum, as in the use of magnetic resonance imaging (MRIs) and X-rays in medical applications; (4) the lack of susceptibility to fatigue and distraction; and (5) in some cases, the use of MVSs is cheaper than the use of human inspectors and it is expected that the cost of MVSs will continue to decrease over time. However, there are several challenges in implementing image monitoring in practical situations. The first challenge is that the number of pixels in a simple cell phone image nowadays is around 4 million pixels and thus, we have to monitor a process with 4 million components over time that faces us to high-dimensional problems. Another challenge is that the neighboring pixels within an image are often highly correlated. This correlation can result in a considerable amount of data redundancy and ignoring the correlation can result in a high level of false alarms as well as poor performance once a fault occurs. In addition, there are several stages for successful impersonation of an image-based monitoring procedure such as the choice of the image-capturing device, the frequency of imaging, the set-up of the imaging to avoid lighting, alignment, the software to use for image analysis, the preliminary image processing and the type of monitoring method to employ. There are no currently existing guidelines for guiding the practitioner through all of these decisions Megahed et al. [115]. Thus, the last challenge is providing easy- and clear-to-used guidelines to applied an efficient image monitoring model in practical applications.

6 Perspectives for Application of Machine Learning in Statistical Process Control Charts in Smart Manufacturing

Making processes smart and digitized, motivate researchers and scientists to develop effective ML strategies for anomaly detection in daily operations. For example, startegies to keep the production systems always dynamic in dealing with unexpected variations and abnormal patterns. Although recent studies have investigated new ML-based techniques in the development, pattern recognition, and interpreting of control charts in manufacturing, there still exists a significant potential for reducing the gap between the theory and application in modern industries. Addressing this gap will ensure that ML tools can be seamlessly integrated into factory operations. The following topics are recommended here for future research.

6.1 Auto-correlated Processes and Non-stationary Processes

Thanks to the rapid evolution of sensor technologies and the velocity of available data in modern industrial processes, a good ability has created to gather observations instantaneously that results in a high degree of autocorrelation within observations.

In fact, the real-world data are in most cases autocorrelated. To deal with such data, most of the existing approaches are not sufficient, and there is an essential need to develop new powerful ML tools to analyze these kinds of datasets. While the effect of autocorrelation on traditional SPC tools has been investigated by several authors, see for example Maragah and Woodall [117] and Alwan [9], a review paper by Apsemidis et al. [35] shows that a few studies have yet been conducted in the area of kernel-based ML methods for such data. In another comprehensive review paper, Weese et al. [34] recommended that although some few ML-based monitoring procedures have been proposed for autocorrelated data considering known time series model by researchers like Arkat et al. [118] , Issam and Mohamed [57], and Kim et al. [119], there is ample potential to develop new algorithms when the time series model is unknown. In literature of ML techniques based CCPR there are also few investigations [120] that recognize relatively simple patterns such as process mean shift [121–123] and process variance shift [37], while more complex patterns including trend, cycle, and systematic patterns only considered in the study of Lin et al. [89]. Concurrently, the existing body of researches needs to be enhanced and improved the existing techniques of monitoring auto-correlated processes. In addition, there are several real-world examples that not only the observations are auto-correlated, but also they have non-stationary behaviors in which they are not oscillating around a common mean or its variance and autocovariance are changing over time. This phenomenon may happen due to several reasons. Researchers such as Ketelaere et al. [107], Chen et al. [106] and Chen et al. [106] presented examples of non-stationary processes in industrial environments. Up to now, most existing researches have focus on developing ML-based control charts and CCPR techniques for monitoring stationary processes and there is a remarkable need for developing such tools for handling time series data from non-stationary processes. This gap should be filled by new researches. Recently, Tran et al. [124] and Nguyen et al. [125] proposed Long Short Term Memory networks (LSTM) and LSTM Autoencoder techniques for monitoring multivariate time series data from non-stationary processes. These techniques can be also employed as efficient solutions for CCPR problems involving auto-correlated non-stationary data for future study. Section 7 of the current chapter provides a good discussion for bearing failure anomaly detection based on the LSTM method. Finally, Explainable AI techniques (see Rudin [126]) should also be used to develop frameworks for interpretation of out-of-control points in this context.

6.2 Big Data Analysis

Nowadays in smart factories, a wide range of sensors have embedded in several devices of production lines as well as are connected to many computers for data analysis, management, and visualization, which brings fruitful business results in the long run. These advanced technologies created the concepts of high-volume, high-dimensional, and high-velocity data that are called in brief Big data. This type of data always has complex natures as well as often has hierarchical or nested structures. In such situations, traditional SPC methods are incapable of monitoring such data and existing methodologies should be stretched to new limits. Although data-driven ML algorithms have a good potential to do this end, there is poor literature about ML-based studies for AD and pattern recognition using data sets that would be considered big data [127–130]. Thus, there is a tremendous opportunity and significant need for the development of advanced ML tools for both AD and CCPR. For example, Qiu [111] provided a comprehensive discussion on some recent SPC methods in the presence of big data and recommended the following research directions as further research. (i) feature selection methods have been suggested by some authors to simplify the computations for monitoring big data sets. In such cases, the key questions that have not been properly addressed yet in the literature and future research is needed in this direction are a) what kind(s) of features are appropriate to use for a specific big data monitoring problem, b) how many features should be extracted for process monitoring, and c) whether the original goals of process monitoring have been substantially compromised by using the selected features. (ii) process observations in SPC literature are widely assumed to be either independent or following some specific parametric time series models such as ARMA models. These assumptions are rarely satisfied in practice, especially when we are dealing with big data sets. More precisely, in the context of big data, process observations have many other complicated structures. Thus, developing SPC tools to properly accommodate such data structures will be an attractive research area. (iii) In practice, the performance of a process is often affected by various covariates that can provide some helpful information to us. Therefore, taking this information into account in developing and designing new SPC tools can improve the efficiency of the monitoring procedures. However, there is no study in the SPC literature yet regarding the proper use of helpful information in covariates. All these topics can also be investigated based on ML methods for both AD, interpreting out-of-control signals, and CCPR as well. More discussions on these topics can be found in Megahed and Jones-Farmer [131] and Reis and Gins [113].

6.3 Real Word Implementation and Hyperparameters Optimization

Scientists and engineers believe that we are at the beginning of the fourth industrial revolution that is being called Industry 4.0. Broadly speaking, this revolution has been happening by decreasing human intervention and adding intelligence into the production processes and service operations. Digitalization and computerization enable manufacturers/managers to make their own smart factories/companies as a unified digital ecosystem of all the different works aspects using advanced technologies to organize and optimize their production/service cycles. In this situation, companies and corporates have begun to adapt and implement the state-of-the-art into daily operations to improve production efficiency, flexibility, and reduce cost (See for example Malaca et al. [132]). On the other hand, Woodall and Montgomery [133] express that "Despite the large number of papers on this topic we have not seen much practical impact on SPC". This is an unacceptable face of SPC literature that may occur because of several reasons. Weese et al. [34] also stated that there are very few discussions in the related literature that, (i) address the step-by-step procedures of selection and operation of algorithm in practical situations, (ii) conduct an illustrative Phase I analysis, and (iii) provide some advice on how to apply the methods in practice, including how to establish an in-control training sample or how large training data size is needed to algorithm learned effectively. As an example of concerns (i) and (iii), one of the most important stages of ML-based algorithms implementation is hyperparameter optimization which is also known as hyperparameter tuning. Hyperparameters are those that lead to the highest accuracy and/or least error in the validation set and provide the best results for the problem they are solving. It is important to note that the hyperparameter is different from the model parameter. Hyperparameters are the model arguments that should be determined before the learning process begins and they are not learned from the training data like model parameters. For example, K in KNN, kernel type and constants in SVMs, number of layers, and neurons in NN are some of the well-known hyperparameters. These hyperparameters can be determined by maximizing (e.g. the accuracy) or minimizing (e.g. the loss function) specified metrics. Although these hyperparameters play a crucial role in utilizing ML algorithms that the effectiveness of the algorithm largely depends on selecting good hyperparameter values, surprisingly, most of the studies applying ML in SPC have not considered hyperparameter optimization in their studies. Bochinski et al. [134] proposed an evolutionary algorithm-based hyper-parameter optimization approach for committees of multiple CNNs. Trittenbach et al. [135] developed a principled approach using the Local Active Min-Max Alignment method to estimate SVDD hyperparameter values by active learning. In a ML-based SPC investigation, Trinh et al. [136] investigated the application of one-class SVM to detect anomalies in wireless sensor networks with data-driven hyperparameter optimization. Also, Wu et al. [137] proposed an effective technique for Hyperparameter tuning using reinforcement learning. Based on the above-mentioned discussions, there is a large gap between the theories and assumptions in literature and real demands in industrial environments that should be reduced through future research. Accordingly, it is recommended to scientists for providing illustrative guidelines for probably non-specialist practitioners to show them clearly how to implement the method in their problems. Moreover, it is also important to prepare the source code of test designs using ML because most of them have no explicit expression of control limits and ARL. This would make the implantation of the proposed methods easy for practitioners.

6.4 Integration of SVM and NN Techniques

It is known that both SVM and NN are powerful ML algorithms in the AD and pattern recognition contexts because of their impressive results which are reported in many references. However, each of them has its advantages and disadvantage. For example, the structural risk minimization of SVMs benefits their performance, in contrast with the empirical risk minimization of NNs, which creates problems. While NNs try to minimize the training error, the SVMs minimize an upper bound of the error, something that enables them to generalize easier even when the dataset is small. Furthermore, SVMs find a global solution and cannot be stuck in local minima, in contrast with the NNs [35]. So, their combination may lead to aggregation of benefits to serve as a unified attractive tool for ML-based SPC activities. For example in an AD problem, one might utilize a deep NN and have the final classification via SVM at the output layer. It is likely to have better classification results compared to ordinary NN. Recently, Hosseini and Zade [138] suggested a new hybrid technique called the MGA-SVM-HGS-PSO-NN model for detection of a malicious attack on computer networks by combining SVM and NN techniques. Their proposed method includes two stages, a feature selection stage and an attack detection stage. The feature selection process was performed using SVM and a Genetic Algorithm (GA). On the other side, the attack detection process was performed using a NN approach. The performance of the MGA-SVM-HGS-PSO-NN method was compared with other popular techniques such as Chi-SVM, NN based on gradient descent and decision tree, and NN based on GA based on performance metrics like classification accuracy, training time, the number of selected features, and testing time on the basis of the well-known NSL-KDD dataset. They showed that the proposed method is the best performing method on all criteria. For example, the proposed MGA-SVM-HGS-PSO-NN method can attain a maximum detection accuracy of 99.3%, dimension reduction of NSL-KDD from 42 to 4 features, and needs only 3 s as maximum training time. In a good review paper on ML Kernel Methods in SPC, Apsemidis et al. [35] showed that while 43% of the papers compare the SVM and NN algorithms and in 51.9% there is no reference of NN in the SVM method, only 5.1% of the cases the SVM and NN are combined to work together in the proposed method. Thus, there is a large room here for developing new methods and improving existing ML-based AD and CCPR models. For instance, investigating the possible design of control charts for monitoring stationary and non-stationary multivariate time series data with LSTM or Autoencoder CNN combined with SVDD technique can be considered as a good research topic (see Tran et al. [124] and Nguyen et al. [125]).

6.5 ML Algorithm in the Presence of Drift

One of the assumptions in supervised learning is that the mapping function f is assumed to be static, meaning that it does not change over time. However, in some problems, but not all problems, this assumption may not hold true. It means that the structure of data can change over time and hence the relation between input and output would be dynamic. This phenomenon in the ML literature known as concept drift and may happen due to several reasons. Ignoring concept drift while we are selecting and learning the predictive model can affect the prediction power of the algorithm. To tackle this problem, many adaptive learning techniques have been proposed by researchers like Žliobaitė et al. [139] and Gama et al. [140]. However, to the best of our knowledge, there is no study for ML-based control charts, pattern recognition, and interpreting out-of-control signals by considering the concept drift. So, there is a large potential here for more researches.

6.6 Data Fusion and Feature Fusion

Data fusion and features are newly developed fields in data science that deal with the problem of the integration of data and knowledge from multiple sources and reducing the features' space of raw data, respectively. This technique can improve available information of data in the sense of decreasing the associated cost, increasing the data quality and veracity, gathering more related information, and increasing the accuracy of ML-based tools. Especially, it can be useful in smart factories with multisensor environments. For example, the main advantages of data fusion are discussed in more detail in the biosurveillance area by Shmueli and Fienberg [141] (pp. 123–133). Castanedo [142] classified the data fusion techniques into three main categories as, (i) data association, (ii) state estimation, and (iii) decision fusion. In addition, feature fusion techniques can improve the ability of mixture CCPRs. Recently, Zhang et al. [143] proposed a CCPR model based on fusion feature reduction (FFR), which makes the features more effective, and fireworks algorithm-optimized MSVM. They showed that the proposed method can significantly improve the recognition accuracy and the recognition rate and the run time of CCPR as well as deliver satisfying prediction results even with relatively small-sized training samples. Another CCPR technique based on the features fusion approach is presented in Zhang et al. [144]. In conclusion, developing new and refining existing ML-based control charts and ML-based CCPR models, as well as interpreting out-of-control techniques based on data fusion and features fusion methods are good directions for future research of the scientist in the field of this chapter [34].

6.7 Control Chart for Complex Data Types

Data in the smart factories are nowadays collected with a high frequency, high dimension, complex structure, and large variety which cannot be treated straightforwardly. These new circumstances create the concept of complex data. Functional data, compositional data, and topological data are some important types of complex data. To handle such data, new data analysis methods have developed or the existing techniques have refined by some researchers. For example, Topological Data Analysis (TDA) has proposed to analyze topological data that emerges as a powerful tool to extract insights from high-dimensional data. The core idea of TDA is to find the shape, the underlying structure of shapes, or relevant low dimensional features of complex structure and massive data. In Umeda et al. [145], the application of TDA is used to describe the time-series DL for analyzing time series data and AD. In particular, two key technologies-Mapper and persistent homology are applied in both supervised learning and unsupervised learning. Mapper presents the distinguishing features of a set of data as an easy-to-understand graph. Persistent homology is a technology that numerically captures a data shape in detail. This paper developed an AD technology for time-series to detect an abnormal state using TDA. Besides that, the data is becoming more and more related to functional data. The studies on monitoring functional data have drawn a lot of attention Colosimo and Pacella [146], Liu et al. [147], and Flores et al. [148]. AD methods for functional data based on functional PCA Yu et al. [149], wavelet functional PCA Liu et al. [147] are developed. However, the application of advanced ML on this type of data for development, pattern recognition, and interpreting of control charts still needs to be discovered. Thus, more efforts are needed to develop tests that use ML to track these types of data, need to find more documentation on ML methods suitable for them in order to write them correctly. For instance, although these studies have eliminated a lot of assumptions about the distribution of data when designing control charts with ML techniques, there are still independent data assumptions that do not exist in the data environment collected from Internet of Things (IoT) sensors. In general, there are still very few studies on this promising approach and further researches needs to be carried out to discover its numerous applications to the smart factory. Accordingly, developing advanced ML techniques to eliminate most of the assumptions of traditional SPC in development, pattern recognition, and interpreting of control charts for monitoring complex data types such as multivariate time series data, image data, and Big Data with complex structures is a high-potential area to carry out more researches. This will be a promising research direction to solve the problem of smart factory SPC implementation with Big Data.

6.8 Monitoring Image Data

Although applications of MVSs in industrial and medical shop floors have been increased dramatically and a huge number of possible applications exist here, but there are only a few papers dealing with image monitoring. Megahed et al. [115] reviewed image-based control charts including univariate, multivariate, profile, spatial, multivariate image analysis, and medical image devices charts and addressed the capability of image-based monitoring in a much wider variety of quality characteristics. They noted that the use of image-based control charts differs from traditional applications of control charts in the SPC area. These differences can be attributed to a number of factors, which include the type of data being monitored, the rationale behind using control charts, and how the control charts are applied. Additionally, preprocessing of image data can also become a factor with 100% inspection since the data preprocessing time can be longer than the production cycle time. Therefore, these factors need to be considered when developing the control charting strategy. He et al. [150] proposed a multivariate control charting method for both single and multiple faults. In their method, each image is divided into non-overlapping regions of equal size, and the mean intensities of these regions are monitored with a multivariate Generalized Likelihood Ratio (GLR)-based statistic. Later, by extending the results of He et al. [151], Stankus and Castillo-Villar [152] developed a multivariate GLR control chart to identify process shifts and locate defects on artifacts by converting 3D point cloud data to a 2D image. They considered the surface dent in addition to two ordinary types of defects, surface curvature, and surface scratch, that does not identify by the existing methodologies. By means of a comparative study, Stankus and Castillo-Villar [152] showed that the new methodology has a significantly shorter out-of-control ARL than the He et al. [151] methodology for the scratch and no significant difference in out-of-control ARL for the incorrect surface curvature. Zuo et al. [116] reported that the existing research in the image-based SPC area has focused on either identification of fault size and/or location or detection of fault occurrence and there is limited research on both fault detection and identification. To handle such situations, they proposed an EWMA and region growing based control chart for monitoring of 8-bit grayscale images of industrial products. The results of the simulation study showed that the new method is not only effective in quick detection of the fault but also accurate in estimating the fault size and location. Recently, Okhrin et al. [153] provided an overview of recent developments on monitoring image processes. While we review some existing literature in this field, there are still some research opportunities in the integration of ML-based control charting methods and pattern recognition models with image data. It is known that with smart manufacturing, the amount of images collected from production lines is very big and each image may include millions of pixels that need ML approaches to develop new control charts and CCPR frameworks. Many authors assume an independent residual process, while there is a natural spatial correlation structure of the pixels in neighborhoods. Therefore, there is a consequent need for some ML-based approaches for the successful monitoring of image processes. The existing methods, for instance, Okhrin et al. [154] and Yuan and Lin [155], can be improved to developing CNN and Transformers control charts to monitoring images in SM.

7 A Case Study: Monitoring and Early Fault Detection in Bearing

In this section, we present an application of ML based control chart for monitoring and early fault detection in bearing. AD in vibration signals is an important technique for monitoring, early detection of the failure, and fault diagnosis for rotating machinery. Very recently, Tran et al. [124], Tran et al. [156] and, Nguyen et al. [125] have developed very efficient methods with LSTM and LSTM Autoencoder techniques in detecting anomalies for multivariate time series data. In this case study, we will combine both of these methods to propose a new ML based control chart that performs AD in an industry context. According to Nguyen et al. [125], we suppose that the autoencoder LSTM has been trained from a normal sequence $\{\mathbf{x}_1, \mathbf{x}_2, \ldots, \mathbf{x}_N\}$, where N is the number of samples and $\mathbf{x}_t = \{x_t^{(1)}, x_t^{(2)}, ..., x_t^{(k)}\}, t = 1, 2, \ldots$ is the value of the multivariate time series at the time t with k number of variables (these notations are from previous section). Using a sliding window of size m, the trained autoencoder LSTM can read the input sequence $\mathbf{X}_i = \mathbf{x}_t, \ldots, \mathbf{x}_{t-m+1}$, encode it and recreate it in the output $\hat{\mathbf{X}}_i = (\hat{\mathbf{x}}_t, \ldots, \hat{\mathbf{x}}_{t-m+1})$, with $i = m + 1, \ldots, N$.. Since these values has been observed from the data, one can calculate the prediction error $e_i = \|\hat{\mathbf{X}}_i - \mathbf{X}_i\|$, $i = m + 1, \ldots, N$. The anomaly detection is then based on these prediction errors. The anomaly scores distribution of the training dataset is shown in Fig. 5. In many studies, these error vectors are supposed that follow a Gaussian distribution and then used the maximum likelihood estimation method to estimate the parameters of this distribution. However, one can argue that the assumption of Gaussian distribution for error vectors may not be true in practice. To overcome the disadvantage of this method, Tran et al. [124] proposed used the kernel quantile estimation (KQE) control chart [157] to automatically determines a threshold for time series AD. In particular, at the new time t, if $e_t > \tau$, x_t is classified as anomaly point and vice versa, see Tran et al. [124] for more details.

The experimental data were generated from a bearing test rig that was able to produce run-to-failure data. These data were downloaded from the Prognostics Center of Excellence (PCoE) through a prognostic data repository contributed by Intelligent Maintenance System (IMS), University of Cincinnati [158]. According to [158], vibrations signals were collected every 10 min with a data sampling rate was 20 kHz and the data length was 20 480 sensor data points.

This ML-based control chart allows for conditional monitoring and prediction of the upcoming bearing malfunction well in advance of the actual physical failure. It allows to automatically define a threshold value for flagging anomalies while avoiding too many false positives during normal operating conditions. The early detection of bearing failure is shown in the Fig. 6, the bearing failure is confirmed at the end of this experiment (Qiu et al. [158]). This promising approach could provide a perfect tool to enable predictive maintenance implementation in SM.

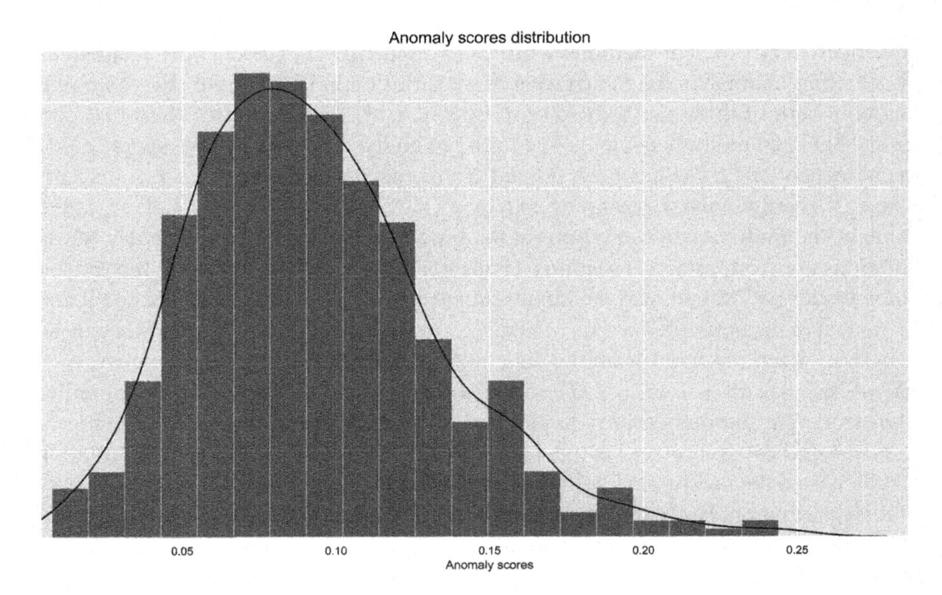

Fig. 5 Anomaly scores distribution of the training dataset

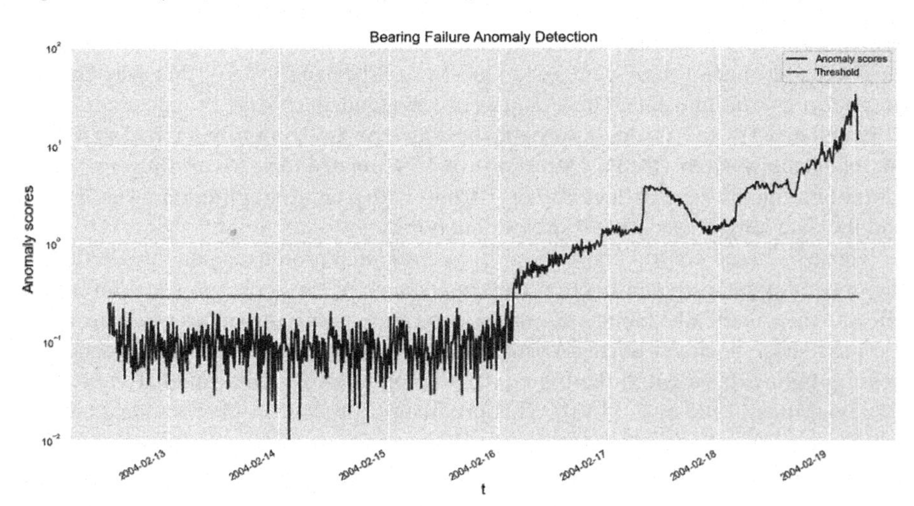

Fig. 6 Bearing failure anomaly detection

8 Conclusion

Along with the development of technologies and AI, leading to production systems become more complex and modern-day by day. Therefore, the application of ML to SPC is an interesting and necessary trend that has been strongly developed in recent years to meet the needs of SM. In this chapter, we have introduced different applications of ML in control chart implementation including designing, recognition trend, and interpreting. A literature review about these issues is discussed. Although there have been many achievements in research in this field, there are still many difficulties and problems that need to be solved in order to be able to apply control charts to SM. There still exists a significant potential for reducing the gap between theory and application in modern industries. A case study is also provided to present a ML-based control chart for monitoring and early fault detection in bearing.

References

1. Kadri F, Harrou F, Chaabane S, Sun Y, Tahon C (2016) Seasonal ARMA-based SPC charts for anomaly detection: application to emergency department systems. Neurocomputing 173:2102–2114
2. Münz G, Carle, G (2008) Application of forecasting techniques and control charts for traffic anomaly detection. In: Proceedings of the 19th ITC specialist seminar on network usage and traffic
3. Tran PH, Tran KP, Truong TH, Heuchenne C, Tran H, Le TMH (2018) Real time data-driven approaches for credit card fraud detection. In: Proceedings of the 2018 international conference on e-business and applications, pp 6–9
4. Tran PH, Heuchenne C, Nguyen HD, Marie H (2020, in press) Monitoring coefficient of variation using one-sided run rules control charts in the presence of measurement errors. J Appl Stat 1–27. https://doi.org/10.1080/02664763.2020.1787356
5. Tran PH, Heuchenne C (2021) Monitoring the coefficient of variation using variable sampling interval CUSUM control charts. J Stat Comput Simul 91(3):501–521
6. Chandola V, Banerjee A, Kumar V (2009) Anomaly detection: a survey. ACM Comput Surv (CSUR) 41(3):1–58
7. Edgeworth FY (1887) XLI. on discordant observations. London Edinburgh Dublin Philos Mag J Sci 23(143):364–375
8. Shewhart WA (1924) Some applications of statistical methods to the analysis of physical and engineering data. Bell Syst Tech J 3(1):43–87
9. Alwan LC (1992) Effects of autocorrelation on control chart performance. Commun Stat Theory Methods 21(4):1025–1049
10. Noorossana R, Vaghefi SJM (2006) Effect of autocorrelation on performance of the MCUSUM control chart. Qual Reliab Eng Int 22(2):191–197
11. Costa AFB, Castagliola P (2011) Effect of measurement error and autocorrelation on the $\bar{X}$ chart. J Appl Stat 38(4):661–673
12. Leoni RC, Costa AFB, Machado MAG (2015) The effect of the autocorrelation on the performance of the T2 chart. Eur J Oper Res 247(1):155–165
13. Vanhatalo E, Kulahci M (2015) The effect of autocorrelation on the hotelling T2 control chart. Qual Reliab Eng Int 31(8):1779–1796
14. Guh RS, Hsieh YC (1999) A neural network based model for abnormal pattern recognition of control charts. Comput Ind Eng 36(1):97–108

15. Swift JA, Mize JH (1995) Out-of-control pattern recognition and analysis for quality control charts using lisp-based systems. Comput Ind Eng 28(1):81–91
16. Guo Y, Dooley KJ (1992) Identification of change structure in statistical process control. Int J Prod Res 30(7):1655–1669
17. Miao Z, Yang M (2019) Control chart pattern recognition based on convolution neural network. In: Smart innovations in communication and computational sciences. Springer, pp 97–104
18. Zan T, Liu Z, Wang H, Wang M, Gao X (2020) Control chart pattern recognition using the convolutional neural network. J Intell Manuf 31(3):703–716
19. Wang TY, Chen LH (2002) Mean shifts detection and classification in multivariate process: a neural-fuzzy approach. J Intell Manuf 13(3):211–221
20. Low C, Hsu CM, Yu FJ (2003) Analysis of variations in a multi-variate process using neural networks. Int J Adv Manuf Technol 22(11):911–921
21. Niaki STA, Abbasi B (2005) Fault diagnosis in multivariate control charts using artificial neural networks. Qual Reliab Eng Int 21(8):825–840
22. Western E (1956) Statistical quality control handbook. Western Electric Co
23. Swift JA (1987) Development of a knowledge based expert system for control chart pattern recognition and analysis. PhD thesis, Oklahoma State University
24. Shewhart M (1992) Interpreting statistical process control (SPC) charts using machine learning and expert system techniques. In: Proceedings of the IEEE 1992 national aerospace and electronics conference@ m_NAECON 1992. IEEE, pp 1001–1006
25. Hotelling H (1947) Multivariate quality control. Techniques of statistical analysis
26. Woodall WH, Ncube MM (1985) Multivariate CUSUM quality-control procedures. Technometrics 27(3):285–292
27. Lowry CA, Woodall WH, Champ CW, Rigdon SE (1992) A multivariate exponentially weighted moving average control chart. Technometrics 34(1):46–53
28. Demircioglu Diren D, Boran S, Cil I (2020) Integration of machine learning techniques and control charts in multivariate processes. Scientia Iranica 27(6):3233–3241
29. Guh RS, Tannock JDT (1999) Recognition of control chart concurrent patterns using a neural network approach. Int J Prod Res 37(8):1743–1765
30. Wu KL, Yang MS (2003) A fuzzy-soft learning vector quantization. Neurocomputing 55(3–4):681–697
31. Cheng CS, Lee HT (2016) Diagnosing the variance shifts signal in multivariate process control using ensemble classifiers. J Chin Inst Eng 39(1):64–73
32. Kang Z, Catal C, Tekinerdogan B (2020) Machine learning applications in production lines: a systematic literature review. Comput Ind Eng 149:106773
33. Qiu P, Xie X (2021, in press) Transparent sequential learning for statistical process control of serially correlated data. Technometrics 1–29. https://doi.org/10.1080/00401706.2021.1929493
34. Weese M, Martinez W, Megahed FM, Jones-Farmer LA (2016) Statistical learning methods applied to process monitoring: an overview and perspective. J Qual Technol 48(1):4–24
35. Apsemidis A, Psarakis S, Moguerza JM (2020) A review of machine learning kernel methods in statistical process monitoring. Comput Ind Eng 142:106376
36. Mashuri M, Haryono H, Ahsan M, Aksioma DF, Wibawati W, Khusna H (2019) Tr r2 control charts based on kernel density estimation for monitoring multivariate variability process. Cogent Eng 6(1):1665949
37. Chinnam RB (2002) Support vector machines for recognizing shifts in correlated and other manufacturing processes. Int J Prod Res 40(17):4449–4466
38. Byvatov E, Sadowski J, Fechner U, Schneider G (2003) Comparison of support vector machine and artificial neural network systems for drug/nondrug classification. J Chem Inf Comput Sci 43(6):1882–1889
39. Li L, Jia H (2013) On fault identification of MEWMA control charts using support vector machine models. In: International Asia conference on industrial engineering and management innovation (IEMI2012) proceedings. Springer, pp 723–730

40. Camci F, Chinnam RB (2008) General support vector representation machine for one-class classification of non-stationary classes. Pattern Recogn 41(10):3021–3034
41. Sun R, Tsung F (2003) A kernel-distance-based multivariate control chart using support vector methods. Int J Prod Res 41(13):2975–2989
42. Ning X, Tsung F (2013) Improved design of kernel distance-based charts using support vector methods. IIE Trans 45(4):464–476
43. Sukchotrat T, Kim SB, Tsung F (2009) One-class classification-based control charts for multivariate process monitoring. IIE Trans 42(2):107–120
44. Kim SB, Jitpitaklert W, Sukchotrat T: One-class classification-based control charts for monitoring autocorrelated multivariate processes. Commun Stat-Simul Comput® 39(3):461–474 (2010)
45. Gani W, Limam M (2013) Performance evaluation of one-class classification-based control charts through an industrial application. Qual Reliab Eng Int 29(6):841–854
46. Gani W, Limam M (2014) A one-class classification-based control chart using the-means data description algorithm. J Qual Reliab Eng 2014. https://www.hindawi.com/journals/jqre/2014/239861/
47. Maboudou-Tchao EM, Silva IR, Diawara N (2018) Monitoring the mean vector with Mahalanobis kernels. Qual Technol Quant Manag 15(4):459–474
48. Zhang J, Li Z, Chen B, Wang Z (2014) A new exponentially weighted moving average control chart for monitoring the coefficient of variation. Comput Ind Eng 78:205–212
49. Wang FK, Bizuneh B, Cheng XB (2019) One-sided control chart based on support vector machines with differential evolution algorithm. Qual Reliab Eng Int 35(6):1634–1645
50. He S, Jiang W, Deng H (2018) A distance-based control chart for monitoring multivariate processes using support vector machines. Ann Oper Res 263(1):191–207
51. Maboudou-Tchao EM (2020) Change detection using least squares one-class classification control chart. Qual Technol Quant Manag 17(5):609–626
52. Salehi M, Bahreininejad A, Nakhai I (2011) On-line analysis of out-of-control signals in multivariate manufacturing processes using a hybrid learning-based model. Neurocomputing 74(12):2083–2095. ISSN 0925-2312
53. Hu S, Zhao L (2015) A support vector machine based multi-kernel method for change point estimation on control chart. In: 2015 IEEE international conference on systems, man, and cybernetics, pp 492–496
54. Gani W, Taleb H, Limam M (2010) Support vector regression based residual control charts. J Appl Stat 37(2):309–324
55. Kakde D, Peredriy S, Chaudhuri A (2017) A non-parametric control chart for high frequency multivariate data. In: 2017 annual reliability and maintainability symposium (RAMS). IEEE, pp 1–6
56. Jang S, Park SH, Baek JG (2017) Real-time contrasts control chart using random forests with weighted voting. Expert Syst Appl 71:358–369. ISSN 0957-4174
57. Issam BK, Mohamed L (2008) Support vector regression based residual MCUSUM control chart for autocorrelated process. Appl Math Comput 201(1):565–574. ISSN 0096-3003
58. Du S, Huang D, Lv J (2013) Recognition of concurrent control chart patterns using wavelet transform decomposition and multiclass support vector machines. Comput Ind Eng 66(4):683–695. ISSN 0360-8352
59. Silva J, Lezama OBP, Varela N, Otero MS, Guiliany JG, Sanabria ES, Rojas VA (2019) U-control chart based differential evolution clustering for determining the number of cluster in k-means. In: International conference on green, pervasive, and cloud computing. Springer, pp 31–41
60. Thirumalai C, SaiSharan GV, Krishna KV, Senapathi KJ (2017) Prediction of diabetes disease using control chart and cost optimization-based decision. In: 2017 International conference on trends in electronics and informatics (ICEI), pp 996–999
61. Stefatos G, Hamza AB (2007) Statistical process control using kernel PCA. In: 2007 Mediterranean conference on control & automation. IEEE, pp 1–6

62. Phaladiganon P, Kim SB, Chen VCP, Jiang W (2013) Principal component analysis-based control charts for multivariate nonnormal distributions. Expert Syst Appl 40(8):3044–3054. ISSN 0957-4174
63. Kullaa J (2003) Damage detection of the z24 bridge using control charts. Mech Syst Signal Process 17(1):163–170. ISSN 0888-3270
64. Lee JM, Yoo CK, Choi SW, Vanrolleghem PA, Lee IB (2004) Nonlinear process monitoring using kernel principal component analysis. Chem Eng Sci 59(1):223–234. ISSN 0009-2509
65. Ahsan M, Khusna H, Mashuri M, Lee MH (2020) Multivariate control chart based on kernel PCA for monitoring mixed variable and attribute quality characteristics. Symmetry 12(11):1838
66. Ahsan M, Prastyo DD, Mashuri M, Kuswanto H, Khusna H (2018) Multivariate control chart based on PCA mix for variable and attribute quality characteristics. Prod Manuf Res 6(1):364–384
67. Mashuri M, Ahsan M, Prastyo DD, Kuswanto H, Khusna H (2021) Comparing the performance of t^2 chart based on PCA mix, kernel PCA mix, and mixed kernel PCA for network anomaly detection. J Phys Conf Ser 1752:012008
68. Lee WJ, Triebe MJ, Mendis GP, Sutherland JW (2020) Monitoring of a machining process using kernel principal component analysis and kernel density estimation. J Intell Manuf 31(5):1175–1189
69. Arkat J, Niaki STA, Abbasi B (2007) Artificial neural networks in applying MCUSUM residuals charts for AR(1) processes. Appl Math Comput 189(2):1889–1901 ISSN 0096-3003
70. Lee S, Kwak M, Tsui KL, Kim SB (2019) Process monitoring using variational autoencoder for high-dimensional nonlinear processes. Eng Appl Artif Intell 83:13–27
71. Chen S, Yu J (2019) Deep recurrent neural network-based residual control chart for autocorrelated processes. Qual Reliab Eng Int 35(8):2687–2708
72. Niaki STA, Abbasi B (2005) Fault diagnosis in multivariate control charts using artificial neural networks. Qual Reliab Eng Int 21(8):825–840
73. Chen P, Li Y, Wang K, Zuo MJ, Heyns PS, Baggerohr S (2021) A threshold self-setting condition monitoring scheme for wind turbine generator bearings based on deep convolutional generative adversarial networks. Measurement 167:108234 ISSN 0263-2241
74. Pugh GA (1989) Synthetic neural networks for process control. Comput Ind Eng 17(1):24–26 ISSN 0360-8352
75. Li TF, Hu S, Wei ZY, Liao ZQ (2013) A framework for diagnosing the out-of-control signals in multivariate process using optimized support vector machines. Math Probl Eng 2013. https://www.hindawi.com/journals/mpe/2013/494626/
76. Guh RS (2008) Real-time recognition of control chart patterns in autocorrelated processes using a learning vector quantization network-based approach. Int J Prod Res 46(14):3959–3991
77. Zaman M, Hassan A (2021) Fuzzy heuristics and decision tree for classification of statistical feature-based control chart patterns. Symmetry 13(1):110 ISSN 2073-8994
78. Hachicha W, Ghorbel A (2012) A survey of control-chart pattern-recognition literature (1991–2010) based on a new conceptual classification scheme. Comput Ind Eng 63(1):204–222 ISSN 0360-8352
79. Pham DT, Oztemel E (1993) Control chart pattern recognition using combinations of multi-layer perceptrons and learning-vector-quantization neural networks. Proc Inst Mech Eng Part I J Syst Control Eng 207(2):113–118
80. Cheng CS (1997) A neural network approach for the analysis of control chart patterns. Int J Prod Res 35(3):667–697
81. Addeh A, Khormali A, Golilarz NA (2018) Control chart pattern recognition using RBF neural network with new training algorithm and practical features. ISA Trans 79:202–216
82. Yu J, Zheng X, Wang S (2019) A deep autoencoder feature learning method for process pattern recognition. J Process Control 79:1–15
83. Xu J, Lv H, Zhuang Z, Lu Z, Zou D, Qin W (2019) Control chart pattern recognition method based on improved one-dimensional convolutional neural network. IFAC-PapersOnLine 52(13):1537–1542

84. Yang WA, Zhou W (2015) Autoregressive coefficient-invariant control chart pattern recognition in autocorrelated manufacturing processes using neural network ensemble. J Intell Manuf 26:1161–1180
85. Fuqua D, Razzaghi T (2020) A cost-sensitive convolution neural network learning for control chart pattern recognition. Expert Syst Appl 150:113275
86. Pham DT, Wani MA (1997) Feature-based control chart pattern recognition. Int J Prod Res 35(7):1875–1890
87. Ranaee V, Ebrahimzadeh A, Ghaderi R (2010) Application of the PSO-SVM model for recognition of control chart patterns. ISA Trans 49(4):577–586
88. Lu CJ, Shao YE, Li, PH (2011) Mixture control chart patterns recognition using independent component analysis and support vector machine. Neurocomputing 74(11):1908–1914. ISSN 0925–2312. Adaptive Incremental Learning in Neural Networks Learning Algorithm and Mathematic Modelling Selected papers from the International Conference on Neural Information Processing 2009 (ICONIP 2009)
89. Lin SY, Guh RS, Shiue YR (2011) Effective recognition of control chart patterns in autocorrelated data using a support vector machine based approach. Comput Ind Eng 61(4):1123–1134
90. Xanthopoulos P, Razzaghi T (2014) A weighted support vector machine method for control chart pattern recognition. Comput Ind Eng 70:134–149 ISSN 0360-8352
91. Wang X (2008) Hybrid abnormal patterns recognition of control chart using support vector machining. In: 2008 international conference on computational intelligence and security, vol 2, pp 238–241
92. Ranaee V, Ebrahimzadeh A (2011) Control chart pattern recognition using a novel hybrid intelligent method. Appl Soft Comput 11(2):2676–2686. ISSN 1568-4946. The Impact of Soft Computing for the Progress of Artificial Intelligence
93. Zhou X, Jiang P, Wang X (2018) Recognition of control chart patterns using fuzzy SVM with a hybrid kernel function. J Intell Manuf 29(1):51–67
94. De la Torre Gutierrez H, Pham DT (2016) Estimation and generation of training patterns for control chart pattern recognition. Comput Ind Eng 95:72–82. ISSN 0360-8352
95. Chen LH, Wang TY (2004) Artificial neural networks to classify mean shifts from multivariate $\chi 2$ chart signals. Comput Ind Eng 47(2–3):195–205
96. Cheng CS, Cheng HP (2008) Identifying the source of variance shifts in the multivariate process using neural networks and support vector machines. Expert Syst Appl 35(1–2):198–206
97. Guh RS, Shiue YR (2008) An effective application of decision tree learning for on-line detection of mean shifts in multivariate control charts. Comput Ind Eng 55(2):475–493
98. Yu J, Xi L, Zhou X (2009) Identifying source (s) of out-of-control signals in multivariate manufacturing processes using selective neural network ensemble. Eng Appl Artif Intell 22(1):141–152
99. Alfaro E, Alfaro JL, Gamez M, Garcia N (2009) A boosting approach for understanding out-of-control signals in multivariate control charts. Int J Prod Res 47(24):6821–6834
100. Verron S, Li J, Tiplica T (2010) Fault detection and isolation of faults in a multivariate process with Bayesian network. J Process Control 20(8):902–911
101. He SG, He Z, Wang GA (2013) Online monitoring and fault identification of mean shifts in bivariate processes using decision tree learning techniques. J Intell Manuf 24(1):25–34
102. Carletti M, Masiero C, Beghi A, Susto GA (2019) Explainable machine learning in industry 4.0: evaluating feature importance in anomaly detection to enable root cause analysis. In: 2019 IEEE international conference on systems, man and cybernetics (SMC). IEEE, pp 21–26
103. Song H, Xu Q, Yang H, Fang J (2017) Interpreting out-of-control signals using instance-based Bayesian classifier in multivariate statistical process control. Commun Stat-Simul Comput 46(1):53–77
104. Salehi M, Bahreininejad A, Nakhai I (2011) On-line analysis of out-of-control signals in multivariate manufacturing processes using a hybrid learning-based model. Neurocomputing 74(12–13):2083–2095

105. Zhao C, Sun H, Tian F (2019) Total variable decomposition based on sparse cointegration analysis for distributed monitoring of nonstationary industrial processes. IEEE Trans Control Syst Technol 28(4):1542–1549
106. Chen Q, Kruger U, Leung AYT (2009) Cointegration testing method for monitoring nonstationary processes. Ind Eng Chem Res 48(7):3533–3543
107. Ketelaere BD, Mertens K, Mathijs F, Diaz DS, Baerdemaeker JD (2011) Nonstationarity in statistical process control–issues, cases, ideas. Appl Stoch Model Bus Ind 27(4):367–376
108. Liu J, Chen DS (2010) Nonstationary fault detection and diagnosis for multimode processes. AIChE J 56(1):207–219
109. Lazariv T, Schmid W (2019) Surveillance of non-stationary processes. AStA Adv Stat Anal 103(3):305–331
110. Lazariv T, Schmid W (2018) Challenges in monitoring non-stationary time series. In: Frontiers in statistical quality control 12. Springer, pp 257–275
111. Qiu P (2020) Big data? Statistical process control can help! Am Stat 74(4):329–344
112. Tuv E, Runger G (2003) Learning patterns through artificial contrasts with application to process control. WIT Trans Inf Commun Technol 29. https://www.witpress.com/elibrary/wit-transactions-on-information-and-communication-technologies/29/1376
113. Reis MS, Gins G (2017) Industrial process monitoring in the big data/industry 4.0 era: from detection, to diagnosis, to prognosis. Processes 5(3):35
114. Capizzi G, Masarotto G (2011) A least angle regression control chart for multidimensional data. Technometrics 53(3):285–296
115. Megahed FM, Woodall WH, Camelio JA (2011) A review and perspective on control charting with image data. J Qual Technol 43(2):83–98
116. Zuo L, He Z, Zhang M (2020) An EWMA and region growing based control chart for monitoring image data. Qual Technol Quant Manag 17(4):470–485
117. Maragah HD, Woodall WH (1992) The effect of autocorrelation on the retrospective x-chart. J Stat Comput Simul 40(1–2):29–42
118. Arkat J, Niaki STA, Abbasi B (2007) Artificial neural networks in applying MCUSUM residuals charts for AR (1) processes. Appl Math Comput 189(2):1889–1901
119. Kim SB, Jitpitaklert W, Park SK, Hwang SJ (2012) Data mining model-based control charts for multivariate and autocorrelated processes. Expert Syst Appl 39(2):2073–2081
120. Cuentas S, Peñabaena-Niebles R, Garcia E (2017) Support vector machine in statistical process monitoring: a methodological and analytical review. Int J Adv Manuf Technol 91(1):485–500
121. Chinnam RB, Kumar VS (2001) Using support vector machines for recognizing shifts in correlated manufacturing processes. In: IJCNN 2001. International joint conference on neural networks. Proceedings (Cat. No. 01CH37222), vol 3. IEEE, pp 2276–2280
122. Hsu CC, Chen MC, Chen LS (2010) Integrating independent component analysis and support vector machine for multivariate process monitoring. Comput Ind Eng 59(1):145–156
123. Hsu CC, Chen MC, Chen LS (2010) Intelligent ICA-SVM fault detector for non-gaussian multivariate process monitoring. Expert Syst Appl 37(4):3264–3273
124. Tran KP, Nguyen HD, Thomassey S (2019) Anomaly detection using long short term memory networks and its applications in supply chain management. IFAC-PapersOnLine 52(13):2408–2412
125. Nguyen HD, Tran KP, Thomassey S, Hamad M (2021) Forecasting and anomaly detection approaches using LSTM and LSTM autoencoder techniques with the applications in supply chain management. Int J Inf Manag 57:102282
126. Rudin C (2019) Stop explaining black box machine learning models for high stakes decisions and use interpretable models instead. Nat Mach Intell 1(5):206–215
127. Wang K, Jiang W (2009) High-dimensional process monitoring and fault isolation via variable selection. J Qual Technol 41(3):247–258
128. Jin Y, Huang S, Wang G, Deng H (2017) Diagnostic monitoring of high-dimensional networked systems via a LASSO-BN formulation. IISE Trans 49(9):874–884
129. Qiu P (2017) Statistical process control charts as a tool for analyzing big data. In: Big and complex data analysis. Springer, pp 123–138

130. Sparks R, Chakraborti S (2017) Detecting changes in location using distribution-free control charts with big data. Qual Reliab Eng Int 33(8):2577–2595
131. Megahed FM, Jones-Farmer LA (2015) Statistical perspectives on "big data". In: Frontiers in statistical quality control 11. Springer, pp 29–47
132. Malaca P, Rocha LF, Gomes D, Silva J, Veiga G (2019) Online inspection system based on machine learning techniques: real case study of fabric textures classification for the automotive industry. J Intell Manuf 30(1):351–361
133. Woodall WH, Montgomery DC (2014) Some current directions in the theory and application of statistical process monitoring. J Qual Technol 46(1):78–94
134. Bochinski E, Senst T, Sikora T (2017) Hyper-parameter optimization for convolutional neural network committees based on evolutionary algorithms. In: 2017 IEEE international conference on image processing (ICIP). IEEE, pp 3924–3928
135. Trittenbach H, Böhm K, Assent I (2020) Active learning of SVDD hyperparameter values. In: 2020 IEEE 7th international conference on data science and advanced analytics (DSAA). IEEE, pp 109–117
136. Trinh VV, Tran KP, Huong TT (2017) Data driven hyperparameter optimization of one-class support vector machines for anomaly detection in wireless sensor networks. In: 2017 international conference on advanced technologies for communications (ATC). IEEE, pp 6–10
137. Wu J, Chen SP, Liu XY (2020) Efficient hyperparameter optimization through model-based reinforcement learning. Neurocomputing 409:381–393
138. Hosseini S, Zade BMH (2020) New hybrid method for attack detection using combination of evolutionary algorithms, SVM, and ANN. Comput Netw 173:107168
139. Žliobaitė I, Pechenizkiy M, Gama J (2016) An overview of concept drift applications. In: Big data analysis: new algorithms for a new society, pp 91–114
140. Gama J, Žliobaitė I, Bifet A, Pechenizkiy M, Bouchachia A (2014) A survey on concept drift adaptation. ACM Comput Surv (CSUR) 46(4):1–37
141. Shmueli G, Fienberg SE (2006) Current and potential statistical methods for monitoring multiple data streams for biosurveillance. In: Statistical methods in counterterrorism. Springer, pp 109–140
142. Castanedo F (2013) A review of data fusion techniques. Sci World J 2013. https://www.hindawi.com/journals/tswj/2013/704504/
143. Zhang M, Yuan Y, Wang R, Cheng W (2020) Recognition of mixture control chart patterns based on fusion feature reduction and fireworks algorithm-optimized MSVM. Pattern Anal Appl 23(1):15–26
144. Zhang M, Zhang X, Wang H, Xiong G, Cheng W (2020) Features fusion exaction and KELM with modified grey wolf optimizer for mixture control chart patterns recognition. IEEE Access 8:42469–42480
145. Umeda Y, Kaneko J, Kikuchi H (2019) Topological data analysis and its application to time-series data analysis. Fujitsu Sci Tech J 55(2):65–71
146. Colosimo BM, Pacella M (2010) A comparison study of control charts for statistical monitoring of functional data. Int J Prod Res 48(6):1575–1601
147. Liu J, Chen J, Wang D (2020) Wavelet functional principal component analysis for batch process monitoring. Chemom Intell Lab Syst 196:103897
148. Flores M, Fernández-Casal R, Naya S, Zaragoza S, Raña P, Tarrío-Saavedra J (2020) Constructing a control chart using functional data. Mathematics 8(1):58
149. Yu G, Zou C, Wang Z (2012) Outlier detection in functional observations with applications to profile monitoring. Technometrics 54(3):308–318
150. He Z, Zuo L, Zhang M, Megahed FM (2016) An image-based multivariate generalized likelihood ratio control chart for detecting and diagnosing multiple faults in manufactured products. Int J Prod Res 54(6):1771–1784
151. He K, Zuo L, Zhang M, Alhwiti T, Megahed FM (2017) Enhancing the monitoring of 3D scanned manufactured parts through projections and spatiotemporal control charts. J Intell Manuf 28(4):899–911

152. Stankus SE, Castillo-Villar KK (2019) An improved multivariate generalised likelihood ratio control chart for the monitoring of point clouds from 3D laser scanners. Int J Prod Res 57(8):2344–2355
153. Okhrin Y, Schmid W, Semeniuk I (2019) Monitoring image processes: overview and comparison study. In: International workshop on intelligent statistical quality control. Springer, pp 143–163
154. Okhrin Y, Schmid W, Semeniuk I (2020) New approaches for monitoring image data. IEEE Trans Image Process 30:921–933
155. Yuan Y, Lin L (2020) Self-supervised pre-training of transformers for satellite image time series classification. IEEE J Sel Top Appl Earth Obs Remote Sens 14:474–487
156. Tran PH, Heuchenne C, Thomassey S (2020) An anomaly detection approach based on the combination of LSTM autoencoder and isolation forest for multivariate time series data. In: Proceedings of the 14th international FLINS conference on robotics and artificial intelligence (FLINS 2020). World Scientific, pp 18–21
157. Sheather SJ, Marron JS (1990) Kernel quantile estimators. J Am Stat Assoc 85(410):410–416
158. Qiu H, Lee J, Lin J, Yu G (2006) Wavelet filter-based weak signature detection method and its application on rolling element bearing prognostics. J Sound Vib 289(4–5):1066–1090
159. Yu J, Zheng X, Wang S (2019) Stacked denoising autoencoder-based feature learning for out-of-control source recognition in multivariate manufacturing process. Q Reliab Eng Int 35(1):204–223

Control Charts for Monitoring Time-Between-Events-and-Amplitude Data

Philippe Castagliola, Giovanni Celano, Dorra Rahali, and Shu Wu

Abstract In recent years, several control charts have been developed for the simultaneous monitoring of the time interval T and the amplitude X of events, denoted as the TBEA (Time Between Events and Amplitude) charts. In general, a decrease in T and/or an increase in X can result in a negative, hazardous or disastrous situation that needs to be efficiently monitored with control charts. The goal of this chapter is to further investigate several TBEA control charts and to hopefully open new research directions. More specifically, this chapter will (1) introduce and compare three different statistics, denoted as Z_1, Z_2 and Z_3, suitable for monitoring TBEA data, in the case of four distributions (gamma, lognormal, normal, and Weibull), when the time T and the amplitude X are considered as *independent*, (2) compare the three statistics introduced in (1) for the same distributions, but considering that the time T and the amplitude X are *dependent* random variables and the joint distribution can be represented using Copulas and (3) introduce a distribution-free approach for TBEA data coupled with an upper-sided EWMA scheme in order to overcome the "distribution choice" dilemma. Two illustrative examples will be presented to clarify the use of the proposed methods.

Keywords Attribute control charts · Binomial AR(1) · Integer-valued time series · Statistical process monitoring

---

P. Castagliola (✉)
Université de Nantes and LS2N UMR CNRS 6004, Nantes, France
e-mail: philippe.castagliola@univ-nantes.fr

G. Celano
Università di Catania, Catania, Italy
e-mail: giovanni.celano@unict.it

D. Rahali
Centre de Recherche en Informatique, Signal et Automatique de Lille, Lille, France
e-mail: dorra.rahali@centralelille.fr

S. Wu
School of Transportation and Logistics Engineering, Wuhan University of Technology, Wuhan, China

K. P. Tran (ed.), *Control Charts and Machine Learning for Anomaly Detection in Manufacturing*, Springer Series in Reliability Engineering,
https://doi.org/10.1007/978-3-030-83819-5_3

1 Introduction

Today, due to the large availability of data, various kinds of processes can (and have to) be monitored using Statistical Process Monitoring (SPM) techniques based on advanced control charts. These kinds of processes can be of course industrial ones, but they can also be non industrial ones like in the biological/health-care (diseases, like the Covid-19 for instance), the geological (earthquakes or volcanic eruptions) or the accidental (traffic accidents, forest fires) fields. In all of these situations, people are usually focusing on two characteristics:

1. the time T between two consecutive specific (usually, adverse) events of interest E,
2. the amplitude X of each of these events.

The characteristics T and X defined above are the key factors to be monitored for an event and they are usually referred to as the TBEA (Time Between Events and Amplitude) characteristics. In general, a decrease in T and/or an increase in X can result in a negative, hazardous or disastrous situation that needs to be efficiently monitored with control charts.

The first TBE (i.e. without taking into account the amplitude characteristic) type of control chart goes back to Calvin [12], who proposed to monitor the cumulative number of conforming items between two non-conforming ones. The initial idea was to find a method to improve the traditional attribute control charts that are known to be ineffective in the case of high-quality processes in which the occurrence of non-conforming products is very rare. This initial idea has then been investigated by Lucas [29] and Vardeman and Ray [42] and, subsequently, many other researchers started to contribute to this area. Radaelli [34] proposed to design and implement one- and two-sided Shewhart-type TBE control charts assuming that the counts can be modeled as a homogeneous Poisson process. Gan [25] developed an EWMA (Exponentially Weighted Moving Average) control chart monitoring the rate of occurrences of rare events based on the inter-arrival times of these events. Benneyan [7] used the geometric ("g" chart) and the negative binomial (called "h" chart) distributions in order to monitor the number of cases between hospital-acquired infections. Xie et al. [45] proposed a control chart for TBE data based on the exponential distribution while Borror et al. [8] extended it using a CUSUM (Cumulative Sum) scheme and evaluated its robustness in the case of the Weibull and lognormal distributions. Liu et al. [28] compared the ATS (Average Time to Signal) performance of several continuous TBE charts including the CQC, CQC-r, exponential EWMA and exponential CUSUM charts. Zhang et al. [46] investigated the case of gamma distributed TBE data and they developed a control chart based on a random-shift model to compute the out-of-control ATS. In the case of multistage manufacturing processes, Shamsuzzaman et al. [39] developed a control chart for TBE data and designed it using a statistical oriented approach while Zhang et al. [47] designed it using a first economic oriented approach and Zhang et al. [48] developed it using a second economic oriented approach assuming random process shifts. The use of supplementary runs rules has also been proposed for monitoring TBE data by Cheng and Chen [17]. Qu et al. [32] studied some TBE control charts that can be used for sampling inspection. Shafae et al. [38] evaluated the performance of three TBE CUSUM charts and Fang et al. [22] proposed a generalized group runs TBE chart for a homogenous Poisson failure process.

The first paper that proposed a combined scheme for monitoring the time interval T of an event E as well as its amplitude X has been introduced by Wu et al. [44] who referred it to as a TBEA (Time Between Events and Amplitude) chart. After this paper, several single TBEA charts have been developed, see for instance Qu et al. [31], Cheng et al. [19], Ali and Pievatolo [6], Qu et al. [33] and, very recently, Sanusi et al. [37].

As it can be noticed, this stream of research is rather recent and few publications have already been devoted to. Therefore, the goal of this chapter is to further investigate it and to hopefully open new research directions. More specifically, this chapter will be splitted into three parts:

1. In Sect. 2 we will introduce and compare three different statistics, denoted as Z_1, Z_2 and Z_3, suitable for monitoring TBEA data, in the case of four distributions (gamma, lognormal, normal and Weibull), when the time T and the amplitude X are considered as *independent* random variables.
2. In Sect. 3, we will compare the three statistics introduced in Sect. 2, for the same distributions, but considering that the time T and the amplitude X are *dependent* random variables. A model based on three types of Copulas will be used to define the dependence between T and X.
3. Finally, in Sect. 4, in order to overcome the "distribution choice" dilemma, we will introduce a distribution-free approach coupled with an upper-sided EWMA scheme. In addition, a specific technique called "continuousify" will be presented in order to compute the Run Length properties of the proposed upper-sided EWMA TBEA control chart in a reliable way.

2 TBEA Charts for Independent Times and Amplitudes

2.1 Model

Let $D_0 = 0, D_1, D_2, \dots$ be the dates of occurrence of a specific negative event E, let $T_1 = D_1 - D_0$, $T_2 = D_2 - D_1, \dots$ be the time intervals between two consecutive occurrences of the event E and let $X_1, X_2, \dots$ be the corresponding magnitudes of this event occurring at times $D_1, D_2, \dots$ (see Fig. 1). It must be noted that $D_0 = 0$ is the date of a "virtual" event which has no amplitude associated with.

In this section, we assume that T and X are two mutually *independent* continuous random variables, both defined on $[0, +\infty)$. Let $F_T(t|\boldsymbol{\theta}_T)$ and $F_X(x|\boldsymbol{\theta}_X)$ be the c.d.f. (cumulative distribution function) of T and X, respectively, and let $f_T(t|\boldsymbol{\theta}_T)$ and $f_X(x|\boldsymbol{\theta}_X)$ be the p.d.f. (probability distribution function) of T and X, respectively,

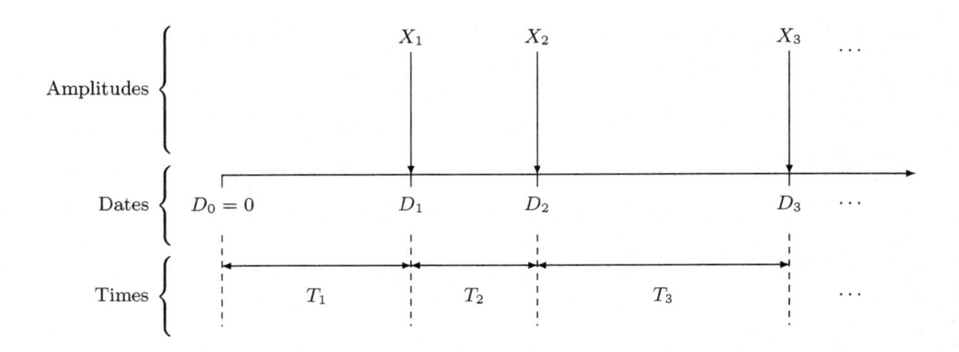

Fig. 1 Dates of occurrence $D_0 = 0, D_1, D_2, \ldots$, time intervals $T_1 = D_1 - D_0, T_2 = D_2 - D_1, \ldots$ and amplitudes $X_1, X_2, \ldots$ of a negative event E

where $\boldsymbol{\theta}_T$ and $\boldsymbol{\theta}_X$ are the corresponding vector of parameters. Let also define $\mu_T = \mathrm{E}(T)$, $\mu_X = \mathrm{E}(X)$, $\sigma_T = \sigma(T)$ and $\sigma_X = \sigma(X)$ be the expectation and standard-deviation of T and X, respectively. By definition, when the process is *in-control*, we have $\boldsymbol{\theta}_T = \boldsymbol{\theta}_{T_0}, \boldsymbol{\theta}_X = \boldsymbol{\theta}_{X_0}, \mu_T = \mu_{T_0}, \mu_X = \mu_{X_0}, \sigma_T = \sigma_{T_0}, \sigma_X = \sigma_{X_0}$ and, when the process is *out-of-control*, we have $\boldsymbol{\theta}_T = \boldsymbol{\theta}_{T_1}$, $\boldsymbol{\theta}_X = \boldsymbol{\theta}_{X_1}$, $\mu_T = \mu_{T_1}$, $\mu_X = \mu_{X_1}$, $\sigma_T = \sigma_{T_1}, \sigma_X = \sigma_{X_1}$.

Because the reference scales for the random variables T and X can be very different and, in order to not favour one random variable over the other one, we suggest to define (and to work with) the "normalized to the mean" new random variables T' and X' as the in-control standardized counterparts of T and X, i.e.

$$T' = \frac{T}{\mu_{T_0}},$$
$$X' = \frac{X}{\mu_{X_0}}.$$

Clearly, when the process is in-control we have $\mathrm{E}(T') = \mathrm{E}(X') = 1$.

2.2 Statistics to Be Monitored

In order to simultaneously monitor the time T between an event E and its amplitude X, we suggest to define several dedicated statistics $Z = Z(T', X')$, functions of the random variables T' and X', satisfying the following two properties:

$$Z \uparrow \text{ if either } T' \downarrow \text{ or } X' \uparrow, \tag{1}$$

$$Z \downarrow \text{ if either } T' \uparrow \text{ or } X' \downarrow . \tag{2}$$

Of course, there are many possible choices for the statistic Z. A first possible choice for the statistic Z (denoted as the Z_1 statistic) satisfying properties (1) and (2) is simply

$$Z_1 = X' - T'. \tag{3}$$

This random variable is defined on $(-\infty, +\infty)$ and its c.d.f. $F_{Z_1}(z|\boldsymbol{\theta}_Z)$ and p.d.f. $f_{Z_1}(z|\boldsymbol{\theta}_Z)$ can be obtained by integrating (see Fig. 2(a) and (b)) over all the couples $(X', T') \in \mathbb{R}^{+2}$ satisfying $Z_1 = X' - T' \leq z$, and they are equal to

$$F_{Z_1}(z|\boldsymbol{\theta}_Z) = 1 - \mu_{X_0} \int_0^{+\infty} F_T((x-z)\mu_{T_0}|\boldsymbol{\theta}_T) f_X(x\mu_{X_0}|\boldsymbol{\theta}_X)\mathrm{d}x, \tag{4}$$

$$f_{Z_1}(z|\boldsymbol{\theta}_Z) = \mu_{T_0}\mu_{X_0} \int_0^{+\infty} f_T((x-z)\mu_{T_0}|\boldsymbol{\theta}_T) f_X(x\mu_{X_0}|\boldsymbol{\theta}_X)\mathrm{d}x, \tag{5}$$

where $\boldsymbol{\theta}_Z = (\boldsymbol{\theta}_T, \boldsymbol{\theta}_X)$ is the corresponding combined vector of parameters.

A second possible choice for the statistic Z (denoted as the Z_2 statistic) satisfying properties (1) and (2) is defining it as the ratio between the two characteristics of an event E:

$$Z_2 = \frac{X'}{T'}. \tag{6}$$

This random variable is defined on $[0, +\infty)$ and its c.d.f. $F_{Z_2}(z|\boldsymbol{\theta}_Z)$ and p.d.f. $f_{Z_2}(z|\boldsymbol{\theta}_Z)$ can be obtained by integrating (see Fig. 2 (c)) over all the couples $(X', T') \in \mathbb{R}^{+2}$ satisfying $Z_2 = \frac{X'}{T'} \leq z$, and they are equal to

$$F_{Z_2}(z|\boldsymbol{\theta}_Z) = 1 - \mu_{X_0} \int_0^{+\infty} F_T\left(\frac{x\mu_{T_0}}{z}|\boldsymbol{\theta}_T\right) f_X(x\mu_{X_0}|\boldsymbol{\theta}_X)\mathrm{d}x, \tag{7}$$

$$f_{Z_2}(z|\boldsymbol{\theta}_Z) = \frac{\mu_{T_0}\mu_{X_0}}{z^2} \int_0^{+\infty} x f_T\left(\frac{x\mu_{T_0}}{z}|\boldsymbol{\theta}_T\right) f_X(x\mu_{X_0}|\boldsymbol{\theta}_X)\mathrm{d}x. \tag{8}$$

Finally, a third possible choice for the statistic Z (denoted as the Z_2 statistic) satisfying properties (1) and (2) is

$$Z_3 = X' + \frac{1}{T'}. \tag{9}$$

This random variable, which should be considered as a hybrid of the two previous ones, is also defined on $[0, +\infty)$ and its c.d.f. $F_{Z_3}(z|\boldsymbol{\theta}_Z)$ and p.d.f. $f_{Z_3}(z|\boldsymbol{\theta}_Z)$ can be obtained by integrating (see Fig. 2 (d)) over all the couples $(X', T') \in \mathbb{R}^{+2}$ satisfying $Z_3 = X' + \frac{1}{T'} \leq z$, and they are equal to

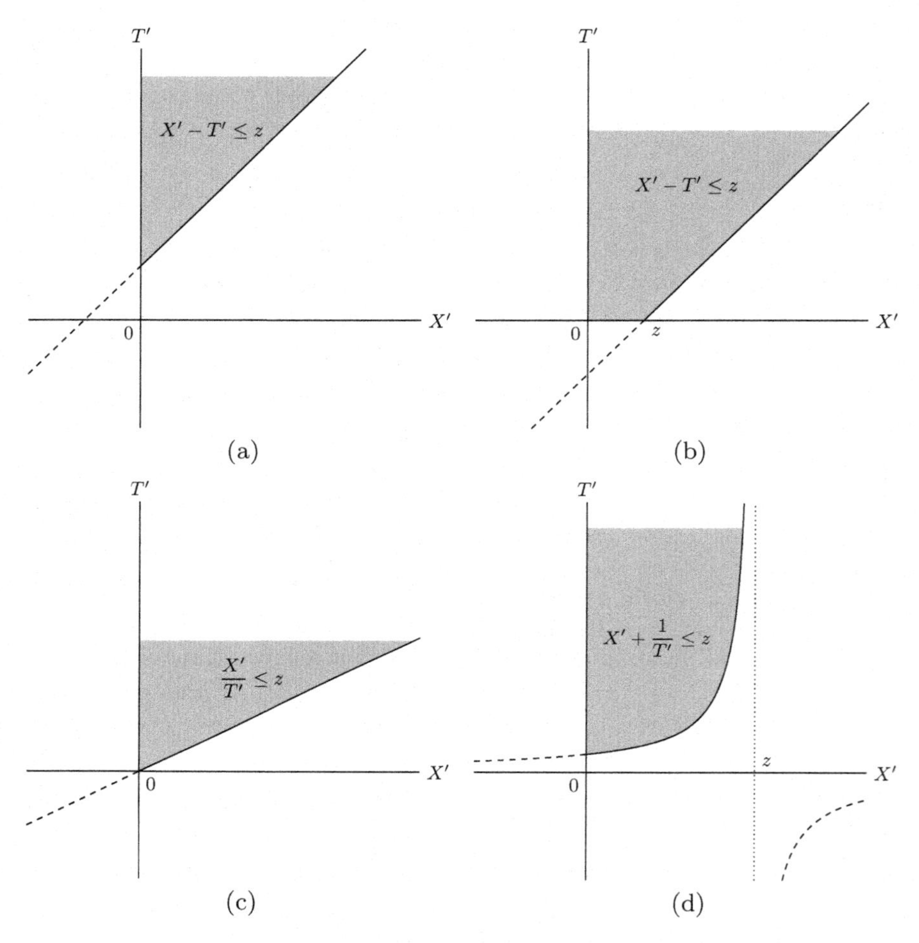

Fig. 2 Integration areas used for statistics (a) and (b) Z_1, (c) Z_2 and (d) Z_3

$$F_{Z_3}(z|\boldsymbol{\theta}_Z) = F_X(z\mu_{X_0}|\boldsymbol{\theta}_X) - \mu_{X_0} \int_0^z F_T\left(\frac{\mu_{T_0}}{z-x}|\boldsymbol{\theta}_T\right) f_X(x\mu_{X_0}|\boldsymbol{\theta}_X)\mathrm{d}x, \quad (10)$$

$$f_{Z_3}(z|\boldsymbol{\theta}_Z) = \mu_{T_0}\mu_{X_0} \int_0^z \frac{1}{(z-x)^2} f_T\left(\frac{\mu_{T_0}}{z-x}|\boldsymbol{\theta}_T\right) f_X(x\mu_{X_0}|\boldsymbol{\theta}_X)\mathrm{d}x. \quad (11)$$

More details on how to derive the c.d.f. and p.d.f. of statistics Z_1, Z_2 and Z_3 provided above can be found in the Appendix section of Rahali et al. [35]. Concerning these c.d.f. and p.d.f. it has to be noted that it is generally not possible to obtain a closed form solution for them and the only solution is to numerically compute these ones by using quadrature techniques.

2.3 Control Limit

As it is more important to detect an increase in Z (in order to avoid more damages or injuries, for instance) rather than a decrease, we suggest to only define an upper control limit UCL_Z for the TBEA charts based on statistics $Z \in \{Z_1, Z_2, Z_3\}$ as

$$\mathrm{UCL}_Z = F_Z^{-1}(1-\alpha|\boldsymbol{\theta}_{Z_0}), \tag{12}$$

where α is the type I error, $\boldsymbol{\theta}_{Z_0} = (\boldsymbol{\theta}_{T_0}, \boldsymbol{\theta}_{X_0})$ and $F_Z^{-1}(\ldots|\boldsymbol{\theta}_{Z_0})$ is the inverse c.d.f. of Z numerically obtained by solving equation $F_Z(z|\boldsymbol{\theta}_{Z_0}) = \alpha$ for z using a one dimension root finder.

2.4 Time to Signal Properties

The type II error β of the upper-sided TBEA charts based on statistic $Z \in \{Z_1, Z_2, Z_3\}$ is equal to

$$\beta = F_Z(\mathrm{UCL}_Z|\boldsymbol{\theta}_{Z_1}), \tag{13}$$

where $\boldsymbol{\theta}_{Z_1} = (\boldsymbol{\theta}_{T_1}, \boldsymbol{\theta}_{X_1})$. The out-of-control ATS (Average Time to Signal) and SDTS (Standard Deviation Time to Signal) of the upper-sided TBEA charts based on statistic $Z \in \{Z_1, Z_2, Z_3\}$ can be obtained using the expectation and variance of compound random variables (see also Rahali et al. [35] for more details) and they are equal to

$$\mathrm{ATS}_1 = \frac{\mu_{T_1}}{1-\beta}, \tag{14}$$

$$\mathrm{SDTS}_1 = \sqrt{\frac{\sigma^2_{T_1}}{1-\beta} + \frac{\mu^2_{T_1}\beta}{(1-\beta)^2}}. \tag{15}$$

when the process is in-control, we have $1-\beta = \alpha$ and, consequently, we have the following equivalence for the in-control ATS

$$\mathrm{ATS}_0 = \frac{\mu_{T_0}}{\alpha} \Leftrightarrow \alpha = \frac{\mu_{T_0}}{\mathrm{ATS}_0}.$$

2.5 Comparative Studies

As in Rahali et al. [35], in order to compare the three TBEA charts defined in Subsect. 2.2 and based on the statistics $Z \in \{Z_1, Z_2, Z_3\}$ we have chosen to investigate

Table 1 Distributions used for the comparison of the 3 TBEA charts

Names	Parameters	$f(x\|a, b)$
Gamma	$a > 0, b > 0$	$\frac{\exp(-\frac{x}{b})x^{a-1}}{b^a \Gamma(a)}$
Lognormal	$a, b > 0$	$\left(\frac{b}{x}\right) f_{\text{Nor}}(a + b\ln(x))$
Normal	$a, b > 0$	$\frac{1}{b} f_{\text{Nor}}\left(\frac{x-a}{b}\right)$
Weibull	$a > 0, b > 0$	$\frac{a}{b}\left(\frac{x}{b}\right)^{a-1} \exp\left(-\left(\frac{x}{b}\right)^a\right)$

Table 2 In-control configurations to be investigated

Distributions	T	X	a_0	b_0	μ_0	σ_0	γ_0
Gamma	•	•	100	0.1	10	1	0.2
	•	•	25	0.4	10	2	0.4
	•	•	4	2.5	10	5	1
Lognormal	•	•	−23.0334	10.0249	10	1	0.3010
	•	•	−11.5277	5.0494	10	2	0.6080
	•	•	−4.6382	2.1169	10	5	1.6250
Normal	○	•	10	1	10	1	0
	○	•	10	2	10	2	0
Weibull	•	•	12.1534	10.4304	10	1	−0.7155
	•	•	5.7974	10.7998	10	2	−0.3519
	•	•	2.1013	11.2906	10	5	0.5664

four different types of distribution that are only dependent on two parameters a and b. The choice of the Gamma, Lognormal, Normal and Weibull distributions is driven by the fact that these ones are very often selected to model time oriented random variables. For this reason, the two parameters beta distribution has been excluded from the benchmark as it is rarely selected for representing time oriented variables. Of course, more complex distributions could have been considered (like the four parameters Beta or Johnson's distributions) but, due to the fact that only the nominal mean μ_0 and standard-deviation σ_0 are assumed to be known, we restricted our choice to a selection of two parameters distributions. These distributions are summarized in Table 1 with their names, parameter settings and p.d.f. $f(x|a, b)$. In this table, $f_{\text{Nor}}(\dots)$ stands for the p.d.f. of the normal (0, 1) distribution.

More specifically, we have selected 9 different in-control configurations to be investigated for T (the normal distribution has not been considered as a possible choice for the *time* between events) and 11 in-control configurations to be investigated for X, i.e. a total of 99 scenarios for (T, X). All of them (see Table 2) are such that the in-control mean is $\mu_0 = 10$ and the in-control standard-deviation is $\sigma_0 \in \{1, 2, 5\}$ (except for the normal distribution, where $\sigma_0 \in \{1, 2\}$ only). In addition, Table 2 also provides the values of the in-control parameters a_0 and b_0 and the corresponding skewness coefficient γ_0.

The upper control limits UCL_Z for the three TBEA charts based on statistics $Z \in \{Z_1, Z_2, Z_3\}$, satisfying $\text{ATS}_0 = 370.4$, have been obtained in Table 3 for the 99 possible scenarios defined in Table 2. As it can be seen, regardless of the statistic Z

Table 3 Upper control limits UCL_Z for the three TBEA charts based on statistics $Z \in \{Z_1, Z_2, Z_3\}$, satisfying $ATS_0 = 370.4$

		Statistic Z_1										
T	$X \rightarrow$	Gamma			Lognormal			Normal		Weibull		
$\downarrow$	σ_0	1	2	5	1	2	5	1	2	1	2	5
Gamma	1	0.273	0.458	1.177	0.275	0.471	1.234	0.268	0.429	0.252	0.400	1.096
	2	0.404	0.547	1.213	0.405	0.556	1.266	0.402	0.527	0.394	0.509	1.138
	5	0.755	0.852	1.395	0.755	0.855	1.431	0.754	0.844	0.752	0.836	1.341
Lognormal	1	0.271	0.457	1.177	0.273	0.471	1.234	0.266	0.428	0.249	0.399	1.096
	2	0.391	0.540	1.212	0.392	0.550	1.264	0.389	0.520	0.380	0.500	1.136
	5	0.682	0.794	1.369	0.682	0.799	1.408	0.681	0.783	0.678	0.772	1.312
Weibull	1	0.293	0.465	1.178	0.295	0.478	1.235	0.289	0.438	0.277	0.410	1.097
	2	0.460	0.578	1.219	0.460	0.587	1.271	0.458	0.562	0.454	0.547	1.145
	5	0.823	0.908	1.418	0.823	0.910	1.452	0.823	0.902	0.822	0.896	1.369
		Statistic Z_2										
T	$X \rightarrow$	Gamma			Lognormal			Normal		Weibull		
$\downarrow$	σ_0	1	2	5	1	2	5	1	2	1	2	5
Gamma	1	1.314	1.500	2.226	1.316	1.513	2.280	1.310	1.474	1.296	1.448	2.148
	2	1.590	1.735	2.422	1.591	1.742	2.463	1.588	1.720	1.583	1.706	2.359
	5	3.615	3.713	4.315	3.615	3.713	4.308	3.615	3.712	3.615	3.711	4.309
Lognormal	1	1.310	1.498	2.225	1.312	1.511	2.279	1.306	1.472	1.291	1.445	2.147
	2	1.555	1.708	2.405	1.556	1.715	2.447	1.553	1.692	1.547	1.677	2.341
	5	2.820	2.941	3.623	2.820	2.942	3.623	2.820	2.937	2.819	2.934	3.603
Weibull	1	1.356	1.524	2.238	1.358	1.537	2.291	1.353	1.500	1.343	1.476	2.160
	2	1.762	1.875	2.507	1.762	1.881	2.550	1.761	1.865	1.758	1.856	2.447
	5	4.934	5.010	5.502	4.934	5.010	5.492	4.934	5.010	4.934	5.010	5.506
		Statistic Z_3										
T	$X \rightarrow$	Gamma			Lognormal			Normal		Weibull		
$\downarrow$	σ_0	1	2	5	1	2	5	1	2	1	2	5
Gamma	1	2.299	2.474	3.188	2.301	2.488	3.245	2.295	2.447	2.282	2.419	3.107
	2	2.566	2.668	3.277	2.567	2.676	3.328	2.565	2.653	2.561	2.640	3.204
	5	4.587	4.604	4.764	4.587	4.605	4.807	4.587	4.604	4.587	4.603	4.738
Lognormal	1	2.295	2.472	3.188	2.297	2.486	3.244	2.291	2.445	2.277	2.416	3.107
	2	2.530	2.641	3.266	2.531	2.649	3.317	2.529	2.625	2.524	2.611	3.193
	5	3.787	3.817	4.084	3.787	3.818	4.127	3.787	3.815	3.787	3.813	4.043
Weibull	1	2.342	2.499	3.196	2.344	2.512	3.252	2.339	2.472	2.329	2.445	3.115
	2	2.742	2.812	3.357	2.743	2.819	3.413	2.742	2.801	2.740	2.793	3.282
	5	5.912	5.921	6.002	5.912	5.922	6.024	5.912	5.921	5.912	5.921	5.994

and the distribution of X, these upper control limits tend to be similar if σ_0 is small (say $\sigma_0 = 1$) and the distribution of T is either gamma or lognormal. But, when T follows a Weibull distribution, the control limits are larger than those of the gamma or lognormal distributions.

When an out-of-control situation occurs in a TBEA process (corresponding to an upper shift in Z), it can be due to (i) a mean shift *only in the time* T from μ_{T_0} to $\mu_{T_1} = \delta_T \mu_{T_0}$, (ii) a mean shift *only in the amplitude* X from μ_{X_0} to $\mu_{X_1} = \delta_X \mu_{X_0}$, or (iii) a combination of the two previous cases, where $\delta_T \leq 1$ and $\delta_X \geq 1$ are the parameters quantifying the change in the time and amplitude, respectively. But, as the actual values of δ_T and δ_X are usually unknown by the practitioner, it is therefore difficult to evaluate the three TBEA charts based on statistics $Z \in \{Z_1, Z_2, Z_3\}$ using the ATS_1 criterion defined in (14) that depends on a specific values for δ_T and/or δ_X. For this reason, it is therefore preferable to use the following more general criterion denoted as EATS_1 (Expected Average Time to Signal) and defined as

$$\text{EATS}_1 = \sum_{\delta_T \in \Omega_T} \sum_{\delta_X \in \Omega_X} f_{\delta_T}(\delta_T) f_{\delta_X}(\delta_X) \text{ATS}_1(\delta_T, \delta_X),$$

where Ω_T and Ω_X are the sets of the potential shifts for δ_T and δ_X, respectively, and $f_{\delta_X}(\delta_X)$ and $f_{\delta_T}(\delta_T)$ are the probability mass functions of the shifts δ_T and δ_X over the sets Ω_T and Ω_X. In this chapter, we adopt the classical assumption that confines $f_{\delta_T}(\delta_T)$ and $f_{\delta_X}(\delta_X)$ to be discrete uniform distributions over Ω_T and Ω_X, respectively. If we want to investigate (i) a mean shift only due to the time T then we suggest to fix $\Omega_T = \{0.5, 0.55, \ldots, 0.9, 0.95\}$ and $\Omega_X = \{1\}$, (ii) a mean shift only due to the amplitude X then we suggest to fix $\Omega_T = \{1\}$ and $\Omega_X = \{1.1, 1.2, \ldots, 1.9, 2\}$ and (iii) a mean shift due to the time T and the amplitude X then we suggest to fix $\Omega_T = \{0.5, 0.55, \ldots, 0.9, 0.95\}$ and $\Omega_X = \{1.1, 1.2, \ldots, 1.9, 2\}$.

For the 99 possible scenarios defined in Table 2, Table 4 gives the EATS_1 values of the three TBEA charts based on statistics $Z \in \{Z_1, Z_2, Z_3\}$ when $\Omega_T = \{0.5, 0.55, \ldots, 0.9, 0.95\}$ and $\Omega_X = \{1.1, 1.2, \ldots, 1.9, 2\}$. Values of EATS_1 in bold characters are the smallest ones among statistics Z_1, Z_2 or Z_3. From Table 4, it can be deduced that when there is a shift in both T and X, the most efficient statistic is Z_1 (in 56% of the cases with an average EATS_1 value $\overline{\text{EATS}_1} = 14.91$), followed by statistic Z_2 (in 35% of the cases with $\overline{\text{EATS}_1} = 29.15$) and, finally, statistic Z_3 (in only 5% of the cases with $\overline{\text{EATS}_1} = 11.74$).

The cases where the mean shift is only due to the time T or the mean shift is only due to the amplitude X have both been investigated in Rahali et al. [35] (see Tables 3 and 4, pages 245–246). In these cases, the conclusions are

- if the mean shift is only due to the time T, then the most efficient statistic is Z_3 (in 71% of the cases with $\overline{\text{EATS}_1} = 59.88$) followed by Z_2 (in 29% of the cases with $\overline{\text{EATS}_1} = 71.82$) while the statistic Z_1 never provides the smallest EATS_1 and should not be considered here as an efficient monitoring statistic.
- if the mean shift is only due to the amplitude X, then the statistic Z_1 is the best option as it always gives the smallest EATS_1 values, regardless of the combination under consideration. In this case, Z_2 and Z_3 should not be considered as potential efficient monitoring statistics.

2.6 *Illustrative Example*

This illustrative example has been detailed for the first time in Rahali et al. [35] and it is based on a real data set concerning the time (T in days) between fires in forests of the region "Provence - Alpes - Côte D'Azur" in the south-east of

Table 4 EATS_1 values when $\Omega_T = \{0.5, 0.55, \ldots, 0.9, 0.95\}$ and $\Omega_X = \{1.1, 1.2, \ldots, 1.9, 2\}$ for the three TBEA charts based on statistics $Z \in \{Z_1, Z_2, Z_3\}$

		Statistic Z_1										
T	$X \rightarrow$	Gamma			Lognormal			Normal		Weibull		
$\downarrow$	σ_0	1	2	5	1	2	5	1	2	1	2	5
Gamma	1	**8.4**	12.4	47.8	**8.5**	13.0	55.9	**8.4**	11.3	**8.1**	10.4	38.6
	2	**10.5**	**14.5**	48.7	**10.5**	**15.0**	56.2	**10.4**	**13.6**	**10.2**	**12.8**	40.1
	5	**16.9**	**21.2**	52.5	**16.9**	**21.5**	58.4	**16.9**	**20.5**	**16.8**	**19.9**	45.6
Lognormal	1	**8.4**	12.4	47.8	**8.5**	13.0	55.9	**8.3**	11.3	**8.1**	10.3	38.6
	2	**10.2**	14.3	48.6	**10.2**	14.8	56.1	**10.1**	**13.3**	**9.9**	**12.5**	39.9
	5	**14.7**	**19.3**	52.1	**14.7**	**19.6**	58.4	**14.7**	**18.5**	**14.5**	**17.8**	44.7
Weibull	1	**8.8**	12.7	47.9	**8.9**	13.3	55.9	**8.8**	11.6	**8.5**	10.7	38.7
	2	**12.4**	**16.0**	49.0	**12.4**	**16.5**	56.4	**12.3**	**15.1**	**12.1**	**14.4**	40.6
	5	**19.5**	**23.7**	54.1	**19.5**	**23.9**	59.6	**19.5**	**23.1**	**19.4**	**22.6**	**47.5**
		Statistic Z_2										
T	$X \rightarrow$	Gamma			Lognormal			Normal		Weibull		
$\downarrow$	σ_0	1	2	5	1	2	5	1	2	1	2	5
Gamma	1	8.5	**11.3**	**30.4**	8.5	**11.6**	**34.4**	8.5	**10.6**	8.3	10.1	**25.7**
	2	12.1	15.0	**32.2**	12.1	15.2	**34.9**	12.1	14.5	12.0	14.1	**28.9**
	5	33.6	34.9	**43.0**	33.6	34.9	**43.1**	33.6	34.9	33.6	34.9	**42.6**
Lognormal	1	8.5	**11.2**	**30.4**	8.5	**11.6**	**34.3**	8.4	**10.6**	8.2	10.0	**25.7**
	2	11.4	14.3	**31.9**	11.4	14.5	**34.7**	11.4	13.8	11.2	13.4	**28.4**
	5	24.5	26.5	**38.2**	24.5	26.5	**38.6**	24.5	26.4	24.4	26.3	**37.1**
Weibull	1	9.2	**11.9**	**30.8**	9.2	**12.3**	**34.7**	9.1	**11.3**	9.0	10.7	**26.3**
	2	16.7	19.1	**34.9**	16.7	19.3	**37.3**	16.6	18.8	16.6	18.5	**32.1**
	5	47.6	48.4	**53.7**	47.6	48.4	**53.6**	47.6	48.4	47.6	48.4	53.6
		Statistic Z_3										
T	$X \rightarrow$	Gamma			Lognormal			Normal		Weibull		
$\downarrow$	σ_0	1	2	5	1	2	5	1	2	1	2	5
Gamma	1	8.5	11.4	38.6	8.5	11.8	44.7	8.4	10.6	8.2	**9.9**	31.7
	2	12.6	14.8	35.7	12.6	15.1	40.0	12.6	14.3	12.5	13.9	30.7
	5	50.9	51.0	52.4	50.9	51.0	53.5	50.9	51.0	50.9	51.0	52.1
Lognormal	1	8.4	11.3	38.7	8.5	11.8	44.8	8.4	10.6	8.2	**9.9**	31.7
	2	11.7	**14.0**	35.7	11.7	**14.3**	40.3	11.7	13.5	11.6	13.1	30.5
	5	35.1	35.5	40.5	35.1	35.5	42.1	35.1	35.4	35.1	35.4	39.4
Weibull	1	9.3	12.1	38.5	9.3	12.6	44.4	9.2	11.3	9.0	**10.6**	31.8
	2	19.0	20.3	38.3	19.0	20.6	42.4	19.0	19.9	18.9	19.7	33.7
	5	73.3	73.3	73.5	73.3	73.3	73.9	73.3	73.3	73.3	73.3	73.5

France and their amplitudes (X measured as the burned surface in $ha = 10000\,\text{m}^2$). This data set reports a total of 92 *significant* fires from 2016/10 to 2017/9: the data set has been split into $m = 47$ fires occurring during the "low season" from 2016/10 to mid 2017/6 (used as Phase 1 data) and $n = 45$ fires occuring during the "high season" from mid 2017/6 to 2017/9 (used as Phase 2 data). The values of T and X are recorded in Table 5 (as well as the values of the statistics Z_1, Z_2 and Z_3) and they are also plotted in Fig. 3 where it is clear that the values of T during the "high season" are smaller than those during the "low season" and the values of X during the "high season" are larger than those during the "low season".

The use of the Kendall's and Spearman's rank correlation tests on the whole data set yields p-values larger than the significance level of 0.05 (0.2 for the Kendall's test and 0.19 for the Spearman's test) validating the fact that the random variables T and X are uncorrelated (a key assumption in this section). Among the four distributions considered in Table 1, the use of the Kolmogorov-Smirnov's test shows that the best fit for both T and X is the lognormal distribution with parameters ($a_0 = -1.2648$, $b_0 = 1.0302$) for T and ($a_0 = -1.6697$, $b_0 = 0.8624$) for X.

The three TBEA charts, corresponding to the statistics $Z \in \{Z_1, Z_2, Z_3\}$ are plotted in Fig. 4 along with their upper control limits $\text{UCL}_{Z_1} = 6.0306$, $\text{UCL}_{Z_2} = 28.1209$ and $\text{UCL}_{Z_3} = 19.3885$ (assuming $\text{ATS}_0 = 730$, i.e. 2 years). As it can be seen, these charts detect several out-of-control situations during the "high season" confirming that a decrease in the time between fires occurred as well as an increase in the amplitude of these fires.

3 TBEA Charts for Dependent Times and Amplitudes

3.1 *Motivation*

In the previous section, the time T between events and their amplitudes X have been considered as *independent* random variables. But, in practice, this is not always the case. For example, there are natural situations for which the amplitudes tend to become larger when the times between events become shorter (i.e. a negative correlation). Such kind of situations is likely to occur for example in the case of earthquakes for which, in a first phase, small amplitude earthquakes may occur with a low frequency (large time between events) and, suddenly, in a second phase, the occurence frequency of these earthquakes may increase (shorter time between events) with a negatively correlated increase in their amplitudes. The same kind of situations may also arise in the case of forest fires occuring, in a first phase during the "humid season", with a low frequency and small amplitudes (surfaces burned) and becoming more disastrous during the "dry season" with shorter time between the occurrence of forest fires and larger amplitudes (see for instance the 2019 forest fires in Amazonia or Siberia or the 2020 forest fires in Australia and USA). Positive correlation between T and X (i.e. the time between events becomes shorter and the amplitude becomes smaller) is also possible as the forthcoming illustrative example in this section will depict it.

Until now, very few research papers have investigated TBEA control charts by considering the potential dependence between the two variables T and X. Cheng and Mukherjee [18] were the first to investigate a T^2 TBEA control chart by using a bivariate SAT (Smith-Adelfang-Tubbs) Gamma distribution to model the joint probability of T and X. This work has been extended later by Cheng et al. [19] who developed a similar approach based on a MEWMA (multivariate exponentially weighted moving average) procedure.

Table 5 Phase 1 and 2 data sets corresponding to time (T in days) between fires, amplitudes (X as the burned surface in ha) and the values of the statistics Z_1, Z_2 and Z_3.

Phase 1

i	T	X	Z_1	Z_2	Z_3
1	9	3.68	-1.37	0.16	0.88
2	17	1.99	-2.96	0.05	0.47
3	34	6.00	-5.78	0.07	0.60
4	7	1.19	-1.19	0.07	0.87
5	3	135.80	**9.45**	18.23	11.82
6	2	14.37	0.69	2.89	3.79
7	14	8.10	-1.96	0.23	0.99
8	2	32.31	2.01	6.51	5.11
9	6	3.07	-0.87	0.21	1.14
10	1	10.03	0.56	4.04	6.21
11	1	7.93	0.40	3.19	6.05
12	1	1.50	-0.07	0.60	5.58
13	6	23.30	0.62	1.56	2.63
14	3	3.73	-0.27	0.50	2.10
15	3	4.73	-0.20	0.63	2.17
16	2	3.19	-0.13	0.64	2.97
17	2	6.25	0.09	1.26	3.19
18	1	3.60	0.08	1.45	5.73
19	1	6.12	0.27	2.46	5.92
20	3	1.50	-0.44	0.20	1.93
21	4	1.33	-0.63	0.13	1.46
22	12	1.42	-2.09	0.05	0.56
23	3	5.75	-0.13	0.77	2.25
24	3	3.47	-0.29	0.47	2.08
25	2	13.31	0.61	2.68	3.71
26	1	26.31	1.75	10.60	7.41
27	1	18.54	1.18	7.47	6.83
28	2	66.17	4.51	13.32	7.61
29	1	9.90	0.55	3.99	6.20
30	3	4.22	-0.24	0.57	2.13
31	7	34.28	1.24	1.97	3.31
32	4	2.23	-0.57	0.22	1.53
33	1	1.84	-0.05	0.74	5.60
34	1	2.88	0.03	1.16	5.68
35	1	21.46	1.40	8.64	7.05
36	1	4.46	0.15	1.80	5.80
37	1	58.27	4.11	23.47	9.76
38	1	8.84	0.47	3.56	6.12
39	13	1.03	-2.30	0.03	0.50
40	7	16.57	-0.06	0.95	2.00
41	14	4.96	-2.20	0.14	0.76
42	1	1.37	-0.08	0.55	5.57
43	3	23.39	1.17	3.14	3.55
44	20	1.70	-3.53	0.03	0.40
45	22	5.30	-3.63	0.10	0.64
46	1	15.64	0.97	6.30	6.62
47	9	5.14	-1.27	0.23	0.99

Phase 2

i	T	X	Z_1	Z_2	Z_3
1	1	1.00	-0.11	0.40	5.54
2	2	3.70	-0.09	0.75	3.01
3	2	3.17	-0.13	0.64	2.97
4	3	18.40	0.81	2.47	3.18
5	3	1.00	-0.47	0.13	1.90
6	1	2.22	-0.02	0.89	5.63
7	2	19.09	1.04	3.84	4.14
8	1	2.00	-0.04	0.81	5.62
9	2	34.28	2.16	6.90	5.26
10	2	3.00	-0.14	0.60	2.95
11	1	6.63	0.31	2.67	5.96
12	1	4.47	0.15	1.80	5.80
13	7	8.24	-0.67	0.47	1.39
14	1	769.45	**56.49**	**309.87**	**62.14**
15	1	4.37	0.14	1.76	5.79
16	1	90.70	**6.50**	**36.53**	12.15
17	1	11.49	0.66	4.63	6.31
18	6	3590.78	**263.36**	**241.01**	**265.37**
19	1	1427.92	**104.98**	**575.04**	**110.63**
20	1	255.96	**18.67**	**103.08**	**24.32**
21	1	1.00	-0.11	0.40	5.54
22	4	13.88	0.29	1.40	2.39
23	1	138.28	**10.00**	**55.69**	15.65
24	2	8.90	0.29	1.79	3.39
25	3	1.50	-0.44	0.20	1.93
26	4	34.63	1.82	3.49	3.92
27	1	82.56	5.90	**33.25**	11.55
28	1	2.00	-0.04	0.81	5.62
29	1	162.08	**11.75**	**65.27**	17.40
30	4	3.26	-0.49	0.33	1.61
31	2	285.91	**20.69**	**57.57**	**23.79**
32	1	2.00	-0.04	0.81	5.62
33	3	11.57	0.30	1.55	2.67
34	9	34.70	0.91	1.55	3.16
35	1	431.00	**31.56**	**173.57**	**37.21**
36	1	10.89	0.62	4.39	6.27
37	4	1.00	-0.66	0.10	1.44
38	6	1.50	-0.99	0.10	1.02
39	1	1.17	-0.10	0.47	5.55
40	2	1.27	-0.27	0.26	2.83
41	1	26.25	1.75	10.57	7.40
42	3	11.66	0.31	1.57	2.68
43	1	3.03	0.04	1.22	5.69
44	1	12.00	0.70	4.83	6.35
45	1	1.00	-0.11	0.40	5.54

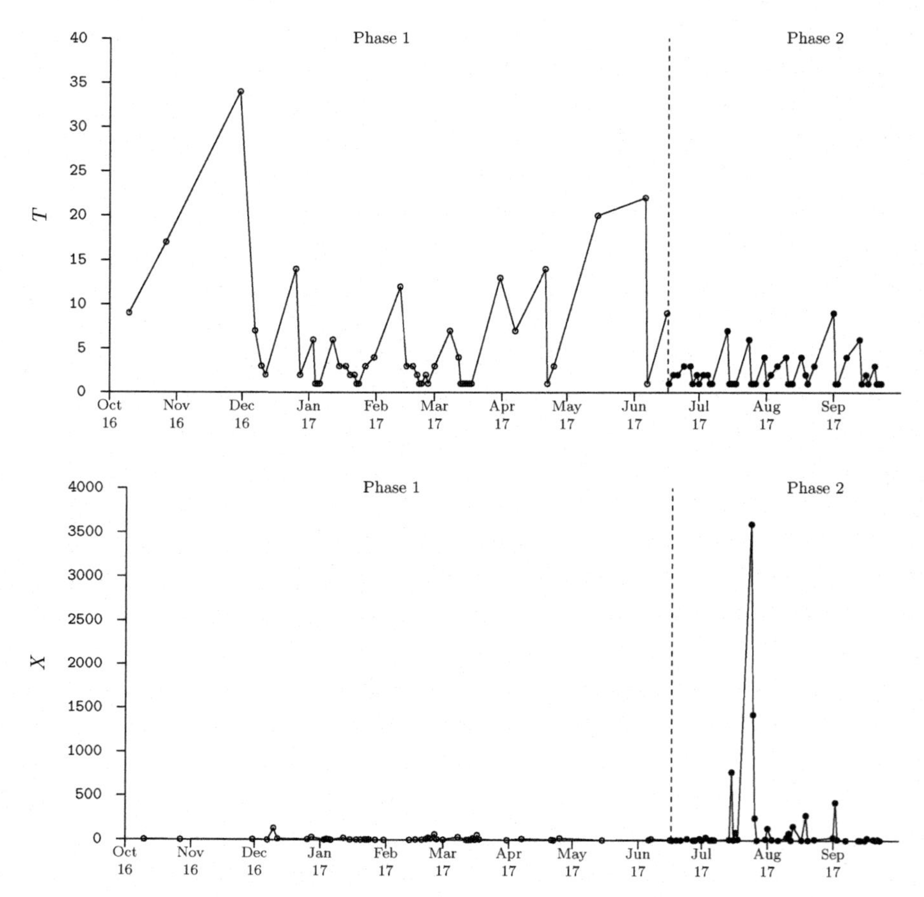

Fig. 3 Time (T in days) between fires and amplitudes (X as the burned surface in ha) corresponding to the data set in Table 5

In this section, instead of specifying a particular bivariate joint distribution for (T, X), like the SAT distribution for instance, we will assume that the marginal distributions of T and X are both known (and they can be almost anything) and we will use the Copulas mechanism (popularized by Sklar [40]) in order to model the dependence between the time T and the amplitude X. The use of Copulas in the Statistical Process Monitoring field is not so common. We can cite for instance Fatahi et al. [23], Dokouhaki and Noorossana [21], Busababodhin and Amphanthong [11] and Sukparungsee et al. [41] who all proposed various types of control charts based on Copulas.

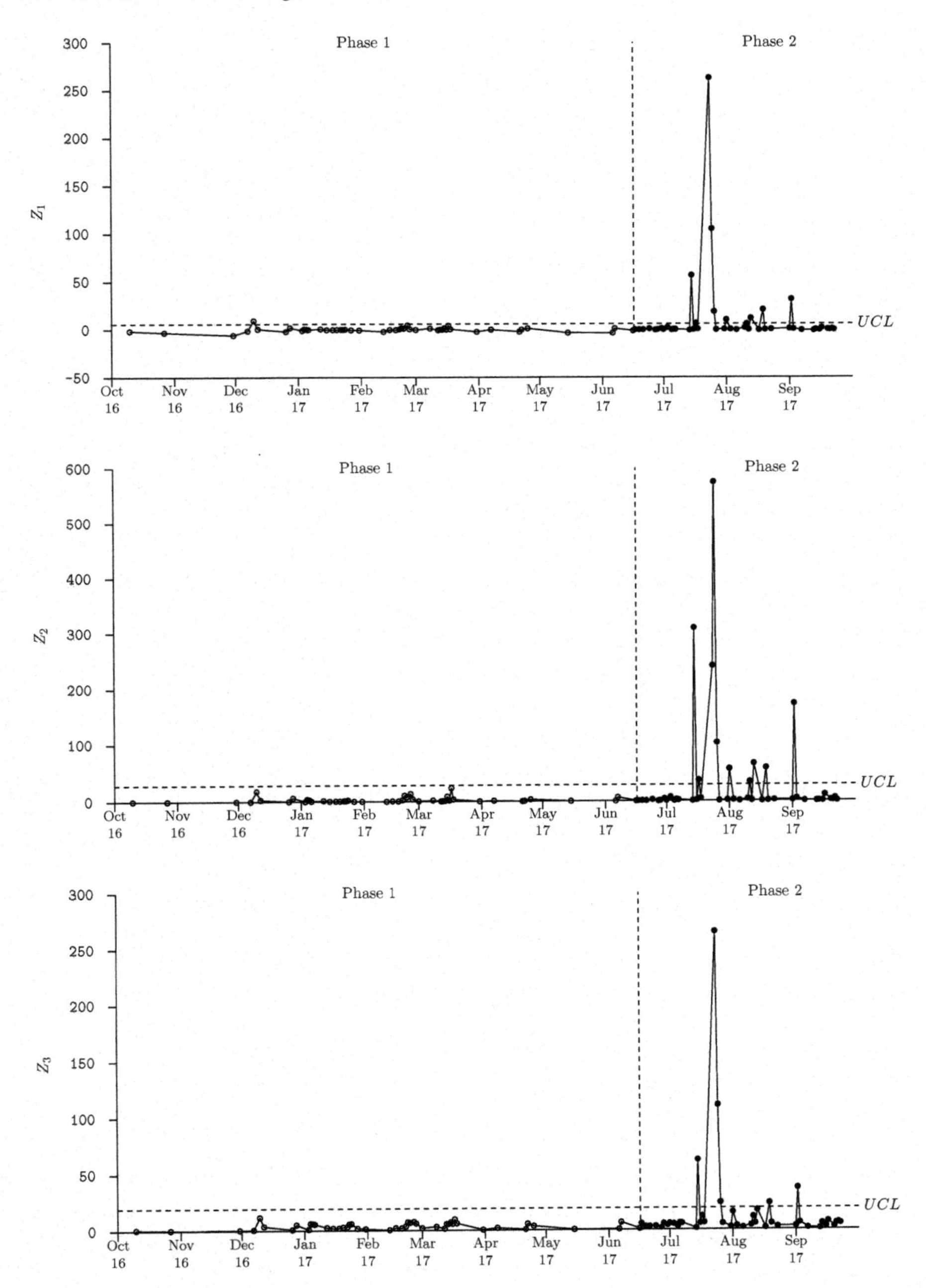

Fig. 4 Statistics Z_1, Z_2 and Z_3 corresponding to the data set in Table 5

3.2 Model

In this section, we assume that $(X, T) \in \mathbb{R}_+^2$ and their joint continuous c.d.f. is equal to

$$F_{(T,X)}(t, x|\boldsymbol{\theta}_T, \boldsymbol{\theta}_X, \theta) = C(F_T(t|\boldsymbol{\theta}_T), F_X(x|\boldsymbol{\theta}_X)|\theta), \tag{16}$$

where $F_T(t|\boldsymbol{\theta}_T)$ and $F_X(x|\boldsymbol{\theta}_X)$, as defined in Sect. 2.1, are the marginal c.d.f. of T and X, respectively, $C(u, v|\theta)$ is a Copula containing all information on the dependence structure between T and X, and θ is a dependence parameter quantifying the dependence between the marginals. In addition, let

$$f_{(T,X)}(t, x|\boldsymbol{\theta}_T, \boldsymbol{\theta}_X, \theta) = c(F_T(t|\boldsymbol{\theta}_T), F_X(x|\boldsymbol{\theta}_X)|\theta) f_T(t|\boldsymbol{\theta}_T) f_X(x|\boldsymbol{\theta}_X) \tag{17}$$

be the joint p.d.f. of (X, T) where $f_T(t|\boldsymbol{\theta}_T)$ and $f_X(x|\boldsymbol{\theta}_X)$ are the marginal p.d.f. of T and X, respectively, and $c(u, v|\theta) = \frac{\partial C(u,v|\theta)}{\partial u \partial v}$ is the Copula density. As explained in Sect. 2.1, in order to not favor one random variable over the other one, the new random variables $T' = \frac{T}{\mu_{T_0}}$ and $X' = \frac{X}{\mu_{X_0}}$ are introduced as the in-control standardized counterparts of T and X, where μ_{T_0} and μ_{X_0} are the (known) in-control expectation of T and X, respectively. It is easy to show that the joint c.d.f. $F_{(T',X')}(t, x|\boldsymbol{\theta}_T, \boldsymbol{\theta}_X, \theta)$ and joint p.d.f. $f_{(T',X')}(t, x|\boldsymbol{\theta}_T, \boldsymbol{\theta}_X, \theta)$ of $(X', T') \in \mathbb{R}_+^2$ are equal to

$$\begin{aligned} F_{(T',X')}(t, x|\boldsymbol{\theta}_T, \boldsymbol{\theta}_X, \theta) =& C(F_T(t\mu_{T_0}|\boldsymbol{\theta}_T), F_X(x\mu_{X_0}|\boldsymbol{\theta}_X)|\theta), \\ f_{(T',X')}(t, x|\boldsymbol{\theta}_T, \boldsymbol{\theta}_X, \theta) =& \mu_{T_0}\mu_{X_0} c(F_T(t\mu_{T_0}|\boldsymbol{\theta}_T), F_X(x\mu_{X_0}|\boldsymbol{\theta}_X)|\theta) \\ & f_T(t\mu_{T_0}|\boldsymbol{\theta}_T) f_X(x\mu_{X_0}|\boldsymbol{\theta}_X). \end{aligned}$$

The closed form formulas for the c.d.f. and p.d.f. of the statistics Z_1, Z_2 and Z_3 defined in Sect. 2.2 are provided in Rahali et al. [36]. The definition of the upper control limit UCL_Z of the TBEA charts with dependent times T and amplitudes X is similar to the one in Eq. (12) and it just requires the addition of the dependence parameter θ, i.e. $\text{UCL}_Z = F_Z^{-1}(1 - \alpha|\boldsymbol{\theta}_{Z_0}, \theta)$. The formulas for computing ATS_1 and SDTS_1 are the same as in Eqs. (14) and (15), respectively.

3.3 Comparative Studies

In order to compare the three TBEA charts for dependent times T and amplitudes X based on statistic $Z \in \{Z_1, Z_2, Z_3\}$, the same distributions listed in Table 1 have been chosen and the same 99 possible scenarios defined in Table 2 have been investigated. The Archimedean bivariate Copulas of Gumbel [26], Clayton [20] and Frank [24] have been chosen in this section to model the dependence between T and X. The Gumbel's (also called Gumbel-Hougaard's) and Clayton's Copulas are two asymmetric Copulas exhibiting a larger dependence in the positive tail than in the negative one (for the Gumbel's Copula) and in the negative tail than in the positive one (for

Table 6 Details concerning Gumbel's, Clayton's and Frank's Copulas

Name	$C(u, v\|\theta)$	Domain for θ	τ and θ
Gumbel	$\exp\left(-\left((-\ln(u))^\theta + (-\ln(v))^\theta\right)^{\frac{1}{\theta}}\right)$	$[1, \infty)$	$\tau = 1 - \frac{1}{\theta} \Leftrightarrow \theta = \frac{1}{1-\tau}$
Clayton	$\max(0, u^{-\theta} + v^{-\theta} - 1)^{-\frac{1}{\theta}}$	$[-1, \infty)\backslash\{0\}$	$\tau = \frac{\theta}{\theta+2} \Leftrightarrow \theta = \frac{2\tau}{1-\tau}$
Frank	$-\frac{1}{\theta}\ln\left(1 + \frac{(e^{-\theta u}-1)(e^{-\theta v}-1)}{e^{-\theta}-1}\right)$	$\mathbb{R}\backslash\{0\}$	$\tau = 1 + \frac{4(D_1(\theta)-1)}{\theta}$

Note: $D_1(\theta)$ is the Debye function of the first kind defined as $D_1(\theta) = \frac{1}{\theta}\int_0^\theta \frac{t}{e^t-1}\mathrm{d}t$

the Clayton's Copulas). The Frank's Copula is a symmetric one that can be used to model dependence structures with either positive or negative correlation. Other Archimedean bivariate Copulas could have also been investigated, like for instance the Ali et al. [5] and Joe [27] Copulas but, for simplicity and also due to their popularity, we only confined our investigations to the Gumbel's, Clayton's and Frank's Copulas. Details concerning the definition of each of these Copulas $C(u, v|\theta)$, the domain of definition for θ and the relationship between the Kendall's rank correlation coefficient τ and the dependence parameter θ are provided in Table 6. In order to facilitate the use of these Copulas, Table 7 simply provides pre-computed values of θ for several selected values of the Kendall's rank correlation coefficient $\tau \in \{0, 0.1, 0.2, \ldots, 0.9\}$.

When it is not possible to model a negative dependence with the Copulas defined above, it is possible to use 90 or 270° rotated versions C_{90} or C_{270} of these Copulas using the following transformations (see Brechmann and Schepsmeier [9])

$$C_{90}(u_1, u_2) = u_2 - C(1 - u_1, u_2),$$
$$C_{270}(u_1, u_2) = u_1 - C(u_1, 1 - u_2).$$

Similarly to Table 3 (for independent times T and amplitudes X) and assuming $\text{ATS}_0 = 370.4$, the upper control limits UCL_Z of the three TBEA charts with dependent times T and amplitudes X are reported in Tables 3–5 of Rahali et al. [36] for the 99 scenarios defined in Table 2 and for the 3 Copulas defined above. From Tables 3–5 in Rahali et al. [36], the following conclusions can be drawn

- For a fixed statistic $Z \in \{Z_1, Z_2, Z_3\}$, scenario in Table 2 and type of Copula, the larger τ, the smaller the control limit UCL_Z.
- For a fixed scenario in Table 2, value of τ and type of Copula, the upper control limits of the statistic $Z \in \{Z_1, Z_2, Z_3\}$ always satisfy $\text{UCL}_{Z_1} < \text{UCL}_{Z_2} < \text{UCL}_{Z_3}$.
- For a fixed scenario in Table 2, value of τ and statistic $Z \in \{Z_1, Z_2, Z_3\}$, the values of UCL_Z are more or less the same no matter the type of Copula considered.

As in Sect. 2.5 for independent times T and amplitudes X, EATS_1 values (for shifts in both T and X) have been reported in Tables 7–9 of Rahali et al. [36] in the case of dependent times T and amplitudes X. These Tables only consider the Frank's Copula and $\tau \in \{0.2, 0.5, 0.8\}$. These results clearly show that, irrespective of the

Table 7 Pre-computed values of θ for several selected values of $\tau \in \{0, 0.1, 0.2, \ldots, 0.9\}$

τ	θ		
	Frank	Clayton	Gumbel
0.0	0.00	0.00	1.00
0.1	0.91	0.22	1.11
0.2	1.86	0.50	1.25
0.3	2.92	0.86	1.43
0.4	4.16	1.33	1.67
0.5	5.74	2.00	2.00
0.6	7.93	3.00	2.50
0.7	11.41	4.67	3.33
0.8	18.19	8.00	5.00
0.9	38.28	18.00	10.00

level of dependence, when a shift in both T and X is likely to occur, the best option is to use the statistics Z_1 or, eventually, Z_2, but not the statistic Z_3 which is never considered as an efficient one. These conclusions are similar to the ones obtained in Sect. 2.5 and they remain identical for other Copulas like the Clayton's or Gumbel's ones. More details concerning these aspects can be found in Rahali et al. [36].

3.4 *Illustrative Example*

The following illustrative example has been detailed for the first time in Rahali et al. [36] and it is related to a company that has recorded for one of its bottleneck machine, during about 6 years (from 08/01/2012 to 27/12/2018) all the breakdown dates (D_i in days) as well as the estimated corresponding incurred costs (X_i, in euros) which include all the repair and restart costs (spare parts, manpower) and the cost of manufacturing disruption, see Table 8. This data set of 44 dates is divided into two parts

- 30 breakdowns recorded during 5 years (2012 to 2016) and used in this example as a Phase 1 data set.
- 14 breakdowns recorded during 2 years (2017 to 2018) and used in this example as a Phase 2 data set.

The times T_i and amplitude X_i, $i = 1, \ldots, 44$ in Table 8 have also been plotted in Fig. 5 with $\circ$ (for Phase 1) and $\bullet$ (for Phase 2), respectively. From the scatterplot shown in Fig. 5, it can be noted that when the time T_i between consecutive breakdowns is smaller (larger), the corresponding cost X_i seems to be also smaller (larger), indicating a potential slight positive correlation between T and X. Investigations (during the period 2012 to 2016) about this phenomena have clarified why such a positive correlation between T and X may occur in this situation. After the occurrence of a breakdown, if the next one occurs after a *short* period of time, it is often due to the same problem occurring to the process: consequently, the time to looking for the breakdown causes is reduced. The spare parts costs are also reduced as they have already been purchased for the previous breakdown. On the contrary, when the next breakdown occurs after a *long* period of time, the causes are usually different from the previous breakdown and need more time to be revealed; new spare parts need to be purchased, thus increasing the cost. In order to evaluate if a positive correlation significantly exists between T and X, the Kendall's and Spearman's rank correlation coefficients have been estimated to $\hat{\tau} = 0.4657$ and $\hat{\rho} = 0.6129$ as well as their correponding p-values 0.00035 and 0.00032, respectively, thus confirming a positive correlation between T and X. In this example, a Frank's Copula has been chosen to model the dependence between T and X. As explained in Rahali et al. [36], if $\hat{\tau} = 0.4657$ then an estimation of the dependence parameter is $\hat{\theta} = 5.14$.

Concerning the marginal distributions of T and X, the best fit using the Kolmogorov -Smirnov's test is to choose a Gamma distribution for the time T with parameters ($a_0 = 11.6488, b_0 = 5.0562$) and the Weibull distribution for the cost X with parameters ($a_0 = 4.8472, b_0 = 5396.4958$).

Assuming an in-control ATS value, $\text{ATS}_0 = 9125$ days (i.e. 25 years), the upper control limits of the three TBEA charts based on statistics Z_1, Z_2 and Z_3 are found to be $\text{UCL}_{Z_1} = 0.57$, $\text{UCL}_{Z_2} = 2.06$ and $\text{UCL}_{Z_3} = 3.18$, respectively, see Rahali et al. [36] for more details. The three TBEA charts, corresponding to the statistics Z_1, Z_2 and Z_3, are plotted in Fig. 6 along with their upper control limits. As it can be seen, the Phase 1 part of these charts seems to confirm the fact that from 2012 to 2016, the time between consecutive breakdowns and their corresponding costs were in stable state. But, from 2017, things seems to have changed as several out-of-control situations (see values in bold in Table 8) have been detected by all the TBEA charts due to more frequent breakdowns and an increasing maintenance cost (also due to an aging machine). Every time an out-of-control situation is detected, the production has been stopped, the root causes of the breakdown have been searched for, analyzed and repaired. Then, the machine has been restarted.

4 A Distribution-Free TBEA Chart

4.1 Motivation

The TBEA control charts developed in the previous sections are *parametric* ones, i.e. they assume that the distributions of the Times X and their Amplitudes X are perfectly known. However, as it is mentioned in Qiu [30], in many practical situations, the distributions of these random variables are unknown (or their parameters cannot be reliably estimated by means of a Phase 1 retrospective study) making the implementation of traditional parametric control charts to be an incorrect approach. For

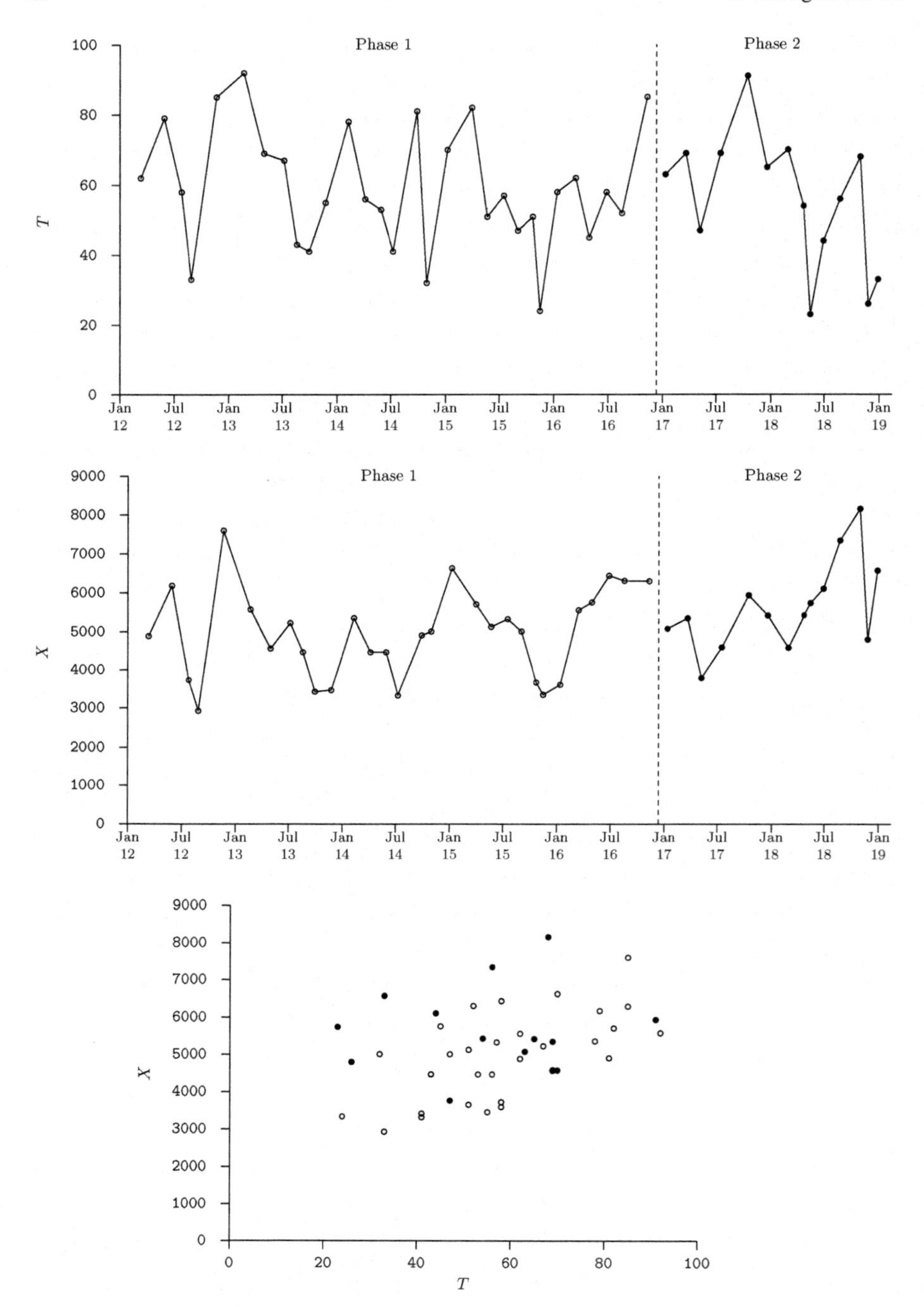

Fig. 5 Phase 1 (∘) and 2 (•) data corresponding to the time (T in days) between breakdowns and amplitudes (X in euros) corresponding to the data set in Table 8.

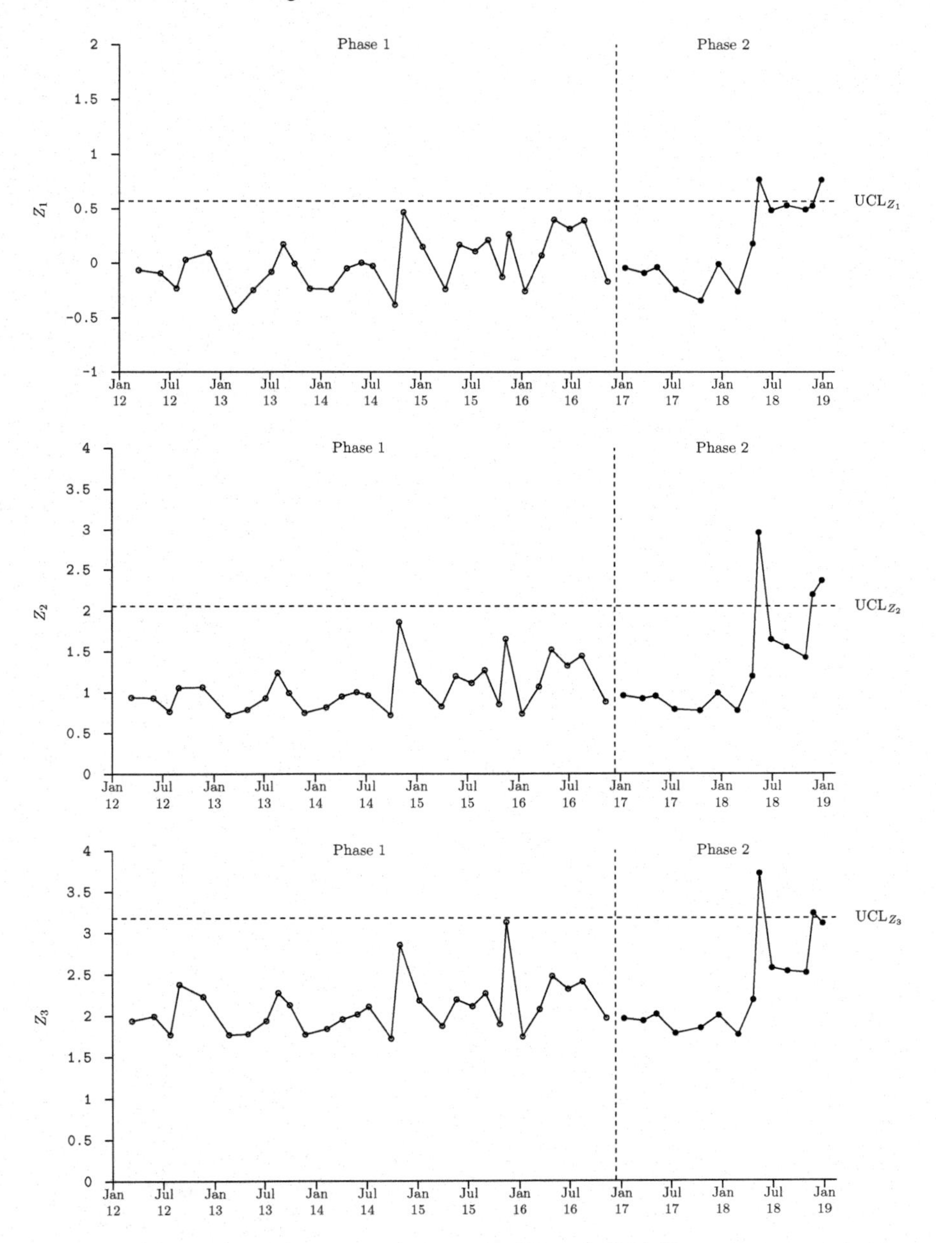

Fig. 6 Statistics Z_1, Z_2 and Z_3 corresponding to the data set in Table 8.

Table 8 Phase 1 and 2 data sets corresponding to the time (T_i in days) between two consecutive breakdowns, amplitudes (X_i cost in euros) and the values of the statistics Z_1, Z_2 and Z_3.

Phase 1						
Date	i	T_i	X_i	$Z_{1,i}$	$Z_{2,i}$	$Z_{3,i}$
10/03/12	1	62	4890	-0.064	0.939	1.939
28/05/12	2	79	6180	-0.092	0.932	1.995
25/07/12	3	58	3730	-0.231	0.766	1.770
27/08/12	4	33	2930	0.032	1.057	2.377
20/11/12	5	85	7600	0.093	1.065	2.230
20/02/13	6	92	5580	-0.434	0.722	1.768
30/04/13	7	69	4570	-0.247	0.789	1.778
06/07/13	8	67	5230	-0.080	0.930	1.937
18/08/13	9	43	4470	0.174	1.238	2.274
28/09/13	10	41	3420	-0.005	0.993	2.128
22/11/13	11	55	3460	-0.234	0.749	1.770
08/02/14	12	78	5360	-0.241	0.818	1.839
05/04/14	13	56	4470	-0.047	0.951	1.956
28/05/14	14	53	4470	0.004	1.004	2.015
08/07/14	15	41	3320	-0.025	0.964	2.108
27/09/14	16	81	4910	-0.382	0.722	1.720
29/10/14	17	32	5010	0.470	1.864	2.854
07/01/15	18	70	6630	0.152	1.128	2.182
30/03/15	19	82	5710	-0.238	0.829	1.873
20/05/15	20	51	5130	0.171	1.198	2.192
16/07/15	21	57	5330	0.110	1.114	2.111
01/09/15	22	47	5010	0.215	1.269	2.266
22/10/15	23	51	3660	-0.126	0.855	1.895
15/11/15	24	24	3340	0.268	1.657	3.129
12/01/16	25	58	3600	-0.257	0.739	1.743
14/03/16	26	62	5560	0.072	1.068	2.074
28/04/16	27	45	5760	0.401	1.524	2.473
25/06/16	28	58	6440	0.317	1.322	2.318
16/08/16	29	52	6310	0.393	1.445	2.408
09/11/16	30	85	6300	-0.169	0.883	1.967

Phase 2						
Date	i	T_i	X_i	$Z_{1,i}$	$Z_{2,i}$	$Z_{3,i}$
11/01/17	1	63	5080	-0.043	0.960	1.962
21/03/17	2	69	5350	-0.090	0.923	1.935
07/05/17	3	47	3770	-0.036	0.955	2.015
15/07/17	4	69	4590	-0.243	0.792	1.782
14/10/17	5	91	5940	-0.344	0.777	1.848
18/12/17	6	65	5420	-0.008	0.993	2.002
26/02/18	7	70	4580	-0.262	0.779	1.767
21/04/18	8	54	5430	0.181	1.197	2.189
14/05/18	9	23	5740	**0.770**	**2.972**	**3.721**
27/06/18	10	44	6110	0.488	1.654	2.574
22/08/18	11	56	7340	0.533	1.561	2.536
29/10/18	12	68	8160	0.495	1.429	2.516
24/11/18	13	26	4800	0.529	**2.199**	**3.236**
27/12/18	14	33	6570	**0.768**	**2.371**	3.113

instance, the parametric distributions that have been investigated in Sects. 2 and 3, are the Gamma, Lognormal, Normal and Weibull, but are these distributions really appropriate in these cases? To overcome this problem, *distribution-free* control charts have been proposed and investigated in the literature. Among the most recent ones, we can cite for instance Celano et al. [14], Abid et al. [1–4], Castagliola et al. [13] and Chakraborti and Graham [16]. Most of these approaches use nonparametric statistics like the Sign or the Wilcoxon signed-rank statistics. Practical guidelines for the implementation of such distribution-free control charts can be found in Chakraborti and Graham [15].

In this Section, we present an upper-sided distribution-free EWMA control chart for monitoring TBEA data. Moreover, as evaluating the Run Length properties of any EWMA scheme based on discrete data is a challenging problem, we will also introduce a specific method called "continuousify" which allows to obtain much better and replicable results.

4.2 Model

In this Section, we assume that $F_T(t|\theta_T)$ and $F_X(x|\theta_X)$ are the *unknown* continuous c.d.f. of T_i and $X_i, i = 1, 2, \ldots$, where θ_T and θ_X are *known* quantiles, respectively. As for the previous sections, when the process is in-control, we have $\theta_T = \theta_{T_0}, \theta_X = \theta_{X_0}$ and, when the process is out-of-control, we have $\theta_T = \theta_{T_1}, \theta_X = \theta_{X_1}$. Without loss of generality, we will consider that θ_T and θ_X are the median values of T_i and X_i, respectively. Of course, if necessary, other quantiles can also be considered.

Let $p_T = \mathrm{P}(T_i > \theta_{T_0}|\theta_T) = 1 - F_T(\theta_{T_0}|\theta_T)$ and $p_X = \mathrm{P}(X_i > \theta_{X_0}|\theta_X) = 1 - F_X(\theta_{X_0}|\theta_X)$, $i = 1, 2, \ldots$, be the probabilities that T_i and X_i are larger than θ_{T_0} and θ_{X_0} assuming that the actual median values are θ_T and θ_X, respectively. Let us also define $q_T = 1 - p_T$ and $q_X = 1 - p_X$. If the process is in-control, we have $p_T = p_{T_0} = 1 - F_T(\theta_{T_0}|\theta_{T_0}) = 0.5, p_X = p_{X_0} = 1 - F_X(\theta_{X_0}|\theta_{X_0}) = 0.5$ and, when the process is out-of-control, we have $p_T = p_{T_1} = 1 - F_T(\theta_{T_0}|\theta_{T_1})$, $p_X = p_{X_1} = 1 - F_X(\theta_{X_0}|\theta_{X_1})$.

The upper-sided distribution-free EWMA TBEA control chart that will be introduced in this section is based on the following sign statistics ST_i and SX_i, for $i = 1, 2, \ldots$ as

$$\mathrm{ST}_i = \mathrm{sign}(T_i - \theta_{T_0}),$$
$$\mathrm{SX}_i = \mathrm{sign}(X_i - \theta_{X_0}),$$

where $\mathrm{sign}(x) = -1$ if $x < 0$ and $\mathrm{sign}(x) = +1$ if $x > 0$. The case $x = 0$ will not be considered here (even if it can happen in practice) because the random variables T_i and X_i are assumed to be continuous ones. In order to simultaneously monitor $(T_i, X_i), i = 1, 2, \ldots$, we suggest to define the statistic S_i, for $i = 1, 2, \ldots$ as

$$S_i = \frac{\mathrm{SX}_i - \mathrm{ST}_i}{2}.$$

Since $\mathrm{ST}_i \in \{-1, +1\}$ and $\mathrm{SX}_i \in \{-1, +1\}$ we have $S_i \in \{-1, 0, +1\}$ and, more precisely, we have:

- $S_i = -1$ when T_i increases ($\mathrm{ST}_i = +1$) and, at the same time, X_i decreases ($\mathrm{SX}_i = -1$). In this case, the process seems to be in an *acceptable* situation.
- $S_i = +1$ when T_i decreases ($\mathrm{ST}_i = -1$) and, at the same time, X_i increases ($\mathrm{SX}_i = +1$). In this case, the process seems to be in an *unacceptable* situation.

- $S_i = 0$ when both T_i and X_i increase or when both T_i and X_i decrease. In this case, the process seems to be in an *intermediate* situation.

It can be easily proven that the p.m.f. (probability mass function) $f_{S_i}(s|p_T, p_X) = \text{P}(S_i = s|p_T, p_X)$ and the c.d.f. $F_{S_i}(s|p_T, p_X) = \text{P}(S_i \leq s|p_T, p_X)$ of S_i are equal to

$$f_{S_i}(s|p_T, p_X) = \begin{cases} p_T q_X & \text{if } s = -1 \\ p_T p_X + q_T q_X & \text{if } s = 0 \\ q_T p_X & \text{if } s = +1 \\ 0 & \text{if } s \notin \{-1, 0, 1\} \end{cases},$$

and

$$F_{S_i}(s|p_T, p_X) = \begin{cases} 0 & \text{if } s \in (-\infty, -1) \\ p_T q_X & \text{if } s \in [-1, 0) \\ p_T + q_T q_X & \text{if } s \in [0, 1) \\ 1 & \text{if } s \in [1, +\infty) \end{cases}.$$

If we define an EWMA TBEA type control chart directly monitoring the statistic S_i, it is known (see Wu et al. [43] for details) that, due to the strong discrete nature of this statistic, an accurate computation using Markov chain or integral equation methods, for instance, of the corresponding Run Length properties (ARL, SDRL, ...) is impossible. Consequently, it is actually possible to define an EWMA TBEA type control chart based on S_i but it is unfortunately impossible to correctly evaluate its properties and, therefore, it is impossible to design it in a reliable way. In order to overcome this problem and before any implementation of an EWMA scheme, Wu et al. [43] suggested to define an extra parameter $\sigma \in [0.1, 0.2]$ and to transform the discrete random variable S_i into a new continuous one, denoted as S_i^* defined as a mixture of 3 normal random variables $Y_{i,-1} \sim \text{Nor}(-1, \sigma)$, $Y_{i,0} \sim \text{Nor}(0, \sigma)$ and $Y_{i,+1} \sim \text{Nor}(+1, \sigma)$, with weights $w_{-1} = p_T q_X$, $w_0 = p_T p_X + q_T q_X$ and $w_{+1} = q_T p_X$ (corresponding to the probabilities $f_{S_i}(s|p_T, p_X)$, $s \in \{-1, 0, +1\}$), respectively, i.e.

$$S_i^* = \begin{cases} Y_{i,-1} & \text{if } S_i = -1, \\ Y_{i,0} & \text{if } S_i = 0, \\ Y_{i,+1} & \text{if } S_i = +1. \end{cases}$$

This strategy has been called *continuousify* by Wu et al. [43]. It is easy to prove that the c.d.f. $F_{S_i^*}(s|p_T, p_X) = \text{P}(S_i^* \leq s|p_T, p_X)$ of S_i^* is equal to

$$\begin{aligned} F_{S_i^*}(s|p_T, p_X) = & p_T q_X F_{\text{Nor}}(s| - 1, \sigma) \\ & + (p_T p_X + q_T q_X) F_{\text{Nor}}(s|0, \sigma) + q_T p_X F_{\text{Nor}}(s| + 1, \sigma), \end{aligned} \tag{18}$$

and its expectation $\text{E}(S_i^*)$ and variance $\text{V}(S_i^*)$ are equal to

$$\begin{aligned} \text{E}(S_i^*) &= p_X - p_T, \\ \text{V}(S_i^*) &= \sigma^2 + p_T q_T + p_X q_X. \end{aligned}$$

when the process is in-control, we have $p_{T_0} = q_{T_0} = 0.5$, $p_{X_0} = q_{X_0} = 0.5$ and the expectation and variance of S_i^* simplify to $\mathrm{E}(S_i^*) = 0$ and $\mathrm{V}(S_i^*) = \sigma^2 + 0.5$. As the main goal is to detect an increase in S_i (in order to avoid, for instance, more damages or injuries/costs) rather than a decrease, the following upper-sided EWMA TBEA control chart based on the statistic Z_i^* is proposed

$$Z_i^* = \max(0, \lambda S_i^* + (1-\lambda) Z_{i-1}^*), \tag{19}$$

with the following upper asymptotic control limit UCL defined as

$$\mathrm{UCL} = \underbrace{\mathrm{E}(S_i^*)}_{=0} + K\sqrt{\frac{\lambda}{2-\lambda}} \times \underbrace{\sqrt{\mathrm{V}(S_i^*)}}_{=\sqrt{\sigma^2+0.5}} = K\sqrt{\frac{\lambda(\sigma^2+0.5)}{2-\lambda}}, \tag{20}$$

where $\lambda \in [0, 1]$ and $K > 0$ are the control chart parameters to be fixed and the initial value $Z_0^* = 0$. The zero-state ARL and SDRL of the proposed distribution-free upper-sided EWMA TBEA control chart can be obtained using the standard approach of Brook and Evans [10] which assumes that the behavior of this control chart can be well represented by a discrete-time Markov chain with $m + 2$ states, where states $i = 0, 1, \ldots, m$ are transient and state $m + 1$ is an absorbing one. The transition probability matrix $\mathbf{P}$ of this discrete-time Markov chain is

$$\mathbf{P} = \begin{pmatrix} \mathbf{Q} & \mathbf{r} \\ \mathbf{0}^\intercal & 1 \end{pmatrix} = \begin{pmatrix} Q_{0,0} & Q_{0,1} & \cdots & Q_{0,m} & r_0 \\ Q_{1,0} & Q_{1,1} & \cdots & Q_{1,m} & r_1 \\ \vdots & \vdots & & \vdots & \vdots \\ Q_{m,0} & Q_{m,1} & \cdots & Q_{m,m} & r_m \\ 0 & 0 & \cdots & 0 & 1 \end{pmatrix},$$

where $\mathbf{Q}$ is the $(m+1, m+1)$ matrix of transient probabilities, where $\mathbf{0} = (0, 0, \ldots, 0)^\intercal$ and where the $(m+1, 1)$ vector $\mathbf{r}$ satisfies $\mathbf{r} = \mathbf{1} - \mathbf{Q}\mathbf{1}$ (i.e. row probabilities must sum to 1) with $\mathbf{1} = (1, 1, \ldots, 1)^\intercal$. The transient states $i = 1, \ldots, m$ are obtained by dividing the interval $[0, \mathrm{UCL}]$ into m subintervals of width 2Δ, where $\Delta = \frac{\mathrm{UCL}}{2m}$. By definition, the midpoint of the i–th subinterval (representing state i) is equal to $H_i = (2i - 1)\Delta$. The transient state $i = 0$ corresponds to the "restart state" feature of our chart and it is represented by the value $H_0 = 0$. Concerning the proposed upper-sided EWMA TBEA control chart, it can be easily shown that the generic element $Q_{i,j}$, $i = 0, 1, \ldots, m$, of the matrix $\mathbf{Q}$ is equal to

- if $j = 0$,

$$Q_{i,0} = F_{S_i^*}\left(-\frac{(1-\lambda)H_i}{\lambda}\middle| p_T, p_X\right), \tag{21}$$

- if $j = 1, 2, \ldots, m$,

$$Q_{i,j} = F_{S_i^*}\left(\frac{H_j + \Delta - (1-\lambda)H_i}{\lambda}\middle| p_T, p_X\right) - F_{S_i^*}\left(\frac{H_j - \Delta - (1-\lambda)H_i}{\lambda}\middle| p_T, p_X\right) \tag{22}$$

Let $\mathbf{q} = (q_0, q_1, \ldots, q_m)^\mathsf{T}$ be the $(m+1, 1)$ vector of initial probabilities associated with the $m+1$ transient states. In our case, we assume $\mathbf{q} = (1, 0, \ldots, 0)^\mathsf{T}$, i.e. the initial state corresponds to the "restart state". When the number m of subintervals is sufficiently large (say $m = 300$), this finite approach provides an effective method that allows the ARL and SDRL to be accurately evaluated using the following classical formulas

$$\text{ARL} = \mathbf{q}^\mathsf{T}(\mathbf{I} - \mathbf{Q})^{-1}\mathbf{1}, \tag{23}$$

$$\text{SDRL} = \sqrt{2\mathbf{q}^\mathsf{T}(\mathbf{I} - \mathbf{Q})^{-2}\mathbf{Q}\mathbf{1} + \text{ARL}(1 - \text{ARL})}. \tag{24}$$

In order to illustrate the "power" of the *continuousify* technique on the upper-sided EWMA TBEA control chart (in the case $K = 3$ and $\lambda = 0.2$), Table 9 presents some ARL values obtained without (left side) and with (right side) this technique, corresponding to 3 combinations for (p_X, p_T), a number of subintervals $m \in \{100, 120, \ldots, 400\}$ and $\sigma = 0.125$. When the *continuousify* technique is not used, the random variable S_i^* in (19) is replaced by S_i or, equivalently, the parameter σ is set to 0. For comparison purpose, the last row of Table 9 also provides the ARL values obtained by simulations. As it can be seen, if the *continuousify* technique is not used (left side), the ARL values obtained using the Markov chain method can have a large variability without any clear monotonic convergence when m increases. It turns out that with these unstable ARL values, it can be really difficult to design and optimize the upper-sided EWMA TBEA control chart. On the other side, if the *continuousify* technique is used (right side), the ARL values obtained using the Markov chain method are clearly very stable, even for small values of m, but (and this is the price to pay for this stability) they are *a bit* larger than those obtained by simulations (compare 26.08, 12.23, 27.88 vs. 24.71, 11.66, 26.46). This property is due to the extra term $\sigma > 0$ in (20).

4.3 Comparative Studies

Since the *continuousify* technique is demonstrated to provide reliable ARL values, it is therefore possible to compute the optimal chart parameters (λ^*, K^*) for the upper-sided EWMA TBEA control chart minimizing the out-of-control $\text{ARL}(\lambda^*, K^*, \sigma, p_T, p_X)$ for $p_T \neq 0.5$ and $p_X \neq 0.5$ under the constraint $\text{ARL}(\lambda^*, K^*, \sigma, 0.5, 0.5) = \text{ARL}_0$, where ARL_0 is a predefined value for the in-control ARL. These optimal values are listed in Table 10 with the corresponding out-of-control (ARL, SDRL) values for $p_T \in \{0.1, 0.2, \ldots, 0.4\}$ (only considering

Table 9 ARL for the distribution-free EWMA TBEA chart computed with and without the *continuousify* technique.

	Without continuousify			**With continuousify** ($\sigma = 0.125$)		
	$p_T = 0.3$	$p_T = 0.2$	$p_T = 0.1$	$p_T = 0.3$	$p_T = 0.2$	$p_T = 0.1$
m	$p_X = 0.8$	$p_X = 0.9$	$p_X = 0.6$	$p_X = 0.8$	$p_X = 0.9$	$p_X = 0.6$
100	32.96	12.18	40.16	26.08	12.23	27.87
120	18.57	10.89	18.55	26.08	12.23	27.87
140	28.88	11.91	31.56	26.08	12.23	27.87
160	20.31	11.08	20.81	26.08	12.23	27.87
180	28.36	11.82	31.36	26.08	12.23	27.87
200	24.47	11.55	26.42	26.08	12.23	27.87
220	17.97	10.05	18.39	26.08	12.23	27.87
240	27.98	11.88	31.36	26.08	12.23	27.87
260	57.68	16.41	77.81	26.08	12.23	27.87
280	21.17	11.41	21.75	26.08	12.23	27.87
300	26.75	11.75	29.94	26.08	12.23	27.88
320	21.69	11.43	22.43	26.08	12.23	27.88
340	26.07	11.93	27.39	26.08	12.23	27.88
360	16.68	10.43	16.5	26.08	12.23	27.88
380	29.2	13.09	29.83	26.08	12.23	27.88
400	20.33	11.02	20.7	26.08	12.23	27.88
Simu	24.71	11.66	26.46	26.09	12.23	27.87

a decrease in T), $p_X \in \{0.5, 0.6, \ldots, 0.9\}$ (only considering an increase in X), for four possible choices for $\sigma \in \{0.1, 0.125, 0.15, 0.2\}$ and assuming $\text{ARL}_0 = 370.4$. These values of (λ^*, K^*) can be freely be used by practitioners who need to optimally detect a specific shift in the times and/or in the amplitudes.

A comparison between the upper-sided EWMA TBEA control chart introduced in this Section and the three parametric TBEA control charts presented in Subsect. 2.1 has been investigated in Wu et al. [43] using the EARL_1 (instead of the EATS_1) for the following two scenarios

- Scenario #1: a Normal distribution for X with in-control mean $\mu_{X_0} = 10$ and standard-deviation $\sigma_{X_0} = 1$ and a gamma distribution for T with in-control mean $\mu_{T_0} = 10$ and standard-deviation $\sigma_{T_0} = 2$, i.e. $X \sim \text{Nor}(10, 1)$ and $T \sim \text{Gam}(25, 0.4)$.
- Scenario #2: a Normal distribution for X with in-control mean $\mu_{X_0} = 10$ and standard-deviation $\sigma_{X_0} = 2$ and a Weibull distribution for T with in-control mean $\mu_{T_0} = 10$ and standard-deviation $\sigma_{T_0} = 1$, i.e. $X \sim \text{Nor}(10, 2)$ and $T \sim \text{Wei}(12.1534, 10.4304)$.

Table 10 Optimal values for (λ^*, K^*) with the corresponding out-of-control values of (ARL, SDRL) for $p_T \in \{0.1, 0.2, \ldots, 0.4\}$, $p_X \in \{0.5, 0.6, \ldots, 0.9\}$ and $\sigma \in \{0.1, 0.125, 0.15, 0.2\}$

	$\sigma = 0.1$				
	p_X				
p_T	0.5	0.6	0.7	0.8	0.9
0.5	(–, –) (370.40, –)				
0.4	(0.010, 1.773) (105.66, 74.04)	(0.025, 2.174) (50.77, 32.32)			
0.3	(0.025, 2.174) (51.54, 32.55)	(0.045, 2.387) (30.55, 18.04)	(0.070, 2.515) (20.50, 11.38)		
0.2	(0.040, 2.348) (31.30, 17.51)	(0.070, 2.515) (20.74, 11.40)	(0.100, 2.591) (14.85, 7.67)	(0.145, 2.639) (11.19, 5.55)	
0.1	(0.060, 2.474) (21.40, 10.76)	(0.090, 2.571) (15.16, 7.37)	(0.135, 2.634) (11.32, 5.40)	(0.180, 2.645) (8.76, 3.84)	(0.240, 2.627) (6.99, 2.74)
	$\sigma = 0.125$				
	p_X				
p_T	0.5	0.6	0.7	0.8	0.9
0.5	(–, –) (370.40, –)				
0.4	(0.010, 1.774) (106.19, 74.55)	(0.025, 2.174) (51.11, 32.63)			
0.3	(0.025, 2.174) (51.88, 32.87)	(0.045, 2.387) (30.79, 18.25)	(0.070, 2.515) (20.68, 11.53)		
0.2	(0.040, 2.348) (31.53, 17.72)	(0.065, 2.496) (20.91, 11.27)	(0.100, 2.592) (14.99, 7.80)	(0.140, 2.638) (11.32, 5.57)	
0.1	(0.060, 2.474) (21.57, 10.92)	(0.090, 2.572) (15.30, 7.50)	(0.135, 2.634) (11.44, 5.49)	(0.175, 2.648) (8.88, 3.89)	(0.225, 2.639) (7.10, 2.75)
	$\sigma = 0.15$				
	p_X				
p_T	0.5	0.6	0.7	0.8	0.9
0.5	(–, –) (370.40, –)				
0.4	(0.010, 1.775) (106.83, 75.16)	(0.025, 2.175) (51.53, 33.01)			
0.3	(0.025, 2.175) (52.30, 33.26)	(0.045, 2.387) (31.08, 18.51)	(0.070, 2.515) (20.90, 11.71)		
0.2	(0.040,2.348) (31.82, 17.97)	(0.065, 2.496) (21.13, 11.45)	(0.095, 2.584) (15.17, 7.81)	(0.135, 2.636) (11.47, 5.61)	
0.1	(0.055, 2.449) (21.79, 10.76)	(0.090, 2.573) (15.47, 7.64)	(0.130, 2.632) (11.59, 5.53)	(0.170, 2.651) (9.02, 3.96)	(0.215, 2.646) (7.23, 2.80)
	$\sigma = 0.2$				
	p_X				
p_T	0.5	0.6	0.7	0.8	0.9
0.5	(–, –) (370.40, –)				
0.4	(0.010, 1.777) (108.43, 76.68)	(0.025, 2.176) (52.57, 33.96)			
0.3	(0.020, 2.085) (53.33, 32.51)	(0.045, 2.387) (31.81, 19.15)	(0.065, 2.496) (21.44, 11.90)		
0.2	(0.040, 2.348) (32.53, 18.61)	(0.065, 2.496) (21.66, 11.92)	(0.090, 2.574) (15.60, 8.01)	(0.125, 2.630) (11.84, 5.74)	
0.1	(0.055, 2.449) (22.31, 11.21)	(0.085, 2.562) (15.89, 7.83)	(0.120, 2.624) (11.95, 5.64)	(0.155, 2.652) (9.34, 4.06)	(0.195, 2.658) (7.53, 2.91)

Based on the results in Tables 3 and 4 in Wu et al. [43], the conclusion is that no matter the scenario (#1 or #2) or the statistic considered $Z \in \{Z_1, Z_2, Z_3\}$, the out-of-control EARL_1 values obtained for the distribution-free upper-sided EWMA TBEA control chart are always smaller than the ones obtained for the parametric Shewhart control charts introduced in Subsect. 2.1, thus showing the advantage of using the proposed distribution-free control chart in situations where the distributions for T and X are unknown.

4.4 Illustrative Example

We consider here the same illustrative example as the one already presented in Sect. 2.6 concerning the days T_i between fires in forests of the French region "Provence - Alpes - Côte D'Azur" and their amplitudes X_i (burned surface in $ha = 10000m^2$). In order to compute the control limit UCL of the distribution-free upper-sided EWMA TBEA chart, the following values have been fixed: $p_T = 0.3$, $p_X = 0.7$, $\sigma = 0.125$ and $\text{ARL}_0 = 370.4$. The corresponding optimal values for λ and K are found to be $\lambda^* = 0.07$ and $K^* = 2.515$ (see results in Table 10) and the upper control limit UCL is equal to

$$\text{UCL} = 2.515 \times \sqrt{\frac{0.07 \times (0.125^2 + 0.5)}{2 - 0.07}} = 0.344.$$

The in-control median values for T and X have been estimated from the Phase 1 data set and they are equal to $\theta_{T_0} = 3$ days and $\theta_{X_0} = 5.3\ ha$. These values are used to compute the values ST_i, SX_i, S_i and S_i^* in Table 11. As it can be noticed, some dates are such that $T_i - \theta_{T_0} = 0$. Of course, this situation is not supposed to happen as the times T_i are supposed to be continuous random variables but, due to the measurement scale (days), this situation actually happens. When this situation occurs, a possible simple strategy consists in assigning $\text{ST}_i = 0$ (instead of -1 or $+1$). For this reason, in Table 11, some values of $S_i = s = \pm 0.5$ and the corresponding values for S_i^* are obtained by randomly generating a $\text{Nor}(s, \sigma)$ random variable, as it is already the case for values $s \in \{-1, 0, +1\}$. For instance, in Table 11, when $D_i = 70$ we have $S_i = 0.5$ and the corresponding value for S_i^* has been randomly generated from a $\text{Nor}(0.5, 0.125)$ distribution ($S_i^* = 0.552$). The values Z_i^* have been computed using Eq. (19), for both Phase 1 and 2 data sets, recorded in Table 11 and plotted in Fig. 7 along with the distribution-free EWMA TBEA upper control limit UCL = 0.344. If the distribution-free upper-sided EWMA TBEA chart does not detect any out-of-control situations during the Phase 1 (validating the in-control state of this phase), it nevertheless detects several out-of-control situations during the period mid-June 2017—end of September 2017, (see also the bold values in Table 11), confirming that a decrease in the time between fires occurred with a concurrent increase in the amplitude of these fires. This conclusion is consistent with the one obtained in

Table 11 Phase 1 and 2 values of D_i, T_i, X_i, ST_i, SX_i, S_i, S_i^* and Z_i^* for the forest fires example.

Phase 1								Phase 2							
D_i	T_i	X_i	ST_i	SX_i	S_i	S_i^*	Z_i^*	D_i	T_i	X_i	ST_i	SX_i	S_i	S_i^*	Z_i^*
9	9	3.68	1	-1	-1.0	-0.917	0.000	258	1	1.00	-1	-1	0.0	-0.078	0.000
26	17	1.99	1	-1	-1.0	-0.802	0.000	260	2	3.70	-1	-1	0.0	0.119	0.008
60	34	6.00	1	1	0.0	-0.081	0.000	262	2	3.17	-1	-1	0.0	-0.063	0.003
67	7	1.19	1	-1	-1.0	-0.901	0.000	265	3	18.40	0	1	0.5	0.333	0.026
70	3	135.80	0	1	0.5	0.552	0.039	268	3	1.00	0	-1	-0.5	-0.145	0.014
72	2	14.37	-1	1	1.0	1.113	0.114	269	1	2.22	-1	-1	0.0	0.208	0.028
86	14	8.10	1	1	0.0	-0.104	0.099	271	2	19.09	-1	1	1.0	1.001	0.096
88	2	32.31	-1	1	1.0	0.892	0.154	272	1	2.00	-1	-1	0.0	0.027	0.091
94	6	3.07	1	-1	-1.0	-1.056	0.069	274	2	34.28	-1	1	1.0	1.086	0.161
95	1	10.03	-1	1	1.0	0.867	0.125	276	2	3.00	-1	-1	0.0	0.070	0.154
96	1	7.93	-1	1	1.0	1.033	0.189	277	1	6.63	-1	1	1.0	0.955	0.210
97	1	1.50	-1	-1	0.0	0.409	0.204	278	1	4.47	-1	-1	0.0	-0.097	0.189
103	6	23.30	1	1	0.0	-0.116	0.182	285	7	8.24	1	1	0.0	0.160	0.187
106	3	3.73	0	-1	-0.5	-0.708	0.120	286	1	769.45	-1	1	1.0	1.024	0.246
109	3	4.73	0	-1	-0.5	-0.677	0.064	287	1	4.37	-1	-1	0.0	-0.144	0.218
111	2	3.19	-1	-1	0.0	0.179	0.072	288	1	90.70	-1	1	1.0	0.961	0.270
113	2	6.25	-1	1	1.0	1.032	0.139	289	1	11.49	-1	1	1.0	1.044	0.324
114	1	3.60	-1	-1	0.0	-0.155	0.118	295	6	3590.78	1	1	0.0	0.033	0.304
115	1	6.12	-1	1	1.0	1.112	0.188	296	1	1427.92	-1	1	1.0	0.949	**0.349**
118	3	1.50	0	-1	-0.5	-0.740	0.123	297	1	255.96	-1	1	1.0	1.054	**0.399**
122	4	1.33	1	-1	-1.0	-1.009	0.044	298	1	1.00	-1	-1	0.0	-0.051	**0.367**
134	12	1.42	1	-1	-1.0	-1.037	0.000	302	4	13.88	1	1	0.0	-0.074	0.336
137	3	5.75	0	1	0.5	0.629	0.044	303	1	138.28	-1	1	1.0	1.117	**0.391**
140	3	3.47	0	-1	-0.5	-0.507	0.005	305	2	8.90	-1	1	1.0	1.153	**0.444**
142	2	13.31	-1	1	1.0	1.217	0.090	308	3	1.50	0	-1	-0.5	-0.342	**0.389**
143	1	26.31	-1	1	1.0	1.041	0.157	312	4	34.63	1	1	0.0	-0.217	**0.347**
144	1	18.54	-1	1	1.0	0.923	0.210	313	1	82.56	-1	1	1.0	0.811	**0.379**
146	2	66.17	-1	1	1.0	1.124	0.274	314	1	2.00	-1	-1	0.0	-0.019	**0.351**
147	1	9.90	-1	1	1.0	0.916	0.319	315	1	162.08	-1	1	1.0	1.071	**0.402**
150	3	4.22	0	-1	-0.5	-0.534	0.260	319	4	3.26	1	-1	-1.0	-1.056	0.300
157	7	34.28	1	1	0.0	-0.110	0.234	321	2	285.91	-1	1	1.0	0.729	0.330
161	4	2.23	1	-1	-1.0	-1.102	0.140	322	1	2.00	-1	-1	0.0	-0.283	0.287
162	1	1.84	-1	-1	0.0	0.152	0.141	325	3	11.57	0	1	0.5	0.347	0.291
163	1	2.88	-1	-1	0.0	-0.018	0.130	334	9	34.70	1	1	0.0	0.068	0.275
164	1	21.46	-1	1	1.0	1.087	0.197	335	1	431.00	-1	1	1.0	1.150	0.337
165	1	4.46	-1	-1	0.0	-0.001	0.183	336	1	10.89	-1	1	1.0	1.003	**0.383**
166	1	58.27	-1	1	1.0	1.034	0.243	340	4	1.00	1	-1	-1.0	-1.004	0.286
167	1	8.84	-1	1	1.0	0.863	0.286	346	6	1.50	1	-1	-1.0	-0.921	0.202
180	13	1.03	1	-1	-1.0	-0.905	0.203	347	1	1.17	-1	-1	0.0	0.100	0.195
187	7	16.57	1	1	0.0	0.156	0.199	349	2	1.27	-1	-1	0.0	0.129	0.190
201	14	4.96	1	-1	-1.0	-1.084	0.110	350	1	26.25	-1	1	1.0	1.098	0.254
202	1	1.37	-1	-1	0.0	-0.087	0.096	353	3	11.66	0	1	0.5	0.332	0.259
205	3	23.39	0	1	0.5	0.498	0.124	354	1	3.03	-1	-1	0.0	0.127	0.250
225	20	1.70	1	-1	-1.0	-1.032	0.043	355	1	12.00	-1	1	1.0	1.130	0.311
247	22	5.30	1	0	-0.5	-0.727	0.000	356	1	1.00	-1	-1	0.0	-0.206	0.275
248	1	15.64	-1	1	1.0	1.161	0.081								
257	9	5.14	1	-1	-1.0	-0.921	0.011								

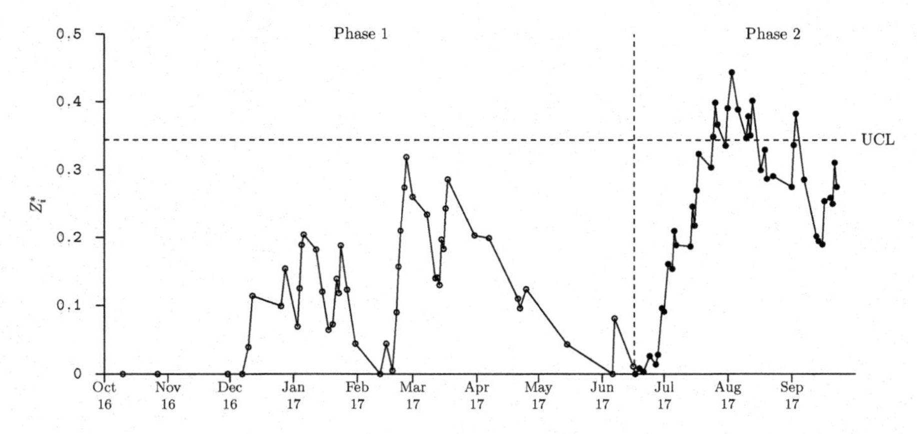

Fig. 7 Distribution-free EWMA TBEA chart with statistic Z_i^* corresponding to the data set in Table 11

Sect. 2.6 in which a parametric approach assuming a lognormal distribution for both T_i and X_i was used.

5 Conclusions

The three contributive Sections of this Chapter have clearly demonstrated that efficient solutions do exist when the aim is to simultaneously monitor the time interval T of an event E as well as its amplitude X. These solutions can be either parametric, for independent or dependent situations, and they can also be distribution-free if there is no a priori knowledge about the distributions associated with T and X.

In our opinion, future researches on the monitoring of TBEA data can be undertaken toward several directions:

- In the proposed parametric approaches, the estimation of the parameters (for the distributions or the Copulas) is not taken into account at all in the design and evaluation of the TBEA control charts. The impact of the parameter estimation is known to strongly influence the efficiency of any control chart and, therefore, researches on this topic should be done.
- Measures like times or amplitudes are obviously subject to measurement errors. These kinds of error are also known to negatively impact the efficiency of any control chart. Reasearch investigating the impact of the measurement errors on T and/or X should also be undertaken in order to evaluate how much they actually impact the performance of parametric TBEA control charts.
- Instead of considering X as a univariate random variable, it could be considered in some cases as a p-variate random vector $\mathbf{X} = (X_1, \ldots, X_p)$ where each X_k is the amplitude of a specific characteristic and the goal would be to simultaneously

monitor $(T, \mathbf{X})$. For instance, in the forest fires example, the amplitude could be considered as a bivariate vector $\mathbf{X} = (X_1, X_2)$ where X_1 would be the burned surface and X_2 would be the cost related to the fires.

- Often, historical data availability in monitoring of adverse events is limited to a few records. Thus, knowledge about the frequency distribution of these events is too restricted to fit a reliable statistical model. With these scenarios, there is room for approaching the monitoring problem with distribution-free approaches, which, therefore, deserve a lot of attention by researchers.

References

1. Abid M, Nazir HZ, Riaz M, Lin Z (2016) Use of ranked set sampling in nonparametric control charts. J Chin Inst Eng 39(5):627–636
2. Abid M, Nazir HZ, Riaz M, Lin Z (2017) An efficient nonparametric EWMA Wilcoxon signed-rank chart for monitoring location. Qual Reliab Eng Int 33(3):669–685
3. Abid M, Nazir HZ, Riaz M, Lin Z (2017) Investigating the impact of ranked set sampling in nonparametric CUSUM control charts. Qual Reliab Eng Int 33(1):203–214
4. Abid M, Nazir HZ, Tahir M, Riaz M (2018) On designing a new cumulative sum Wilcoxon signed rank chart for monitoring process location. PloS One 13(4):e0195762
5. Ali MM, Mikhail NN, Haq MS (1978) A class of bivariate distributions including the bivariate logistic. J Multivar Anal 8(3):405–412
6. Ali S, Pievatolo A (2018) Time and magnitude monitoring based on the renewal reward process. Reliab Eng Syst Saf 179:97–107
7. Benneyan JC (2001) Number-between g-type statistical quality control charts for monitoring adverse events. Health Care Manag Sci 4(4):305–318
8. Borror CM, Keats JB, Montgomery DC (2003) Robustness of the time between events CUSUM. Int J Prod Res 41(15):3435–3444
9. Brechmann EC, Schepsmeier U (2013) Modeling dependence with C- and D-vine copulas: the R package CDVine. J Stat Softw 52(3):1–27
10. Brook D, Evans DA (1972) An approach to the probability distribution of CUSUM run length. Biometrika 59(3):539–549
11. Busababodhin P, Amphanthong P (2016) Copula modelling for multivariate statistical process control: a review. Commun Stat Appl Methods 23(6):497–515
12. Calvin T (1983) Quality control techniques for "zero defects". IEEE Trans Compon Hybrids Manuf Technol 6(3):323–328
13. Castagliola P, Tran KP, Celano G, Rakitzis AC, Maravelakis PE (2019) An EWMA-type sign chart with exact run length properties. J Qual Technol 51(1):51–63
14. Celano G, Castagliola P, Chakraborti S, Nenes G (2016) The performance of the Shewhart sign control chart for finite horizon processes. Int J Adv Manuf Technol 84(5):1497–1512
15. Chakraborti S, Graham MA (2019a) Nonparametric statistical process control. Wiley, Hoboken
16. Chakraborti S, Graham MA (2019b) Nonparametric (distribution-free) control charts: an updated overview and some results. Qual Eng 31(4):523–544
17. Cheng CS, Chen PW (2011) An ARL-unbiased design of time-between-events control charts with runs rules. J Stat Comput Simul 81(7):857–871
18. Cheng Y, Mukherjee A (2014) One hotelling T^2 chart based on transformed data for simultaneous monitoring the frequency and magnitude of an event. In: 2014 IEEE international conference on industrial engineering and engineering management, pp 764–768
19. Cheng Y, Mukherjee A, Xie M (2017) Simultaneously monitoring frequency and magnitude of events based on bivariate gamma distribution. J Stat Comput Simul 87(9):1723–1741

20. Clayton DG (1978) A model for association in bivariate life tables and its application in epidemiological studies of familial tendency in chronic disease incidence. Biometrika 65(1):141–151
21. Dokouhaki P, Noorossana R (2013) A Copula Markov CUSUM chart for monitoring the bivariate auto-correlated binary observations. Qual Reliab Eng Int 29(6):911–919
22. Fang YY, Khoo MBC, Teh SY, Xie M (2016) Monitoring of time-between-events with a generalized group runs control chart. Qual Reliab Eng Int 32(3):767–781
23. Fatahi AA, Dokouhak P, Moghaddam BF (2011) A bivariate control chart based on copula function. In: 2011 IEEE international conference on quality and reliability (ICQR). IEEE, pp 292–296
24. Frank MJ (1979) On the simultaneous associativity of $F(x, y)$ and $x + y - F(x, y)$. Aequationes Math 19(1):194–226
25. Gan FF (1998) Designs of one- and two-sided exponential EWMA charts. J Qual Technol 30(1):55
26. Gumbel EJ (1960) Distributions des Valeurs Extremes en Plusieurs Dimensions. Publications de l'Institut de Statistique de l'Université de Paris 9:171–173
27. Joe H (1993) Parametric families of multivariate distributions with given margins. J Multivar Anal 46(2):262–282
28. Liu JY, Xie M, Goh TN, Sharma PR (2006) A comparative study of exponential time between events charts. Qual Technol Quant Manag 3(3):347–359
29. Lucas JM (1985) Counted data CUSUM's. Technometrics 27(2):129–144
30. Qiu P (2014) Introduction to statistical process control. Chapman and Hall, Boca Raton
31. Qu L, Wu Z, Khoo MBC, Castagliola P (2013) A CUSUM scheme for event monitoring. Int J Prod Econ 145(1):268–280
32. Qu L, Wu Z, Khoo MBC, Rahim A (2014) Time-between-event control charts for sampling inspection. Technometrics 56(3):336–346
33. Qu L, He S, Khoo MBC, Castagliola P (2018) A CUSUM chart for detecting the intensity ratio of negative events. Int J Prod Res 56(19):6553–6567
34. Radaelli G (1998) Planning time-between-events Shewhart control charts. Total Qual Manag 9(1):133–140
35. Rahali D, Castagliola P, Taleb H, Khoo MBC (2019) Evaluation of Shewhart time-between-events-and-amplitude control charts for several distributions. Qual Eng 31(2):240–254. https://doi.org/10.1080/08982112.2018.1479036
36. Rahali D, Castagliola P, Taleb H, Khoo MBC (2021) Evaluation of Shewhart time-between-events-and-amplitude control charts for correlated data. Qual Reliab Eng Int 37(1):219–241. https://doi.org/10.1002/qre.2731
37. Sanusi RA, Teh SY, Khoo MBC (2020) Simultaneous monitoring of magnitude and time-between-events data with a max-EWMA control chart. Comput Ind Eng 142:106378
38. Shafae MS, Dickinson RM, Woodall WH, Camelio JA (2015) Cumulative sum control charts for monitoring Weibull-distributed time between events. Qual Reliab Eng Int 31(5):839–849
39. Shamsuzzaman M, Xie M, Goh TN, Zhang HY (2009) Integrated control chart system for time-between-events monitoring in a multistage manufacturing system. Int J Adv Manuf Technol 40(3):373–381
40. Sklar A (1959) Fonctions de Répartition à n Dimensions et leurs Marges. Publications de l'Institut de Statistique de l'Université de Paris 8:229–231
41. Sukparungsee S, Kuvattana S, Busababodhin P, Areepong Y (2018) Bivariate copulas on the Hotelling's T^2 control chart. Commun Stat-Simul Comput 47(2):413–419
42. Vardeman S, Ray DO (1985) Average run lengths for CUSUM schemes when observations are exponentially distributed. Technometrics 27(2):145–150
43. Wu S, Castagliola P, Celano G (2021) A distribution-free EWMA control chart for monitoring time-between-events-and-amplitude data. J Appl Stat 48(3):434–454. https://doi.org/10.1080/02664763.2020.1729347
44. Wu Z, Jiao J, Zhen H (2009) A control scheme for monitoring the frequency and magnitude of an event. Int J Prod Res 47(11):2887–2902

45. Xie M, Goh TN, Ranjan P (2002) Some effective control chart procedures for reliability monitoring. Reliab Eng Syst Saf 77(2):143–150
46. Zhang CW, Xie M, Liu JY, Goh TN (2007) A control chart for the Gamma distribution as a model of time between events. Int J Prod Res 45(23):5649–5666
47. Zhang HY, Shamsuzzaman M, Xie M, Goh TN (2011) Design and application of exponential chart for monitoring time-between-events data under random process shift. Int J Adv Manuf Technol 57(9):849–857
48. Zhang HY, Xie M, Goh TN, Shamsuzzaman M (2011) Economic design of time-between-events control chart system. Comput Ind Eng 60(4):485–492

Monitoring a BAR(1) Process with EWMA and DEWMA Control Charts

Maria Anastasopoulou and Athanasios C. Rakitzis

Abstract In this work we study one-sided and two-sided EWMA and Double EWMA control charts for monitoring an integer-valued autocorrelated process with a bounded support. The performance of the proposed charts is studied via simulation. We compare the performance of the proposed charts and provide aspects for the statistical design and practical implementation. The results of an extensive numerical study, that consists of the examination of a wide variety of out-of-control situations, show that none of the chart outperforms the other uniformly. Specifically, both charts have a difficulty in detecting decreasing shifts in the autocorrelation parameter. An illustrative example based on real data is also provided.

Keywords Attributes control charts · Binomail AR(1) · Integer-valued time series · Statistical process monitoring

1 Introduction

Statistical process monitoring (SPM) is a collection of tools that allows the monitoring of a process. Among them the control chart is the most widely used SPM tool. Initially, its main use was in the monitoring of manufacturing processes, aiming at the detection of abnormal (usually unwanted) situations such as an increase in the percentage of defective items that are produced by the process. However, nowadays, since processes become more and more complex, their use is not restricted in industry but also on several other areas of applied science like public health, environment and social networks (see, for example, Bersimis et al. [1], Woodall et. al. [2], Aykroyd et al. [3]).

M. Anastasopoulou · A. C. Rakitzis (✉)
Department of Statistics and Actuarial Financial-Mathematics, Laboratory of Statistics and Data Analysis, University of the Aegean, 83200 Karlovasi, Samos, Greece
e-mail: arakitz@aegean.gr

M. Anastasopoulou
e-mail: anastasopoulou@aegean.gr

K. P. Tran (ed.), *Control Charts and Machine Learning for Anomaly Detection in Manufacturing*, Springer Series in Reliability Engineering,
https://doi.org/10.1007/978-3-030-83819-5_4

Popular charts to monitor the proportion and the number of nonconforming units, respectively, within a sample of finite size are the Shewhart p and np charts (Montgomery [4]). These monitoring schemes are developed under the assumption that the number of nonconforming units follows a binomial distribution $B(n, \pi)$, where n is the sample size and π is the success probability (i.e. the probability for an item or a unit to be nonconforming). Moreover, a common assumption when the p and the np charts are applied in practice is that the successive counts are independent and identically distributed (iid) binomial random variables (rv).

It is well known that Shewhart charts are control charts without memory, since they make use of only the value of the most recent observation. Consequently, they are not very sensitive in the detection of small and moderate changes in the values of process parameters. On the other hand, the cumulative sum (CUSUM) and exponentially weighted moving average (EWMA) control charts, as control charts with memory, detect this type of changes more quickly than the Shewhart charts (Montgomery [4]). Efficient alternative control charting procedures for monitoring binomial counts have been proposed and studied by Gan [5], Gan [6], Chang and Gan [7], Wu et al. [8], Yeh et al. [9], Haridy et al. [10] and Haridy et al. [11]. All the above mentioned control charts are based on the assumption of iid binomial rv. Even though the iid assumption is a common assumption in practice, observations on a process will be autocorrelated when the sampling rate is very high, which, in turn, commonly happens because of the technological progress in automated sampling (Psarakis and Papaleonida [12], Kim and Lee [13]). Therefore, in a variety of real-life problems, the iid assumption is violated.

In that case, the previously mentioned control charts cannot be used because they demonstrate an increased false alaram rate (FAR). This means that there are more frequent (than expected) signals that the process is out-of-control, when actually it is in-control and nothing has changed.

In the case of variables control charts, there are several approaches to deal with autocorrelated data. In the case of attribute (or count) data, there has been an increasing interest in the recent years, to deal with this problem. However, the methods and techniques that are used in the case of variables data, need first an appropriate adjustment due to the discrete nature of the count data.

One solution to this problem is to select first an appropriate model of integer-valued time series, and then to develop control charts based on this model. Weiß [14] provided a literature review for the available SPC methods that are used in the monitoring of a process that it is modelled as an integer-valued time-series model. In particular, if it is of interest to monitor the number X of defects in a sample of n objects, then there is a finite number of possible values for X. Therefore, the appropriate integer-valued time series model must be such that X takes a finite number of possible values. Consequently, an appropriate model for correlated binomial counts needs to be selected first.

The monitoring of correlated binomial counts has been considered by Weiß [15] who developed and studied Shewhart and Moving Average (MA) control charts for monitoring a process that is properly described by the first-order binomial autoregressive model (binomial AR(1) or BAR(1) model) of McKenzie [16] and Al-Osh and Alzaid [17]. Apart from this work, Shewhart, CUSUM and EWMA control charts have been proposed and studied by Rakitzis et al. [18] and Anastasopoulou and Rakitzis [19] in the case of monitoring a BAR(1) process

It is well-known that although the Shewhart-type charts are easy to use and effective in detecting large, sudden and sustained shifts in the process parameters, they are not very sensitive in the detection of small and moderate shifts.

Even though CUSUM and EWMA control charts are better than Shewhart charts in the detection of small and moderate shifts in process parameters, there is an increasing interest in improving further their performance. Sometimes this can be achieved by developing more "sophisticated" charts which are control charts with memory and are defined by mixing different (or the same) schemes. A method that belongs to this class of control charts is the double EWMA (DEWMA) chart. Shamma et al. [20] and Shamma and Shamma [21] developed the DEWMA control chart in an attempt to improve the performance of usual EWMA chart in the detection of small shifts in process mean. The idea behind the DEWMA is on the method of double exponentially weighted moving average, which is a common forecasting method in time series analysis. The DEWMA chart has been studied by many authors (see, for example, Mahmoud and Woodall [22], Khoo et al. [23], Adeoti and Malela-Majika [24], Raza et al. [25] and references therein). Zhang et al. [26] studied the DEWMA chart in the case of monitoring and detecting changes in a Poisson process while its performance has been also studied in the case of zero-inflated Poisson (Alevizakos and Koukouvinos [27]), a zero-inflated binomial (Alevizakos and Koukouvinos [28]) and Conway-Maxwell Poisson (Alevizakos and Koukouvinos [29]) process.

Motivated by the previously mentioned works, in this work we study, via Monte Carlo simulation, the performance of one- and two-sided EWMA and double EWMA (DEWMA) control charts in the monitoring of BAR(1) process. To the best of our knowledge, the performance of the DEWMA chart has not been investigated in the case of serially dependent count data. Moreover, in order to highlight the usefulness and the applicability of the proposed EWMA and DEWMA schemes in anomaly detection, we consider various types of shifts as possible out-of-control cases (anomalies or abnormalities). The aim is to assess how much both schemes are affected by the different types of shifts that may occur in the values of process parameters and also, how possible is to detect these anomalies.

The rest of this work is organized as follows: In Sect. 2 we briefly present the main properties of the BAR(1) model. In Sect. 3 we demonstrate the methodology for the proposed one- and two-sided EWMA and DEWMA control charts in the case of monitoring a BAR(1) process (Sects. 3.1 and 3.2) as well as the measures of performance for each chart and their statistical design (Sect. 3.3). Section 4 consists of the results of an extensive numerical study on the performance of the proposed charts. In Sect. 5, we provide an example for the practical implementation of the proposed charts by using a real dataset of correlated counts with bounded support. Finally, conclusions and topics for future research are summarized in Sect. 6.

2 The Binomial AR(1) Model

The BAR(1) model (McKenzie [16]) is a simple model for autocorrelated processes of counts with a finite range. This model is based on the binomial thinning operator "$\circ$" (Steutel and van Harn [30]). More specifically, if X is a non-negative discrete rv and $\alpha \in (0, 1)$ then, by using the binomial thinning operator, it is possible to define the rv $\alpha \circ X = \sum_{i=1}^{X} Y_i$, as an alternative to the usual multiplication $\alpha \cdot X$. However, the result of $\alpha \circ X$ will always be an integer value. The rv Y_i, $i = 1, 2, \ldots$, are iid Bernoulli rv with success probability α, independent also of the count data rv X. Therefore, the conditional distribution of $\alpha \circ X$, given $X = x$, is the binomial distribution $B(x, \alpha)$. We will refer to a process X_t, $t \in \mathbb{N} = \{1, 2, \ldots\}$, as a BAR(1) process if it is of the form

$$X_t = \alpha \circ X_{t-1} + \beta \circ (n - X_{t-1}), \tag{1}$$

where $\beta = \pi \cdot (1 - \rho)$, $\alpha = \beta + \rho$, $\pi \in (0, 1)$, $\rho \in (\max\{-\pi/(1-\pi), -(1-\pi)/\pi\}, 1)$ and $n \in \mathbb{N}$ is fixed. The condition on ρ guarantees that $\alpha, \beta \in (0, 1)$. Moreover, all thinnings are performed independently of each other and the thinnings at time t are independent of X_s, $s < t$, as well.

It is known (see, for example, Weiß [15]) that the process X_t, $t \in \mathbb{N}_0 = \{0, 1, 2, \ldots\}$, is a stationary Markov chain with marginal distribution $B(n, \pi)$. Clearly, the marginal mean and variance are, respectively, equal to

$$\mathbb{E}(X_t) = n\pi, \quad \mathbb{V}(X_t) = n\pi(1 - \pi). \tag{2}$$

Moreover, the transition probabilities are

$$\begin{aligned} p_{k|l} &= P(X_t = k \,|X_{t-1} = l) \\ &= \sum_{m=max\{0,k+l-n\}}^{min\{k,l\}} \binom{l}{m}\binom{n-l}{k-m} \alpha^m (1-\alpha)^{l-m} \beta^{k-m} (1-\beta)^{n-l+m-k}, \end{aligned} \tag{3}$$

for $k, l \in \{0, 1, 2, \ldots, n\}$.

The conditional mean and variance, respectively, are equal to (see Weiß and Kim [31])

$$\begin{aligned} \mathbb{E}(X_t \,|X_{t-1}) &= \rho \cdot X_{t-1} + n\beta, \\ \mathbb{V}(X_t \,|X_{t-1}) &= \rho(1-\rho)(1-2\pi) \cdot X_{t-1} + n\beta(1-\beta), \end{aligned} \tag{4}$$

while the autocorrelation function is given by $\rho(k) = \rho^k$ for $k = 0, 1, \ldots$.

Parameters π and ρ of the BAR(1) model can be estimated via the method of Conditional Maximum Likelihood (CML, see Weiß and Kim [31]), when time series data are available. Let us assume that $x_1, \ldots, x_T$, $T \in \mathbb{N}$, is a segment from a stationary BAR(1) process. Then the conditional likelihood function equals

$$L(\pi, \rho) = \binom{n}{x_1} \pi^{x_1} (1 - \pi)^{n - x_1} \prod_{t=2}^{T} p_{x_{t-1}|x_t}, \tag{5}$$

where the probabilities $p_{x_{t-1}|x_t}$ are given in Eq. (3). There is no closed-form formula for the maximum likelihood (ML) estimators $\hat{\pi}_{ML}$, $\hat{\rho}_{ML}$ of π, ρ and therefore, they are obtained by maximizing numerically the log-likelihood function $l(\pi, \rho) = \log L(\pi, \rho)$. The corresponding standard errors can be computed from the observed Fisher's Information matrix.

3 Methods

In this section, we introduce the proposed one-sided and two-sided EWMA and DEWMA control charts for monitoring a BAR(1) process. The aim is to detect quickly and accuratelly a change in the either parameters of the process. When the process is in-control (IC), we will denote its IC process mean level as $\mu_{0,X}$ while in the out-of-control state (OoC), it is denoted as $\mu_{1,X}$. In a similar manner, the IC (OoC) parameter values of the BAR(1) model are denoted as π_0 and ρ_0 (π_1 and ρ_1).

Usually, practitioners focus on changes in the mean level $\mu \equiv \mu_X = E(X_t) = n\pi$ of the process. Specifically, in several applications, practitioners are interested in detecting increases in the process mean level, from an IC value $\mu_{0,X}$ to an OoC value $\mu_{1,X} > \mu_{0,X}$. For example, if X is the number of non-conforming items produced by a manufacturing process, then the presence of assignable causes might affect (increase) the average number of the produced nonconforming items. This is also an indication of process deterioration. On the contrary, when there is a decrease in the mean level of the process, i.e. when $\mu_{1,X} < \mu_{0,X}$, then less non-conforming items are produced, which is an indication of process improvement. In this work we consider both cases. Moreover, under certain circumstances, the presence of assignable causes has an effect on the IC value ρ_0 of ρ. Note also, that a change in ρ_0 does not affect directly the value of $\mu_{0,X}$ (see Eq. 2). Generally speaking, the presence of assignable causes might affect both μ and ρ or exactly one of them. In this work, we consider a wide variety of possible OoC situations.

3.1 EWMA Control Charts

The EWMA control chart was introduced by Roberts [32]. For $t = 1, 2, \ldots$, the vaules of the following statistic are plotted on the chart

$$Z_t = \lambda X_t + (1 - \lambda) Z_{t-1}, \quad Z_0 = z_0, \tag{6}$$

where λ is a smoothing parameter such that $0 < \lambda \leq 1$ and the initial value Z_0 equals $\mu_{0,X} = n\pi_0$. For small values of λ, is given less weight to the most recent observation X_t and more weight is given to all the available observations since the beginning of process monitoring. Usually, λ takes values in the interval $[0.05, 0.30]$. This is a control chart with memory and it is more capable of than a Shewhart control chart in detecting shifts of small or medium magnitude in the mean level of the process. For $\lambda = 1$, the EWMA chart coincides with the usual Shewhart chart.

The two-sided EWMA chart for a BAR(1) process gives an OoC signal when for the first time $Z_t \notin [LCL_{\text{ewma}}, UCL_{\text{ewma}}]$, where LCL_{ewma}, UCL_{ewma} are the control limits of the chart. The values of these limits are determined so as the two-sided EWMA chart has the desired performance.

One-sided control charts are recommended when the direction of the shift in known and/or predetermined. In public-health surveillance, sometimes we are interested in monitoring the effects that has "corrective action" (or intervention), like a vaccination programme, in the weekly number of new cases from a disease. In this case we would like to detect a decrease in weekly number of new cases and therefore, the use of a lower one-sided chart is recommended. On the contrary, for the detection of an increase in the weekly number of new cases, the use of an upper one-sided chart is more appropriate.

Therefore, in the case of an upper (resp. lower) one-sided EWMA chart, only an upper (lower) control limit UCL_{ewma} (LCL_{ewma}) needs to be determined, for a given λ value. When the value of the EWMA statistic crosses for the first the control limit, then the one-sided EWMA chart gives an OoC signal.

3.2 DEWMA Control Charts

The DEWMA control chart was introduced by Shamma and Shamma [21] and for $t = 1, 2, \ldots$, the values of the following statistic are plotted on it:

$$Y_t = \lambda Z_t + (1 - \lambda)Y_0, \quad t = 1, 2, \ldots, \tag{7}$$

where Z_t is given in Eq. (6), $0 < \lambda \leq 1$ is the smoothing parameter and the initial value $Y_0 = \mu_{0,X} = n\pi_0$. Therefore, the exponential smoothing is performed twice and the Y_t values are extra smoothed (compared to the Z_t). Similar to the case of the EWMA chart, popular values for λ are in the interval $[0.05, 0.30]$. The two-sided DEWMA chart for a BAR(1) process gives an out-of-control signal when for the first time $Y_t \notin [LCL_{\text{dewma}}, UCL_{\text{dewma}}]$, where LCL_{dewma}, UCL_{dewma} are the control limits of the chart. The values of these limits are determined so as the two-sided DEWMA chart has the desired performance. The development and implementation of the corresponding one-sided DEWMA charts is made similar to the one-sided EWMA charts.

3.3 Performance Measures

In order to evaluate the performance of the EWMA and DEWMA charts, it is necessary to determine their run length distribution. For the case of a two-sided EWMA control chart, with control limits LCL_e, UCL_e, the run length distribution is defined as the distribution of the rv

$$RL = \min\{t : Z_t \notin [LCL_{\text{ewma}}, UCL_{\text{ewma}}]\}$$

and expresses the number of points plotted on the chart until it gives for the first time an OoC signal. In a similar manner, we define the RL distribution in the case of the two-sided DEWMA chart, as well as for each of the one-sided schemes. In this work we use the method of Monte Carlo simulation since the values of charting statistics (in (6) and (7)) are not integers, they can take a variable number of possible different values. Also, it is worth mentioning that in almost all of the works related to these "mixed" charts, like the DEWMA and its extensions, Monte Carlo simulation is used to evaluate their performance, mainly due to their complexity. On the contrary, in the case of the EWMA chart, it is possible to use the Markov chain method (see, for example, Weiß [33]) and evaluate its exact performance. However, before applying the Markov chain method, a modification is needed for Eq. (6). The common modification is to apply a rounding function in order to have integer values for Z_t. In this work we make use of the increased computational power that it is available nowadays. Thus, we use the usual EWMA statistic, without any modification, and evaluate its performance.

The most common performance measure of a control chart is the expected value $\mathbb{E}(RL)$, the well-known average run length (ARL). The ARL expresses the average number of points to be plotted on the chart until it gives for the first time an OoC signal. In this work, the IC performance of the proposed schemes is evaluated in terms of the zero-state ARL ($zsARL$) which is the expected number of points plotted on the chart until the first (false) alarm is given.

For an OoC process, the performance of the proposed schemes is evaluated in terms of the steady-state ARL ($ssARL$) which gives an approximation of the true mean delay for detection after a change in the process, from the IC state to the OoC state. We assume that a change in process happens at an (unknown) change-point $\tau \in \{1, 2, \ldots\}$. Specifically, for $t < \tau$, the process is in the IC state while for $t \geq \tau$, the process has shifted to the OoC state. Therefore, the $ssARL$ expresses the expected number of points to be plotted on the chart until it gives for the first time an indication of an OoC process, given that the process has been operated for"sufficient time" in control. According to Weiß and Testik [34], the $zsARL$ and the $ssARL$ are substantially different in the case of monitoring processes with correlated counts.

The statistical design of the two-sided EWMA (or DEWMA) control chart requires the determination of the values for the triple $(\lambda, LCL_{\text{ewma}}, UCL_{\text{ewma}})$ (or $(\lambda, LCL_{\text{dewma}}, UCL_{\text{dewma}})$). Next, we provide the steps of the algorithmic procedure that is followed in order to determine the values of the design parameters in the case of the two-sided EWMA chart with the desired IC ARL performance.

Step 1 Choose the IC values of the process parameters n, π_0, ρ_0 and the desired in-control ARL_0 value for the $zsARL$.
Step 2 Choose the value for λ
Step 3 Set the control limits of the chart equal to $LCL_{\text{ewma}} = CL - K, UCL_{\text{ewma}} = CL + K$, where $CL = \lceil n\pi_0 \rceil$ and $CL = \lceil x \rceil$ denotes the minimum integer that it is greater than or equal to x. Use as starting value $K = 0.001$.
Step 4 Simulate 50000 BAR(1) processes with parameters (n, π_0, ρ_0) and for each sequence record the number of samples until the first false alarm is triggered.
Step 5 Estimate the $zsARL$ as the sample mean of the 50000 RL values obtained in Step 4. If $zsARL \notin (ARL_0 - 1, ARL_0 + 1)$, increase K by 0.001 and go back to Step 4. Otherwise, go to Step 6.
Step 6 Use the value for K that has been obtained in the previous step, set up the control limits for the two-sided chart and declare the process as OoC at sample t if $Z_t \notin [LCL_{\text{ewma}}, UCL_{\text{ewma}}]$.

We mention that Steps 1–6 apply for a pre-specified λ value in [0, 1]. In this work we considered several values for λ. However, our numerical analysis is focused on the most popular values for λ that are used in practice, such as $\lambda \in \{0.05, 0.10, 0.20, 0.30\}$. Once the triple $(\lambda, LCL_{\text{ewma}}, UCL_{\text{ewma}})$ has been determined, we evaluated the OoC performance of this scheme, for various shifts in process parameters. Below are the steps that have been used in order to determine the OoC $ssARL$ values for the two-sided EWMA control chart (see also Weiß and Testik [34]).

Step 1 Choose the IC values of the process parameters n, π_0, ρ_0 and the desired in-control ARL_0 value for the $zsARL$.
Step 2 Choose the shifts in process parameters or, equivalently, the OoC values π_1, ρ_1.
Step 3 Set up a two-sided EWMA control chart by using the values (λ, K) that have been obtained during the design phase of the chart.
Step 4 Simulate 50000 BAR(1) processes as follows: For each simulation run, generate first an IC BAR(1) process with $\pi = \pi_0$, $\rho = \rho_0$, until the $t = 199$ observation. Then, the process "shifts" to the OoC state, where $\pi = \pi_1$, $\rho = \rho_1$ and the simulation run continues. Now, the observations from $t \geq 200$ are generated from the OoC model.
Step 5 For each of the 50000 sequences, record the number of samples until the first (true) alarm is triggered. Use all the available data but, if an alarm is triggered on or before $t = 199$, then this simulation run is skipped. If the first alarm is triggered at some $t \geq 200$, then compute the $RL - 199$, which gives the (conditional) delay in the detection of the OoC situation.
Step 6 Average all the (conditional) delays and estimate the expected conditional delay $\mathbb{E}(RL - 200 + 1\,|\,RL \geq 200)$ which serves as an estimation of the $ssARL$, since the $ssARL$ is defined as

$$ssARL = \lim_{\tau\to\infty} \mathbb{E}(RL - \tau + 1 \,|\, RL \geq \tau)$$

After some direct (but necessary) modifications in the above steps, for both the IC and the OoC case, we may design and evaluate the performance of the two-sided DEWMA chart as well as the performance of the upper and the lower one-sided EWMA and DEWMA charts.

4 Numerical Analysis

In this section we present the results of an extensive numerical study on the performance of the proposed one- and two-sided EWMA and DEWMA control charts in the monitoring of a BAR(1) process. For the IC design parameters we assume that $\mu_0 \in \{4, 8, 12\}$, $\rho_0 \in \{0.25, 0.50, 0.75\}$ and $n \in \{20, 50\}$. The desired IC $zsARL$ equals 200. In addition, when the process is OoC, we assume the following OoC scenarios:

- A shift only in $\mu_{0,X}$ with $\mu_{1,X} = n(\delta \cdot \pi_0)$.
- A shift only in ρ_0 with $\rho_1 = \rho_0 + \tau$.
- A simultaneous shift in both $\mu_{0,X}$, ρ_0.

When $\delta > 1$ then μ_0 has been increased whereas a decreasing shift occurs for $0 < \delta < 1$. Also, when $\tau > 0$ there is an increase in the correlation structure of the process while a $\tau < 0$ denotes the case of a decreasing shift in ρ_0. Tables 1, 2, 3, 4, 5, 6, 7 and 8 provide the chart with the best performance (minimum $ssARL$ value) in the detection of a specific OoC case. Due to space economy we do not provide all the available results (the $ssARL$ profiles for all the examined charts) from the complete study. However, it is available from the authors upon request.

More specifically, Tables 1 and 2 consist of the best upper one-sided chart (between the EWMA and DEWMA) in the detection of shifts only in μ_0 (Table 1) or only in ρ_0 (Table 2). The values of the design parameters (λ, UCL) chart are given in the respective columns. For λ we pre-specified one value in $\{0.05, 0.10, 0.20, 0.30\}$. Then, following the steps of the algorithmic procedure described in Sect. 3.3, we determined first the value for K (either in the two-sided or the one-sided case) and then the control limits of the charts. For each pair (λ, K), we determined next the $ssARL$ value for the given shift in process parameter(s). The column "δ" (Table 1) provides the shifts in μ_0 while the column "τ" (Table 2) gives the shifts in ρ_0. Also, the column "chart" gives the appropriate chart (EWMA or DEWMA) that has to be used for the detection of the specific OoC case. The IC parameter values of the process are given in the column entitled "Process".

Table 1 reveals that, in general, the upper one-sided EWMA chart has better performance than the DEWMA chart. In almost all cases, at a given increasing shift δ, the EWMA chart attains a $ssARL$ value that it is lower than the one that it is attained by the DEWMA chart. For small increasing shifts (e.g. $\delta = 1.1$ or 1.2) we

Table 1 Suggested upper one-sided charts, shifts only in μ_0

Process	λ	UCL	δ	$ssARL$	Chart	Process	λ	UCL	δ	$ssARL$	Chart
$\mu_0 = 4$	0.05	4.267	1.1	54.12	DEWMA	$\mu_0 = 4$	0.05	4.284	1.1	57.49	DEWMA
$\rho_0 = 0.25$	0.05	4.651	1.2	26.69	EWMA	$\rho_0 = 0.25$	0.05	4.702	1.2	28.99	EWMA
$n = 20$	0.10	5.092	1.3	16.71	EWMA	$n = 50$	0.10	4.702	1.3	18.60	EWMA
	0.10	5.092	1.4	11.78	EWMA		0.10	5.186	1.4	13.13	EWMA
	0.20	5.757	1.5	8.93	EWMA		0.20	5.186	1.5	10.03	EWMA
$\mu_0 = 8$	0.05	8.783	1.1	32.77	EWMA	$\mu_0 = 8$	0.05	8.936	1.1	39.71	EWMA
$\rho_0 = 0.25$	0.10	9.302	1.2	14.48	EWMA	$\rho_0 = 0.25$	0.10	8.936	1.2	18.57	EWMA
$n = 20$	0.20	10.056	1.3	8.56	EWMA	$n = 50$	0.20	9.574	1.3	11.13	EWMA
	0.30	10.654	1.4	5.81	EWMA		0.30	10.512	1.4	7.72	EWMA
	0.30	10.654	1.5	4.26	EWMA		0.30	11.268	1.5	5.74	EWMA
$\mu_0 = 12$	0.05	12.772	1.1	20.26	EWMA	$\mu_0 = 12$	0.05	13.084	1.1	29.87	EWMA
$\rho_0 = 0.25$	0.20	13.983	1.2	8.10	EWMA	$\rho_0 = 0.25$	0.20	13.810	1.2	13.27	EWMA
$n = 20$	0.30	14.541	1.3	4.55	EWMA	$n = 50$	0.30	14.884	1.3	7.89	EWMA
	0.30	14.541	1.4	3.13	EWMA		0.30	15.732	1.4	5.36	EWMA
	0.30	14.541	1.5	2.45	EWMA		0.30	15.732	1.5	3.99	EWMA
$\mu_0 = 4$	0.05	4.352	1.1	69.36	DEWMA	$\mu_0 = 4$	0.05	4.375	1.1	72.18	DEWMA
$\rho_0 = 0.50$	0.05	4.819	1.2	37.17	EWMA	$\rho_0 = 0.50$	0.05	4.882	1.2	40.38	EWMA
$n = 20$	0.05	4.819	1.3	23.99	EWMA	$n = 50$	0.05	4.882	1.3	26.22	EWMA
	0.10	5.354	1.4	17.31	EWMA		0.10	5.472	1.4	19.13	EWMA
	0.10	5.354	1.5	13.08	EWMA		0.10	5.472	1.5	14.63	EWMA
$\mu_0 = 8$	0.05	8.980	1.1	44.42	EWMA	$\mu_0 = 8$	0.05	8.510	1.1	52.71	DEWMA
$\rho_0 = 0.50$	0.05	8.980	1.2	21.03	EWMA	$\rho_0 = 0.50$	0.05	9.177	1.2	25.88	EWMA
$n = 20$	0.10	9.607	1.3	12.73	EWMA	$n = 50$	0.10	9.941	1.3	16.37	EWMA
	0.20	10.458	1.4	8.69	EWMA		0.20	9.941	1.4	11.40	EWMA
	0.30	11.066	1.5	6.38	EWMA		0.30	11.003	1.5	8.58	EWMA
$\mu_0 = 12$	0.05	12.965	1.1	28.24	EWMA	$\mu_0 = 12$	0.05	13.362	1.1	41.20	EWMA
$\rho_0 = 0.50$	0.10	13.562	1.2	12.20	EWMA	$\rho_0 = 0.50$	0.10	13.362	1.2	19.47	EWMA
$n = 20$	0.30	14.928	1.3	6.89	EWMA	$n = 50$	0.30	14.237	1.3	11.74	EWMA
	0.30	14.928	1.4	4.60	EWMA		0.30	15.444	1.4	8.05	EWMA
	0.30	14.928	1.5	3.55	EWMA		0.30	16.332	1.5	5.98	EWMA
$\mu_0 = 4$	0.05	4.512	1.1	92.25	DEWMA	$\mu_0 = 4$	0.05	4.548	1.1	97.58	DEWMA
$\rho_0 = 0.75$	0.05	5.080	1.2	55.85	EWMA	$\rho_0 = 0.75$	0.05	5.168	1.2	60.05	EWMA
$n = 20$	0.05	5.080	1.3	38.36	EWMA	$n = 50$	0.05	5.168	1.3	41.34	EWMA
	0.05	5.080	1.4	28.13	EWMA		0.05	5.168	1.4	30.81	EWMA
	0.10	5.730	1.5	22.28	EWMA		0.10	5.168	1.5	24.47	EWMA
$\mu_0 = 8$	0.05	8.632	1.1	65.90	DEWMA	$\mu_0 = 8$	0.05	8.742	1.1	74.70	DEWMA
$\rho_0 = 0.75$	0.05	9.304	1.2	33.92	EWMA	$\rho_0 = 0.75$	0.05	9.557	1.2	41.18	EWMA
$n = 20$	0.10	10.046	1.3	21.59	EWMA	$n = 50$	0.10	9.557	1.3	26.89	EWMA
	0.20	10.946	1.4	15.28	EWMA		0.20	10.479	1.4	19.49	EWMA
	0.30	11.529	1.5	11.44	EWMA		0.30	10.479	1.5	15.03	EWMA
$\mu_0 = 12$	0.05	13.280	1.1	44.65	EWMA	$\mu_0 = 12$	0.05	13.800	1.1	61.81	EWMA
$\rho_0 = 0.75$	0.10	13.990	1.2	20.64	EWMA	$\rho_0 = 0.75$	0.10	13.800	1.2	31.42	EWMA
$n = 20$	0.20	14.844	1.3	12.27	EWMA	$n = 50$	0.20	14.850	1.3	19.88	EWMA
	0.30	15.386	1.4	8.22	EWMA		0.30	16.126	1.4	14.18	EWMA
	0.30	15.386	1.5	6.17	EWMA		0.30	16.942	1.5	10.60	EWMA

Table 2 Suggested upper one-sided charts, shifts only in ρ_0

Process	λ	UCL	τ	$ssARL$	Chart	Process	λ	UCL	τ	$ssARL$	Chart
$\mu_0 = 4$	0.30	5.500	0.15	119.41	DEWMA	$\mu_0 = 4$	0.30	5.63	0.15	118.51	DEWMA
$\rho_0 = 0.25$	0.30	5.500	0.10	76.39	DEWMA	$\rho_0 = 0.25$	0.30	5.63	0.10	76.63	DEWMA
$n = 20$						$n = 50$					
$\mu_0 = 8$	0.30	9.78	0.15	117.47	DEWMA	$\mu_0 = 8$	0.30	10.15	0.15	118.20	DEWMA
$\rho_0 = 0.25$	0.30	9.78	0.10	72.56	DEWMA	$\rho_0 = 0.25$	0.30	10.15	0.10	75.88	DEWMA
$n = 20$						$n = 50$					
$\mu_0 = 12$	0.30	13.73	0.15	117.85	DEWMA	$\mu_0 = 12$	0.30	14.48	0.15	120.49	DEWMA
$\rho_0 = 0.25$	0.30	13.73	0.10	71.51	DEWMA	$\rho_0 = 0.25$	0.30	14.48	0.10	73.65	DEWMA
$n = 20$						$n = 50$					
$\mu_0 = 4$	0.20	5.37	0.15	123.29	DEWMA	$\mu_0 = 4$	0.20	5.48	0.15	125.17	DEWMA
$\rho_0 = 0.50$	0.20	5.37	0.35	83.49	DEWMA	$\rho_0 = 0.50$	0.20	5.48	0.35	87.17	DEWMA
$n = 20$						$n = 50$					
$\mu_0 = 8$	0.20	9.63	0.15	118.83	DEWMA	$\mu_0 = 8$	0.20	9.97	0.15	122.03	DEWMA
$\rho_0 = 0.50$	0.20	9.63	0.35	81.56	DEWMA	$\rho_0 = 0.50$	0.20	9.97	0.35	84.42	DEWMA
$n = 20$						$n = 50$					
$\mu_0 = 12$	0.30	14.15	0.15	117.60	DEWMA	$\mu_0 = 12$	0.20	14.26	0.15	121.17	DEWMA
$\rho_0 = 0.50$	0.20	13.59	0.35	79.47	DEWMA	$\rho_0 = 0.50$	0.20	14.26	0.35	81.88	DEWMA
$n = 20$						$n = 50$					
$\mu_0 = 4$	0.10	5.05	0.10	140.16	DEWMA	$\mu_0 = 4$	0.10	5.14	0.10	144.90	DEWMA
$\rho_0 = 0.75$	0.10	5.05	0.20	134.74	DEWMA	$\rho_0 = 0.75$	0.10	5.14	0.20	135.85	DEWMA
$n = 20$						$n = 50$					
$\mu_0 = 8$	0.20	10.168	0.10	137.87	DEWMA	$\mu_0 = 8$	0.10	9.52	0.10	141.61	DEWMA
$\rho_0 = 0.75$	0.10	9.28	0.20	129.66	DEWMA	$\rho_0 = 0.75$	0.10	9.52	0.20	133.00	DEWMA
$n = 20$						$n = 50$					
$\mu_0 = 12$	0.10	13.26	0.10	136.24	DEWMA	$\mu_0 = 12$	0.10	13.76	0.10	139.86	DEWMA
$\rho_0 = 0.75$	0.10	13.26	0.20	123.08	DEWMA	$\rho_0 = 0.75$	0.10	13.76	0.20	134.34	DEWMA
$n = 20$						$n = 50$					

recommend a small value forλ (e.g. 0.05 or 0.10), while for larger shifts (e.g. $\delta \geq 1.5$) we suggest $\lambda = 0.30$. Note also that the difference in the $ssARL$ values, between the EWMA and the DEWMA charts can be up to a 35% difference, especially for moderate to large shifts.

Table 2 gives the results for the upper one-sided charts in the case of increasing shifts only in ρ_0. Contrary to the results in Table 1, we notice that now the DEWMA chart has better performance than the EWMA chart. Our numerical analysis reveals that when the IC autocorrelation is of low or medium size (e.g. $\rho_0 = 0.25$ or 0.5), then the recommended value for λ is 0.3. As ρ_0 increases, a smaller value for λ are recommended in order to achieve an increased detection ability. For $\rho_0 = 0.75$, we suggest $\lambda = 0.05$. It should be also noticed that the difference in using the suggested chart for detecting the specific shift is about 10–20%, depending on shift and the IC process parameters.

In Table 3 we provide the suggested chart for the detection of simultaneous shifts in μ_0 and ρ_0. The shifts in both parameters are given in the form (δ, τ). A 20%

Table 3 Suggested upper one-sided charts, simultaneous shifts in μ_0, ρ_0

Process	λ	UCL	(δ, τ)	$ssARL$	Chart	Process	λ	UCL	(δ, τ)	$ssARL$	Chart
$\mu_0 = 4$	0.30	5.500	(1.2,0.35)	26.66	DEWMA	$\mu_0 = 4$	0.30	5.629	(1.2,0.35)	28.31	DEWMA
$\rho_0 = 0.25$	0.10	4.562	(1.2,−0.10)	26.51	DEWMA	$\rho_0 = 0.25$	0.05	4.702	(1.2,−0.10)	28.62	EWMA
$n = 20$						$n = 50$					
$\mu_0 = 8$	0.30	9.778	(1.2,0.35)	16.30	DEWMA	$\mu_0 = 8$	0.30	10.152	(1.2,0.35)	19.65	DEWMA
$\rho_0 = 0.25$	0.10	9.302	(1.2,−0.10)	14.35	EWMA	$\rho_0 = 0.25$	0.05	8.936	(1.2,−0.10)	18.29	EWMA
$n = 20$						$n = 50$					
$\mu_0 = 12$	0.20	13.983	(1.2,0.35)	10.22	EWMA	$\mu_0 = 12$	0.30	14.476	(1.2,0.35)	15.35	DEWMA
$\rho_0 = 0.25$	0.20	13.983	(1.2,−0.10)	7.79	EWMA	$\rho_0 = 0.25$	0.10	13.810	(1.2,−0.10)	12.85	EWMA
$n = 20$						$n = 50$					
$\mu_0 = 4$	0.20	5.366	(1.2,0.35)	40.55	DEWMA	$\mu_0 = 4$	0.20	5.484	(1.2,0.35)	43.64	DEWMA
$\rho_0 = 0.50$	0.05	4.352	(1.2,−0.10)	37.84	DEWMA	$\rho_0 = 0.50$	0.05	4.375	(1.2,−0.10)	40.13	DEWMA
$n = 20$						$n = 50$					
$\mu_0 = 8$	0.20	9.627	(1.2,0.35)	27.93	DEWMA	$\mu_0 = 8$	0.20	9.965	(1.2,0.35)	32.97	DEWMA
$\rho_0 = 0.50$	0.10	9.607	(1.2,−0.10)	21.06	EWMA	$\rho_0 = 0.50$	0.05	9.177	(1.2,−0.10)	25.09	EWMA
$n = 20$						$n = 50$					
$\mu_0 = 12$	0.20	13.090	(1.2,0.35)	19.12	DEWMA	$\mu_0 = 12$	0.20	14.264	(1.2,0.35)	26.38	DEWMA
$\rho_0 = 0.50$	0.10	13.562	(1.2,−0.10)	11.58	EWMA	$\rho_0 = 0.50$	0.05	13.362	(1.2,−0.10)	18.75	EWMA
$n = 20$						$n = 50$					
$\mu_0 = 4$	0.10	5.053	(1.2,0.20)	74.99	DEWMA	$\mu_0 = 4$	0.10	5.136	(1.2,0.25)	78.84	DEWMA
$\rho_0 = 0.75$	0.05	4.512	(1.2,−0.25)	52.91	DEWMA	$\rho_0 = 0.75$	0.05	4.548	(1.2,−0.25)	58.00	DEWMA
$n = 20$						$n = 50$					
$\mu_0 = 8$	0.10	9.276	(1.2,0.20)	55.17	DEWMA	$\mu_0 = 8$	0.10	9.518	(1.2,0.25)	63.60	DEWMA
$\rho_0 = 0.75$	0.05	9.304	(1.2,−0.25)	31.03	EWMA	$\rho_0 = 0.75$	0.05	8.742	(1.2,−0.25)	38.54	DEWMA
$n = 20$						$n = 50$					
$\mu_0 = 12$	0.10	13.259	(1.2,0.20)	40.64	DEWMA	$\mu_0 = 12$	0.10	13.764	(1.2,0.25)	52.89	DEWMA
$\rho_0 = 0.75$	0.10	13.990	(1.2,−0.25)	17.45	EWMA	$\rho_0 = 0.75$	0.05	13.800	(1.2,−0.25)	27.93	EWMA
$n = 20$						$n = 50$					

increase is assumed for μ_0 while an increasing as well as a decreasing shift in ρ_0 are also considered. For a small μ_0, the DEWMA chart has a better performance than the EWMA chart while the latter is better when μ_0 increases and ρ_0 decreases. When both shifts are on the same direction (increase), the DEWMA chart attains a lower $ssARL$ value than the EWMA chart, in almost all cases. The difference in the $ssARL$ between the two charts is 5–20%, depending on the shifts in μ_0, ρ_0.

Similarly, Table 4 provides the best lower one-sided chart, between the EWMA and DEWMA, in the detection of downward shifts only in μ_0. The results reveal that the EWMA chart has the best performance in almost all cases. We suggest a λ value equal to 0.20 or 0.30 for moderate to large decreasing shifts while for small decreasing shifts (up to a 20% decrease), the recommended value is $\lambda = 0.10$. The DEWMA chart outperforms the EWMA chart only in case of a shift $\delta = 0.9$ (a 10% decrease in μ_0), for processes with a weak or moderate correlation structure ($\rho_0 = 0.25$ or 0.5). Using the EWMA chart instead of the DEWMA chart in the detection of decreasing shifts only in μ_0 can result even in a 50% decrease in the $ssARL$ value, especially for large decreasing shifts,

Table 4 Suggested lower one-sided charts, shifts only in μ_0

Process	λ	LCL	δ	$ssARL$	Chart	Process	λ	LCL	δ	$ssARL$	Chart
$\mu_0 = 4$	0.20	2.453	0.5	7.29	EWMA	$\mu_0 = 4$	0.20	2.375	0.5	7.90	EWMA
$\rho_0 = 0.25$	0.30	2.640	0.6	10.07	DEWMA	$\rho_0 = 0.25$	0.30	2.563	0.6	11.00	DEWMA
$n = 20$	0.20	3.008	0.7	15.14	DEWMA	$n = 50$	0.20	2.947	0.7	16.36	DEWMA
	0.05	3.380	0.8	25.00	EWMA		0.20	2.947	0.8	26.87	DEWMA
	0.05	3.732	0.9	52.20	DEWMA		0.05	3.712	0.9	55.50	DEWMA
$\mu_0 = 8$	0.30	5.458	0.5	3.94	EWMA	$\mu_0 = 8$	0.30	5.119	0.5	4.65	EWMA
$\rho_0 = 0.25$	0.30	5.458	0.6	5.43	EWMA	$\rho_0 = 0.25$	0.30	5.119	0.6	6.55	EWMA
$n = 20$	0.20	6.013	0.7	8.18	EWMA	$n = 50$	0.20	5.719	0.7	9.96	EWMA
	0.10	6.728	0.8	14.02	EWMA		0.10	6.529	0.8	16.82	EWMA
	0.05	7.227	0.9	32.51	EWMA		0.10	7.204	0.9	37.78	DEWMA
$\mu_0 = 12$	0.30	9.346	0.5	2.65	EWMA	$\mu_0 = 12$	0.30	8.554	0.5	3.46	EWMA
$\rho_0 = 0.25$	0.30	9.346	0.6	3.41	EWMA	$\rho_0 = 0.25$	0.30	8.554	0.6	4.66	EWMA
$n = 20$	0.30	9.346	0.7	4.91	EWMA	$n = 50$	0.20	9.303	0.7	7.05	EWMA
	0.20	9.944	0.8	8.56	EWMA		0.10	10.267	0.8	12.41	EWMA
	0.10	10.698	0.9	20.72	EWMA		0.05	10.944	0.9	29.13	EWMA
$\mu_0 = 4$	0.20	2.178	0.5	10.79	EWMA	$\mu_0 = 4$	0.20	2.093	0.5	11.47	EWMA
$\rho_0 = 0.50$	0.10	2.771	0.6	14.76	EWMA	$\rho_0 = 0.50$	0.10	2.703	0.6	15.88	EWMA
$n = 20$	0.10	2.771	0.7	21.48	EWMA	$n = 50$	0.10	2.703	0.7	23.10	EWMA
	0.05	3.228	0.8	34.39	EWMA		0.05	3.181	0.8	36.42	EWMA
	0.05	3.643	0.9	67.40	DEWMA		0.05	3.620	0.9	70.45	DEWMA
$\mu_0 = 8$	0.30	5.070	0.5	5.87	EWMA	$\mu_0 = 8$	0.30	4.668	0.5	6.98	EWMA
$\rho_0 = 0.50$	0.20	5.632	0.6	8.07	EWMA	$\rho_0 = 0.50$	0.20	5.307	0.6	9.66	EWMA
$n = 20$	0.10	6.438	0.7	12.20	EWMA	$n = 50$	0.10	6.196	0.7	14.57	EWMA
	0.05	7.035	0.8	20.21	EWMA		0.05	6.878	0.8	24.09	EWMA
	0.05	7.035	0.9	44.17	EWMA		0.05	6.878	0.9	50.19	EWMA
$\mu_0 = 12$	0.30	8.931	0.5	3.85	EWMA	$\mu_0 = 12$	0.30	8.031	0.5	5.12	EWMA
$\rho_0 = 0.50$	0.30	8.931	0.6	5.02	EWMA	$\rho_0 = 0.50$	0.20	8.800	0.6	6.99	EWMA
$n = 20$	0.30	8.931	0.7	7.42	EWMA	$n = 50$	0.20	8.800	0.7	10.54	EWMA
	0.10	10.396	0.8	12.74	EWMA		0.10	9.880	0.8	17.98	EWMA
	0.05	11.020	0.9	28.68	EWMA		0.10	10.781	0.9	39.64	DEWMA
$\mu_0 = 4$	0.10	2.434	0.5	18.59	EWMA	$\mu_0 = 4$	0.30	8.031	0.5	19.88	EWMA
$\rho_0 = 0.75$	0.10	2.434	0.6	25.01	EWMA	$\rho_0 = 0.75$	0.20	8.800	0.6	26.51	EWMA
$n = 20$	0.05	2.981	0.7	34.78	EWMA	$n = 50$	0.20	8.800	0.7	36.44	EWMA
	0.05	2.981	0.8	52.63	EWMA		0.10	9.880	0.8	55.16	EWMA
	0.05	3.480	0.9	88.89	DEWMA		0.05	3.445	0.9	93.75	DEWMA
$\mu_0 = 8$	0.30	4.618	0.5	10.53	EWMA	$\mu_0 = 8$	0.20	4.756	0.5	12.50	EWMA
$\rho_0 = 0.75$	0.20	5.157	0.6	14.24	EWMA	$\rho_0 = 0.75$	0.10	5.704	0.6	16.94	EWMA
$n = 20$	0.10	6.010	0.7	20.63	EWMA	$n = 50$	0.05	6.509	0.7	24.61	EWMA
	0.05	6.718	0.8	33.15	EWMA		0.05	6.509	0.8	38.38	EWMA
	0.05	7.370	0.9	64.58	DEWMA		0.05	7.250	0.9	71.92	DEWMA
$\mu_0 = 12$	0.30	8.468	0.5	6.77	EWMA	$\mu_0 = 12$	0.30	7.427	0.5	9.24	EWMA
$\rho_0 = 0.75$	0.30	8.468	0.6	8.85	EWMA	$\rho_0 = 0.75$	0.20	8.147	0.6	12.62	EWMA
$n = 20$	0.20	9.060	0.7	13.16	EWMA	$n = 50$	0.10	9.284	0.7	17.94	EWMA
	0.10	9.957	0.8	21.68	EWMA		0.05	10.248	0.8	29.47	EWMA
	0.05	10.701	0.9	45.67	EWMA		0.05	11.126	0.9	59.53	DEWMA

Table 5 Suggested lower one-sided charts, simultaneous shifts in μ_0, ρ_0

Process	λ	LCL	(δ, τ)	$ssARL$	Chart	Process	λ	LCL	(δ, τ)	$ssARL$	Chart
$\mu_0 = 4$	0.05	3.380	(0.8,−0.10)	25.06	EWMA	$\mu_0 = 4$	0.10	3.413	(0.8,−0.10)	26.81	DEWMA
$\rho_0 = 0.25$	0.30	2.640	(0.8,0.35)	23.71	DEWMA	$\rho_0 = 0.25$	0.30	2.563	(0.8,0.35)	24.71	DEWMA
$n = 20$						$n = 50$					
$\mu_0 = 8$	0.10	6.728	(0.8,−0.10)	13.82	EWMA	$\mu_0 = 8$	0.10	6.529	(0.8,−0.10)	16.97	EWMA
$\rho_0 = 0.25$	0.30	6.266	(0.8,0.35)	15.75	DEWMA	$\rho_0 = 0.25$	0.30	6.005	(0.8,0.35)	17.84	DEWMA
$n = 20$						$n = 50$					
$\mu_0 = 12$	0.20	9.944	(0.8,−0.10)	8.42	EWMA	$\mu_0 = 12$	0.10	10.267	(0.8,−0.10)	12.27	EWMA
$\rho_0 = 0.25$	0.30	10.221	(0.8,0.35)	10.71	DEWMA	$\rho_0 = 0.25$	0.30	9.639	(0.8,0.35)	14.09	DEWMA
$n = 20$						$n = 50$					
$\mu_0 = 4$	0.05	3.228	(0.8,−0.10)	35.06	EWMA	$\mu_0 = 4$	0.05	3.181	(0.8,−0.10)	37.48	EWMA
$\rho_0 = 0.50$	0.20	2.743	(0.8,0.35)	34.49	DEWMA	$\rho_0 = 0.50$	0.10	2.703	(0.8,0.35)	35.75	EWMA
$n = 20$						$n = 50$					
$\mu_0 = 8$	0.10	6.438	(0.8,−0.10)	19.88	EWMA	$\mu_0 = 8$	0.05	6.878	(0.8,−0.10)	23.80	EWMA
$\rho_0 = 0.50$	0.10	6.438	(0.8,0.35)	21.52	EWMA	$\rho_0 = 0.50$	0.10	6.196	(0.8,0.35)	24.99	EWMA
$n = 20$						$n = 50$					
$\mu_0 = 12$	0.10	10.396	(0.8,−0.10)	12.07	EWMA	$\mu_0 = 12$	0.10	9.880	(0.8,−0.10)	17.45	EWMA
$\rho_0 = 0.50$	0.10	10.396	(0.8,0.35)	14.19	EWMA	$\rho_0 = 0.50$	0.10	9.880	(0.8,0.35)	19.27	EWMA
$n = 20$						$n = 50$					
$\mu_0 = 40$	0.05	3.480	(0.8,−0.25)	51.49	DEWMA	$\mu_0 = 4$	0.05	3.445	(0.8,−0.25)	54.16	DEWMA
$\rho_0 = 0.75$	0.10	2.998	(0.8,0.20)	67.63	DEWMA	$\rho_0 = 0.75$	0.10	2.922	(0.8,0.20)	68.91	DEWMA
$n = 20$						$n = 50$					
$\mu_0 = 8$	0.05	6.718	(0.8,−0.25)	29.94	EWMA	$\mu_0 = 8$	0.05	6.509	(0.8,−0.25)	37.11	EWMA
$\rho_0 = 0.75$	0.10	6.739	(0.8,0.20)	54.10	DEWMA	$\rho_0 = 0.75$	0.10	6.525	(0.8,0.20)	58.86	DEWMA
$n = 20$						$n = 50$					
$\mu_0 = 12$	0.10	9.957	(0.8,−0.25)	17.94	EWMA	$\mu_0 = 12$	0.05	10.248	(0.8,−0.25)	26.66	EWMA
$\rho_0 = 0.75$	0.05	10.701	(0.8,0.20)	41.75	EWMA	$\rho_0 = 0.75$	0.10	10.273	(0.8,0.20)	50.97	DEWMA
$n = 20$						$n = 50$					

Table 5 provides the suggested chart for the detection of simultaneous shifts in μ_0 and ρ_0. An 20% decrease is assumed for μ_0 while an increasing as well as a decreasing shift in ρ_0 are also considered. Similar to case of the upper one-sided charts, when μ_0 is small (e.g. $\mu_0 = 4$), the DEWMA chart has a better performance than the EWMA chart while the latter is better when both μ_0 and ρ_0 decrease. Thus, when both shifts are on the same direction (decrease), the EWMA chart attains a lower $ssARL$ value than the EWMA chart, in almost all cases. The DEWMA chart outperforms the EWMA chart when shifts are on the opposite direction, i.e. μ_0 decreases and ρ_0 increases. Similar to the case of upper one-sided charts, the difference in the $ssARL$ between the two charts is at most 20%, depending on the shifts in μ_0, ρ_0.

It should be also mentioned that our numerical analysis showed that both lower one-sided EWMA and DEWMA charts are not able to detect a decrease only in ρ_0. Specifically, we considered $\tau = -0.10$ (for $\rho_0 = 0.25$ or 0.50) and $\tau = -0.2$ (for $\rho_0 = 0.75$) and our simulation results showed that the $ssARL$ values are larger than the $zsARL$ value.

The performance of the two-sided EWMA and DEWMA charts is presented in Tables 6, 7 and 8. Specifically, from the results in Table 6 we deduce that the EWMA chart outperforms the DEWMA in the detection of shifts only in μ_0, especially for moderate to large shifts, either decreasing or increasing. The DEWMA outperforms the EWMA chart when the IC μ_0 is small (e.g. $\mu_0 = 4$) and there is small decreasing or increasing shift in it (e.g. a 10% decrease or a 10% increase). The suggested λ value for the DEWMA is 0.05. It should be also noted that $\lambda = 0.05$ is a good choice for the most of the OoC cases. Thus, in practice and depending on the shift we want to detect, we recommend the use of an EWMA (or a DEWMA) chart with $\lambda = 0.05$, because it seems to have the best performance for a range shifts.

In the case of shifts only in ρ_0 (Table 7), the DEWMA chart outperforms the EWMA chart, in almost all of the considered cases. Specifically, the EWMA chart has a better performance than the EWMA chart, only in the case of strong dependence ($\rho_0 = 0.75$) and large sample size ($n = 50$). The suggested value for λ in the DEWMA chart is 0.20 or 0.30 (for increasing shifts in ρ_0) or 0.05 for decreasing shifts.

Finally, Table 8 provides the best two-sided chart, when there is a simultaneous change in both parameters μ_0 and ρ_0. The DEWMA chart outperforms the EWMA chart in the most of the examined cases, especially when there is an increase in ρ_0. When both μ_0 and ρ_0 decrease, the DEWMA chart has also better performance than the EWMA chart while when μ_0 increases and ρ_0 decreases, we recommend the EWMA chart.

As a general conclusion from Tables 1, 2, 3, 4, 5, 6, 7 and 8 we state that when there is a shift only in μ_0 either increasing or decreasing, the recommended chart is the EWMA chart. The λ value depends on the size of shift and the general rule of a "small λ for small shift" applies here, as well. The DEWMA chart is recommended when we are interested in detecting an increasing shift in ρ_0. A λ equal to 0.10 or 0.20 is suggested. When both parameters shift, there is not a clear pattern on the λ value and depends on the shift we want to detect. For the two-sided charts, the DEWMA chart has the best performance in the most of the examined cases and we recommend its use, especially when there is an increase in ρ_0, no matter to which direction is the change in μ_0. Finally, both charts have a difficulty to detect a downward shift only in ρ_0.

5 A Real-Data Example

In this section, we present an example with real data, in order to demonstrate the usefulness and practical implementation of the EWMA and DEWMA control charts. The example is from the area of network monitoring and the data are about the number of log-ins in the 15 available workstations. The data have been collected per minute from the computer centre of the University of Würzburg (Weiß [15]). Clearly, the available data are counts and they constitute a time serie that can be modelled via an appropriate integer-valued time series model.

Table 6 Suggested two-sided charts, shifts only in μ_0

Process	λ	LCL	UCL	δ	$ssARL$	Chart	Process	λ	LCL	UCL	δ	$ssARL$	Chart
$\mu_0 = 4$	0.30	2.359	5.641	0.5	9.66	DEWMA	$\mu_0 = 4$	0.20	2.704	5.296	0.5	10.60	DEWMA
$\rho_0 = 0.25$	0.10	3.289	4.711	0.8	33.87	DEWMA	$\rho_0 = 0.25$	0.05	3.571	4.429	0.8	37.12	DEWMA
$n = 20$	0.05	3.599	4.401	0.9	72.96	DEWMA	$n = 50$	0.05	3.571	4.429	0.9	77.97	DEWMA
	0.20	2.789	5.110	1.1	67.21	DEWMA		0.05	3.571	4.429	1.1	78.90	DEWMA
	0.20	2.789	5.110	1.2	31.19	DEWMA		0.05	3.158	4.842	1.2	36.73	EWMA
	0.20	2.789	5.110	1.5	9.78	DEWMA		0.20	1.997	6.003	1.5	10.94	EWMA
$\mu_0 = 8$	0.30	5.109	10.891	0.5	4.71	EWMA	$\mu_0 = 8$	0.20	5.286	10.714	0.5	6.06	EWMA
$\rho_0 = 0.25$	0.05	7.039	8.961	0.8	18.01	EWMA	$\rho_0 = 0.25$	0.05	6.861	9.139	0.8	22.26	EWMA
$n = 20$	0.05	7.509	8.491	0.9	43.73	DEWMA	$n = 50$	0.05	7.420	8.580	0.9	52.12	DEWMA
	0.05	7.509	8.491	1.1	44.14	DEWMA		0.05	7.420	8.580	1.1	53.08	DEWMA
	0.10	6.495	9.505	1.2	17.86	EWMA		0.05	6.861	9.139	1.2	22.67	EWMA
	0.30	5.109	10.891	1.5	4.79	EWMA		0.30	4.554	11.446	1.5	6.22	EWMA
$\mu_0 = 12$	0.30	9.107	14.893	0.5	2.86	EWMA	$\mu_0 = 12$	0.30	7.994	16.006	0.5	4.16	EWMA
$\rho_0 = 0.25$	0.20	9.713	14.287	0.8	10.30	EWMA	$\rho_0 = 0.25$	0.10	9.926	14.074	0.8	16.08	EWMA
$n = 20$	0.05	11.038	12.962	0.9	26.11	EWMA	$n = 50$	0.05	11.324	12.676	0.9	39.76	DEWMA
	0.05	11.038	12.962	1.1	26.21	EWMA		0.05	10.672	13.328	1.1	39.20	EWMA
	0.30	9.986	14.014	1.2	10.19	DEWMA		0.10	9.926	14.074	1.2	16.00	EWMA
	0.30	9.107	14.893	1.5	2.76	EWMA		0.30	7.994	16.006	1.5	4.35	EWMA
$\mu_0 = 4$	0.10	2.479	5.521	0.5	14.53	EWMA	$\mu_0 = 4$	0.05	2.937	5.063	0.5	16.04	EWMA
$\rho_0 = 0.50$	0.05	3.467	4.533	0.8	47.86	DEWMA	$\rho_0 = 0.50$	0.05	3.429	4.571	0.8	50.32	DEWMA
$n = 20$	0.05	3.467	4.533	0.9	96.53	DEWMA	$n = 50$	0.05	3.429	4.571	0.9	101.76	DEWMA
	0.05	3.467	4.533	1.1	97.08	DEWMA		0.30	1.038	6.962	1.1	101.73	EWMA
	0.05	3.007	4.993	1.2	48.61	EWMA		0.05	2.937	5.063	1.2	51.94	EWMA
	0.10	2.479	5.521	1.5	15.01	EWMA		0.20	1.595	6.405	1.5	16.43	EWMA
$\mu_0 = 8$	0.20	5.255	10.745	0.5	7.21	EWMA	$\mu_0 = 8$	0.30	5.019	10.981	0.5	9.26	DEWMA
$\rho_0 = 0.50$	0.05	6.780	9.220	0.8	26.27	EWMA	$\rho_0 = 0.50$	0.05	6.556	9.444	0.8	32.72	EWMA
$n = 20$	0.05	7.346	8.654	0.9	60.36	EWMA	$n = 50$	0.05	7.228	8.772	0.9	70.84	DEWMA
	0.05	7.346	8.654	1.1	60.83	EWMA		0.05	7.228	8.772	1.1	72.94	DEWMA
	0.05	6.780	9.220	1.2	26.30	EWMA		0.05	6.556	9.444	1.2	32.94	EWMA
	0.30	4.617	11.383	1.5	7.36	EWMA		0.20	4.744	11.256	1.5	9.56	EWMA

(continued)

Table 6 (continued)

Process	λ	LCL	UCL	δ	$ssARL$	Chart	Process	λ	LCL	UCL	δ	$ssARL$	Chart
$\mu_0 = 12$	0.30	8.618	15.382	0.5	4.23	EWMA	$\mu_0 = 12$	0.30	7.322	16.678	0.5	6.39	EWMA
$\rho_0 = 0.50$	0.10	10.131	13.869	0.8	15.40	DEWMA	$\rho_0 = 0.50$	50.00	0.05	10.319	13.681	23.42	EWMA
$n = 20$	0.05	10.779	13.221	0.9	38.26	DEWMA	$n = 50$	0.05	11.101	12.899	0.9	55.01	EWMA
	0.05	10.779	13.221	1.1	38.53	EWMA		0.05	11.101	12.899	1.1	55.89	DEWMA
	0.10	10.131	13.869	1.2	15.23	EWMA		0.05	10.319	13.681	1.2	23.88	DEWMA
	0.30	8.618	15.382	1.5	4.06	EWMA		0.30	7.322	16.678	1.5	6.60	EWMA
$\mu_0 = 4$	0.05	2.652	5.348	0.5	25.00	EWMA	$\mu_0 = 4$	0.05	2.558	5.442	0.5	27.30	EWMA
$\rho_0 = 0.75$	0.05	3.213	4.787	0.8	73.67	DEWMA	$\rho_0 = 0.75$	0.05	3.157	4.843	0.8	78.91	DEWMA
$n = 20$	0.05	3.213	4.787	0.9	128.09	DEWMA	$n = 50$	0.05	3.157	4.843	0.9	132.77	DEWMA
	0.30	0.806	7.194	1.1	125.91	EWMA		0.30	0.574	7.426	1.1	122.61	EWMA
	0.05	2.652	5.348	1.2	76.78	EWMA		0.30	0.574	7.426	1.2	77.96	EWMA
	0.10	2.028	5.972	1.5	26.33	EWMA		0.30	0.574	7.426	1.5	28.19	EWMA
$\mu_0 = 8$	0.20	4.628	11.372	0.5	13.47	EWMA	$\mu_0 = 8$	0.10	5.122	10.878	0.5	16.55	EWMA
$\rho_0 = 0.75$	0.05	6.338	9.662	0.8	45.15	EWMA	$\rho_0 = 0.75$	0.05	6.859	9.141	0.8	53.67	DEWMA
$n = 20$	0.05	7.031	8.969	0.9	91.40	DEWMA	$n = 50$	0.05	6.859	9.141	0.9	104.54	DEWMA
	0.05	7.031	8.969	1.1	91.02	DEWMA		0.05	6.859	9.141	1.1	104.86	DEWMA
	0.05	6.338	9.662	1.2	45.58	EWMA		0.05	6.042	9.958	1.2	55.21	EWMA
	0.20	4.628	11.372	1.5	13.61	EWMA		0.20	4.028	11.972	1.5	17.45	EWMA
$\mu_0 = 12$	0.30	8.058	15.942	0.5	7.65	EWMA	$\mu_0 = 12$	0.20	7.348	16.652	0.5	11.82	EWMA
$\rho_0 = 0.75$	0.05	10.340	13.660	0.8	27.33	EWMA	$\rho_0 = 0.75$	0.05	9.715	14.285	0.8	40.41	EWMA
$n = 20$	0.05	11.030	12.970	0.9	62.71	DEWMA	$n = 50$	0.05	10.667	13.333	0.9	84.16	DEWMA
	0.05	11.030	12.970	1.1	61.88	DEWMA		0.05	10.667	13.333	1.1	86.13	DEWMA
	0.05	10.340	13.660	1.2	26.74	EWMA		0.05	9.715	14.285	1.2	40.99	EWMA
	0.30	8.058	15.942	1.5	7.30	EWMA		0.30	6.532	17.468	1.5	12.27	EWMA

Table 7 Suggested two-sided charts, shifts only in ρ_0

Process	λ	LCL	UCL	τ	$ssARL$	Chart	Process	λ	LCL	UCL	τ	$ssARL$	Chart
$\mu_0 = 4$	0.20	2.789	5.110	0.15	85.40	DEWMA	$\mu_0 = 4$	0.30	2.240	5.760	0.15	99.26	DEWMA
$\rho_0 = 0.25$	0.20	2.789	5.110	0.35	45.05	DEWMA	$\rho_0 = 0.25$	0.30	2.240	5.760	0.35	49.58	DEWMA
$n = 20$	0.20	2.789	5.110	−0.10	246.08	DEWMA	$n = 50$	0.05	3.571	4.429	−0.10	244.21	DEWMA
$\mu_0 = 8$	0.30	5.988	10.012	0.15	96.00	DEWMA	$\mu_0 = 8$	0.30	5.618	10.382	0.15	97.85	DEWMA
$\rho_0 = 0.25$	0.30	5.988	10.012	0.35	48.38	DEWMA	$\rho_0 = 0.25$	0.20	6.244	9.756	0.35	49.17	DEWMA
$n = 20$	0.05	7.509	8.491	−0.10	249.25	DEWMA	$n = 50$	0.05	7.420	8.580	−0.10	239.72	DEWMA
$\mu_0 = 12$	0.30	9.986	14.014	0.15	95.97	DEWMA	$\mu_0 = 12$	0.30	9.223	14.777	0.15	97.61	DEWMA
$\rho_0 = 0.25$	0.30	9.986	14.014	0.35	47.92	DEWMA	$\rho_0 = 0.25$	0.30	9.223	14.777	0.35	49.07	DEWMA
$n = 20$	0.05	11.510	12.490	−0.10	240.59	DEWMA	$n = 50$	0.05	11.324	12.676	−0.10	247.42	DEWMA
$\mu_0 = 4$	0.20	2.446	5.554	0.15	92.79	DEWMA	$\mu_0 = 4$	0.20	2.338	5.662	0.15	94.46	DEWMA
$\rho_0 = 0.50$	0.20	2.446	5.554	0.35	45.71	DEWMA	$\rho_0 = 0.50$	0.20	2.338	5.662	0.35	46.75	DEWMA
$n = 20$	0.05	3.467	4.533	−0.10	259.95	DEWMA	$n = 50$	0.05	3.429	4.571	−0.10	263.65	DEWMA
$\mu_0 = 8$	0.20	6.090	9.910	0.15	92.24	DEWMA	$\mu_0 = 8$	0.20	5.742	10.258	0.15	94.63	DEWMA
$\rho_0 = 0.50$	0.20	6.090	9.910	0.35	45.77	DEWMA	$\rho_0 = 0.50$	0.20	5.742	10.258	0.35	46.29	DEWMA
$n = 20$	0.05	7.346	8.654	−0.10	258.58	DEWMA	$n = 50$	0.05	7.228	8.772	−0.10	231.94	DEWMA
$\mu_0 = 12$	0.20	10.090	13.910	0.15	92.49	DEWMA	$\mu_0 = 12$	0.20	9.367	14.633	0.15	92.10	DEWMA
$\rho_0 = 0.50$	0.20	10.090	13.910	0.35	46.25	DEWMA	$\rho_0 = 0.50$	0.20	9.367	14.633	0.35	46.16	DEWMA
$n = 20$	0.05	11.346	12.654	−0.10	259.63	DEWMA	$n = 50$	0.05	11.101	12.899	−0.10	258.34	DEWMA
$\mu_0 = 4$	0.10	2.659	5.341	0.10	99.77	DEWMA	$\mu_0 = 4$	0.10	2.568	5.432	0.10	100.45	DEWMA
$\rho_0 = 0.75$	0.05	3.213	4.787	0.20	64.00	DEWMA	$\rho_0 = 0.75$	0.05	2.937	5.063	0.20	49.59	EWMA
$n = 20$	0.30	0.806	7.194	−0.25	541.67	EWMA	$n = 50$	0.30	1.038	6.962	−0.25	272.93	EWMA
$\mu_0 = 8$	0.10	6.347	9.653	0.10	100.02	DEWMA	$\mu_0 = 8$	0.10	5.793	10.207	0.10	98.08	EWMA
$\rho_0 = 0.75$	0.05	7.031	8.969	0.20	63.38	DEWMA	$\rho_0 = 0.75$	0.05	6.556	9.444	0.20	49.61	EWMA
$n = 20$	0.30	4.060	11.940	−0.25	684.11	EWMA	$n = 50$	0.30	3.995	12.005	−0.25	303.14	EWMA
$\mu_0 = 12$	0.10	10.347	13.653	0.10	102.33	DEWMA	$\mu_0 = 12$	0.10	9.421	14.579	0.10	99.96	EWMA
$\rho_0 = 0.75$	0.05	11.030	12.970	0.20	64.79	DEWMA	$\rho_0 = 0.75$	0.05	10.319	13.681	0.20	50.14	EWMA
$n = 20$	0.30	8.058	15.942	−0.25	711.77	EWMA	$n = 50$	0.30	7.322	16.678	−0.25	308.65	EWMA

Table 8 Suggested two-sided charts, simultaneous shifts in μ_0, ρ_0

Process	λ	LCL	UCL	(δ, τ)	$ssARL$	Chart	Process	λ	LCL	UCL	(δ, τ)	$ssARL$	Chart
$\mu_0 = 4$	0.20	2.789	5.110	(0.8, 0.35)	28.87	DEWMA	$\mu_0 = 4$	0.20	2.704	5.296	(0.8, 0.35)	31.86	DEWMA
$\rho_0 = 0.25$	0.20	2.789	5.110	(1.2, 0.35)	26.07	DEWMA	$\rho_0 = 0.25$	0.30	2.240	5.760	(1.2, 0.35)	30.13	DEWMA
$n = 20$	0.10	3.289	4.711	(0.8,–0.10)	34.15	DEWMA	$n = 50$	0.05	3.571	4.429	(0.8,–0.10)	36.71	DEWMA
	0.20	2.789	5.110	(1.2,–0.10)	32.48	DEWMA		0.05	3.158	4.842	(1.2,–0.10)	37.28	EWMA
$\mu_0 = 8$	0.10	6.495	9.505	(0.8, 0.35)	18.97	EWMA	$\mu_0 = 8$	0.20	6.244	9.756	(0.8, 0.35)	22.44	DEWMA
$\rho_0 = 0.25$	0.30	5.988	10.012	(1.2, 0.35)	18.85	DEWMA	$\rho_0 = 0.25$	0.30	5.618	10.382	(1.2, 0.35)	22.49	DEWMA
$n = 20$	0.05	7.039	8.961	(0.8,–0.10)	17.68	EWMA	$n = 50$	0.10	6.968	9.032	(0.8,–0.10)	22.03	DEWMA
	0.05	7.039	8.961	(1.2,–0.10)	17.86	EWMA		0.05	6.861	9.139	(1.2,–0.10)	22.48	EWMA
$\mu_0 = 12$	0.30	9.986	14.014	(0.8, 0.35)	11.91	DEWMA	$\mu_0 = 12$	0.20	9.954	14.046	(0.8, 0.35)	17.16	DEWMA
$\rho_0 = 0.25$	0.30	9.986	14.014	(1.2, 0.35)	11.88	DEWMA	$\rho_0 = 0.25$	0.30	9.223	14.777	(1.2, 0.35)	17.24	DEWMA
$n = 20$	0.30	9.986	14.014	(0.8,–0.10)	10.02	DEWMA	$n = 50$	0.10	9.926	14.074	(0.8,-0.10)	15.88	EWMA
	0.10	10.494	13.506	(1.2,–0.10)	9.90	EWMA		0.10	9.926	14.074	(1.2,–0.10)	15.70	EWMA
$\mu_0 = 4$	0.10	3.065	4.935	(0.8, 0.35)	37.43	DEWMA	$\mu_0 = 4$	0.10	2.998	5.002	(0.8, 0.35)	38.63	DEWMA
$\rho_0 = 0.50$	0.20	2.446	5.554	(1.2, 0.35)	36.49	DEWMA	$\rho_0 = 0.50$	0.20	2.338	5.662	(1.2, 0.35)	38.04	DEWMA
$n = 20$	0.05	3.467	4.533	(0.8,–0.10)	46.92	DEWMA	$n = 50$	0.05	3.429	4.571	(0.8,–0.10)	50.61	DEWMA
	0.05	3.467	4.533	(1.2,–0.10)	49.78	DEWMA		0.05	2.937	5.063	(1.2,–0.10)	54.17	EWMA
$\mu_0 = 8$	0.20	6.090	9.910	(0.8, 0.35)	29.30	DEWMA	$\mu_0 = 8$	0.20	5.742	10.258	(0.8, 0.35)	33.04	DEWMA
$\rho_0 = 0.50$	0.20	6.090	9.910	(1.2, 0.35)	28.86	DEWMA	$\rho_0 = 0.50$	0.20	5.742	10.258	(1.2, 0.35)	32.57	DEWMA
$n = 20$	0.05	6.780	9.220	(0.8,–0.10)	26.07	EWMA	$n = 50$	0.05	6.556	9.444	(0.8,–0.10)	32.48	EWMA
	0.05	6.780	9.220	(1.2,–0.10)	25.96	EWMA		0.05	6.556	9.444	(1.2,–0.10)	32.87	EWMA
$\mu_0 = 12$	0.20	10.090	13.910	(0.8, 0.35)	21.86	DEWMA	$\mu_0 = 12$	0.20	9.367	14.633	(0.8, 0.35)	27.66	DEWMA
$\rho_0 = 0.50$	0.20	10.090	13.910	(1.2, 0.35)	21.02	DEWMA	$\rho_0 = 0.50$	0.20	9.367	14.633	(1.2, 0.35)	27.50	DEWMA
$n = 20$	0.10	10.131	13.869	(0.8,–0.10)	14.87	EWMA	$n = 50$	0.05	6.556	9.444	(0.8,–0.10)	23.03	EWMA
	0.10	10.131	13.869	(1.2,–0.10)	14.65	EWMA		0.05	6.556	9.444	(1.2,–0.10)	23.62	EWMA

(continued)

Table 8 (continued)

Process	λ	LCL	UCL	(δ, τ)	$ssARL$	Chart	Process	λ	LCL	UCL	(δ, τ)	$ssARL$	Chart
$\mu_0 = 4$	0.05	3.213	4.787	(0.8, 0.20)	58.02	DEWMA	$\mu_0 = 4$	0.20	1.096	6.904	(0.8, 0.20)	50.03	EWMA
$\rho_0 = 0.75$	0.10	2.659	5.341	(1.2, 0.20)	58.91	DEWMA	$\rho_0 = 0.75$	0.20	1.096	6.904	(1.2, 0.20)	40.73	EWMA
$n = 20$	0.05	3.213	4.787	(0.8,–0.25)	81.88	DEWMA	$n = 50$	0.20	1.096	6.904	(0.8,–0.25)	61.49	EWMA
	0.05	3.213	4.787	(1.2,–0.25)	83.84	DEWMA		0.20	1.096	6.904	(1.2,–0.25)	36.57	EWMA
$\mu_0 = 8$	0.05	7.031	8.969	(0.8, 0.20)	51.41	DEWMA	$\mu_0 = 8$	0.05	6.859	9.141	(0.8, 0.20)	54.66	DEWMA
$\rho_0 = 0.75$	0.10	6.347	9.653	(1.2, 0.20)	51.76	DEWMA	$\rho_0 = 0.75$	0.05	6.859	9.141	(1.2, 0.20)	54.95	DEWMA
$n = 20$	0.05	7.031	8.969	(0.8,–0.25)	42.44	DEWMA	$n = 50$	0.05	6.859	9.141	(0.8,–0.25)	50.65	DEWMA
	0.05	7.031	8.969	(1.2,–0.25)	43.36	DEWMA		0.05	6.859	9.141	(1.2,–0.25)	54.24	DEWMA
$\mu_0 = 12$	0.10	10.347	13.653	(0.8, 0.20)	42.74	DEWMA	$\mu_0 = 12$	0.05	10.667	13.333	(0.8, 0.20)	49.56	DEWMA
$\rho_0 = 0.75$	0.10	10.347	13.653	(1.2, 0.20)	41.90	DEWMA	$\rho_0 = 0.75$	0.10	9.722	14.278	(1.2, 0.20)	51.03	DEWMA
$n = 20$	0.05	10.340	13.660	(0.8,–0.25)	23.97	EWMA	$n = 50$	0.05	10.667	13.333	(0.8,–0.25)	37.62	DEWMA
	0.05	10.340	13.660	(1.2,–0.25)	23.57	EWMA		0.05	9.715	14.285	(1.2,–0.25)	38.83	EWMA

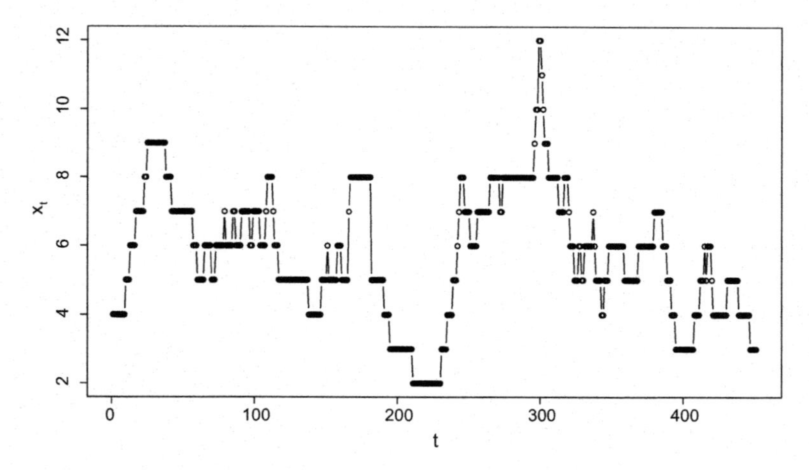

Fig. 1 Log-ins data, May 3rd, 2005

We start with a time series plot with the available data on the 3rd May 2005 (Fig. 1). At this day, it is available at each minute t the number of log-ins in 15 workstations in the computer centre of the University of Würzburg, from 10:00 to 17:30. The total number of observations equals 451. The possible values for the number X_t of log-ins at minute t are in $\{0, 1, \ldots, 15\}$ which also supports the BAR(1) as a candidate model for capturing their stochastic behavior.

We will use these 451 values as a Phase I sample and estimate the proceed parameters as well as the control limits for the EWMA and DEWMA two-sided charts. By using the function `optim` in R (R Core Team [35]), we estimate the parameters π and ρ via the method of maximum likelihood. The results (which verify those in Weiß [15]) are $\hat{\pi} = 0.36482$ (0.04306) and $\hat{\rho} = 0.96822$ (0.00355). In the parentheses we provide the standard errors of the estimates. Therefore, the process is modelled as a BAR(1) process with $(n, \pi, \rho) = (15, 0.36482, 0.96822)$.

Next, we apply the two-sided *np* chart with control limits $LCL = 2$ and $UCL = 9$, as a Phase-I method, on the data from May 3rd. The $zsARL$ is around 360. The chart is given in Fig. 2. There are 6 values beyond the UCL, observations 297–302. Further investigation is needed in order to verify that these signals are due to the presence of assignable causes in the process or that they are false alarms. Here, if we assume that these are true alarms, then we have to remove them. Therefore, in order to estimate process parameters with incomplete data, we have to use the modified maximum likelihood estimation method, provided by Weiß and Testik [36].

Below, in Table 9, we provide the estimates for π and ρ from the complete data and from the data without the observations 297–302. The difference in estimates cannot be considered as big. Therefore, we proceed with the estimates from the complete data set and control limits at $LCL = 2$, $UCL = 9$.

Next, we use first the log-in count data on May 10th 2005 as the Phase II data and construct again the two-sided *np* chart with control limits at 2 and 9 as well as the

Table 9 Estimates of process parameters π, ρ from the Phase I data

Estimates	Complete data	Without Obs. 297–302
$\hat{\pi}_{ML}$	0.36482 (0.04306)	0.36254 (0.04420)
$\hat{\rho}_{ML}$	0.96822 (0.00355)	0.97013 (0.00345)

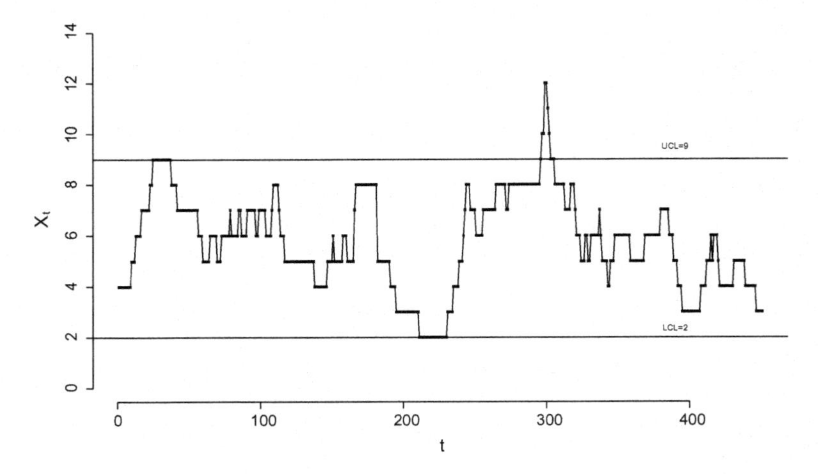

Fig. 2 Log-in count data on May 3rd 2005, Phase I analysis

two-sided EWMA and DEWMA charts. We assume that the estimated values for π and ρ from the Phase I analysis are the true values for the process parameters. For illustrative purposes we use $\lambda = 0.10$ and by applying the procedure for the statistical of the EWMA and DEWMA charts (described in Sect. 3) we determine their control limits so as their IC performance is comparable to that of the *np* chart. Thus, for the two-sided EWMA chart, the control limits are $LCL_{\text{ewma}} = 2.30215$, $UCL_{\text{ewma}} = 8.57414$ while for the two-sided DEWMA chart, they are $LCL_{\text{dewma}} = 2.73414$, $UCL_{\text{dewma}} = 8.14215$. We notice that the control chart limits for the DEWMA are narrower than the EWMA limits. This result holds in general.

The *np* chart is provided in Fig. 3 while in Fig. 4 we provide both EWMA and DEWMA charts. The *np* chart gives for the first time an OoC signal at time $t = 15$, indicating a possible increase in the mean number of log-ins, compared to the IC baseline model (during the 3rd of May). Also, from Fig. 4 we notice that the DEWMA chart gives an OoC for the first time at $t = 35$ (about 20 min later than the *np* chart) while the EWMA chart gives an OoC signal for the first time at sample 161. At the same time, an OoC signal is given by the *np* chart, as well. It is worth mentioning that Weiß [15] in his analysis, concluded that there are not statistically significant indications that the process has changed from the IC model. However, it seems that since the estimated IC process mean level is $15 \cdot 0.36482 \approx 5.44$, there are indications of increased process variation, especially whithin the first 3-3.5 h of the day.

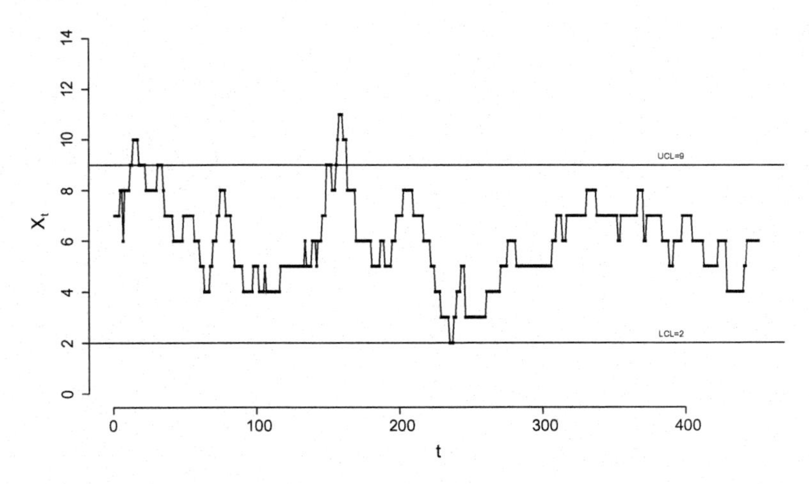

Fig. 3 Two-sided *np* chars for the log-ins data, May 10th, 2005

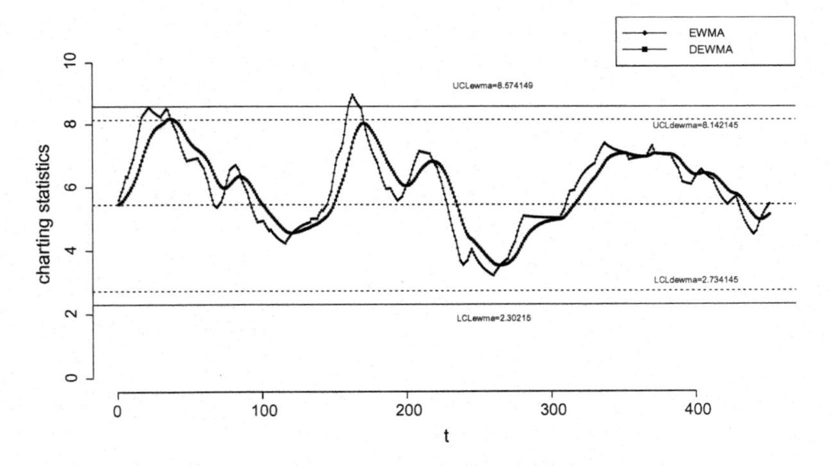

Fig. 4 Two-sided EWMA and DEWMA charts for the log-ins data, May 10th, 2005

The three charts are also applied in the log-in count data that have been obtained a week later, on May 17 2005. The *np* chart is provided in Fig. 5 while both the EWMA and the DEWMA charts are provided in Fig. 6. Clearly, the *np* chart gives an OoC signal even from the first minute while almost all points are below the process mean level. There is a clear indication that the actual process mean level has been decreased (compared to the one under the IC model). This is also confirmed by the EWMA and DEWMA charts. However, the EWMA chart signals for the first time at time $t = 306$ (almost 5 h after the beginning of process monitoring) whereas the DEWMA signals for the first time at $t = 272$, about 35 min earlier than the EWMA chart (but still, about 4.5 h since the beginning of process monitoring). According to

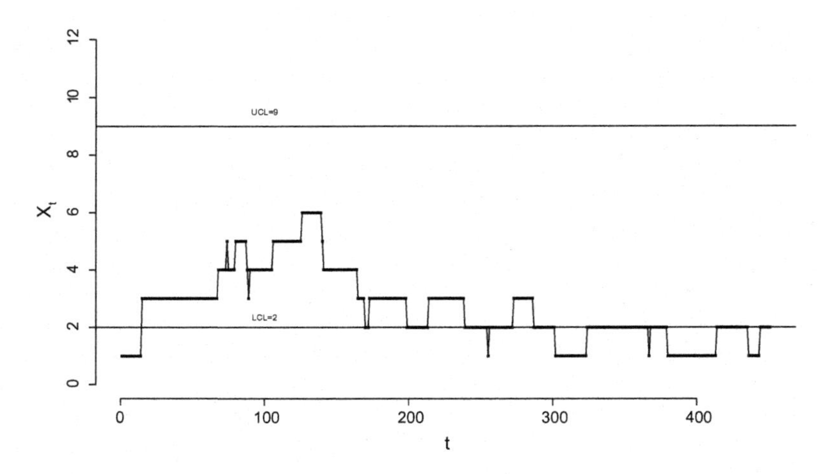

Fig. 5 Two-sided *np* chars for the log-ins data, May 17th, 2005

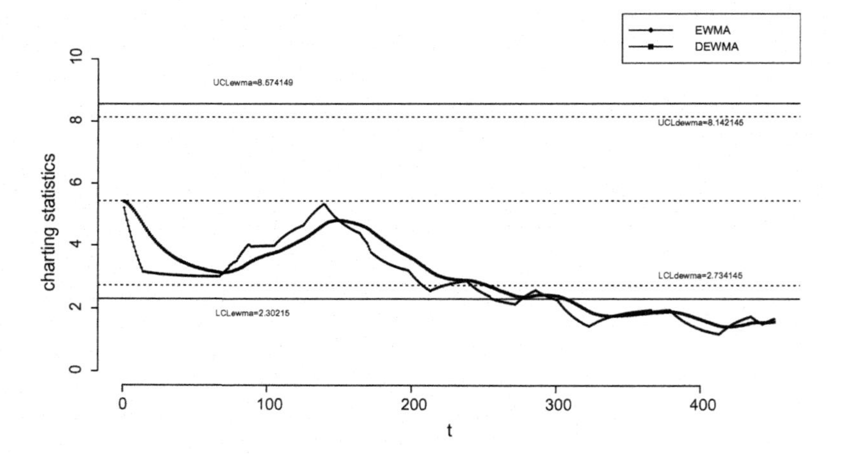

Fig. 6 Two-sided EWMA and DEWMA charts for the log-ins data, May 17th, 2005

Weiß [15], this (unusual) behavior is attributed to the fact that this day was the first day after a long-weekend (a public holiday after a weekend), and traditionally, there were no lectures at that day. Note also that the shift here is sudden, sustained and of large magnitude. Therefore, it is not surprising that both EWMA and DEWMA charts do not react immediately on this change in the process.

6 Conclusion and Future Work

In this work, we developed and studied one-sided and two-sided EWMA and DEWMA control charts that are suitable for the detection of upward and downward shifts in the parameters of a BAR(1) process. Both charts have been frequently applied in the monitoring of count (or attributes) data, but in the case of serially dependent observation, their performance was not investigated previously. The results of an extensive simulation study regarding the statistical design and the performance of the proposed EWMA charts revealed that in the case of one-sided charts, the EWMA chart is preferable than the DEWMA chart, when the interest is on detecting shifts only in parameter μ_0. When we are interested in detecting shifts only in ρ_0 or in detecting a simultaneous shift in μ_0 and ρ_0, the recommended chart is the DEWMA. Also, both charts are not capable of detecting decreasing shifts only in ρ_0. In the two-sided case, our numerical analysis revealed that for small shifts in exactly one of the process parameters, the recommended chart is the DEWMA whereas for larger shifts, the EWMA has better performance.

Finally, the practical application of the proposed schemes was illustrated via a real-data example. For all calculations, the R statistical software R Core Team [35] was used and the programs are available from the authors upon request.

Topics for future research consist of the development and study of other types of control charts, such as combined or composite charts, which are able to detect shifts in either direction (upward or downward) in any of the process parameters. Specifically, instead of detecting the shift, it is also important to provide some information about the parameter(s) that has been changed. Moreover, the application of the proposed schemes needs to be investigated in the monitoring of processes that exhibit overdispersion. It is expected that apart from autocorrelation on the data, the overdispersion will also affect the performance of the usual EWMA and DEWMA charts. Therefore, proper adjustments are necessary. Finally, the performance of other mixed-type control charts like the generally weighted moving average (GWMA) and the double GWMA (DGWMA) charts needs to be investigated in the case of serially dependend count data.

Acknowledgments We would like to thank one anonymous reviewer for his/her comments. Also, the authors would like to thank Prof. Christian Weiß for providing the login data.

References

1. Bersimis S, Sgora A, Psarakis S (2018) The application of multivariate statistical process monitoring in non-industrial processes. Qual Technol Quant Manage 15(4):526–549
2. Woodall WH, Zhao MJ, Paynabar K, Sparks R, Wilson JD (2017) An overview and perspective on social network monitoring. IISE Trans 49(3):354–365
3. Aykroyd RG, Leiva V, Ruggeri F (2019) Recent developments of control charts, identification of big data sources and future trends of current research. Technol Forecast Soc Change 144:221–232

4. Montgomery DC (2009) Introduction to Statistical Quality Control, 6th edn. John Wiley & Sons Inc, New York
5. Gan FF (1990) Monitoring observations generated from a binomial distribution using modified exponential weighted moving average control chart. J Stat Comput Simul 37:45–60
6. Gan FF (1993) An optimal design of CUSUM control charts for binomial counts. J Appl Stat 20(4):445–460
7. Chang TC, Gan FF (2001) Cumulative sum charts for high yield processes. Statistica Sinica 11:791–805
8. Wu Z, Jiao J, Liu Y (2008) A binomial CUSUM chart for detecting large shifts in fraction nonconforming. J Appl Stat 35(11):1267–1276
9. Yeh AB, Mcgrath RN, Sembower MA, Shen Q (2008) EWMA control charts for monitoring high-yield processes based on non-transformed observations. Int J Prod Res 46(20):5679–5699
10. Haridy S, Wu Z, Yu F-J, Shamsuzzaman M (2013) An optimisation design of the combined np-CUSUM scheme for attributes. Eur J Ind Eng 7(1):16–37
11. Haridy S, Wu Z, Chen S, Knoth S (2014) Binomial CUSUM chart with curtailment. Int J Prod Res 52(15):4646–4659
12. Psarakis S, Papaleonida GEA (2007) SPC procedures for monitoring autocorrelated processes. Qual Technol Quant Manage 4(4):501–540
13. Kim H, Lee S (2019) Improved CUSUM monitoring of Markov counting process with frequent zeros. Qual Reliab Eng Int 35(7):2371–2394
14. Weiß CH (2015) SPC methods for time-dependent processes of counts-a literature review. Cogent Math 2(1):1111116
15. Weiß CH (2009) Monitoring correlated processes with binomial marginals. J Appl Stat 36(4):399–414
16. McKenzie E (1985) Some simple models for discrete variate time series. Water Res Bull 21:645–650
17. Al-Osh MA, Alzaid AA (1987) First-order integer-valued autoregressive (INAR(1)) process. J Time Ser Anal 8:261–275
18. Rakitzis AC, Weiß CH, Castagliola P (2017) Control charts for monitoring correlated counts with a finite range. Appl Stoch Models Bus Ind 33(6):733–749
19. Anastasopoulou M, Rakitzis AC (2020) EWMA control charts for monitoring correlated counts with finite range. J Appl Stat 1–21. https://doi.org/10.1080/02664763.2020.1820959
20. Shamma SE, Amin RW, Shamma AK (1991) A double exponentially weighted moving average control procedure with variable sampling intervals. Commun Stat Simul Comput 20(2–3):511–528
21. Shamma SE, Shamma AK (1992) Development and evaluation of control charts using double exponentially weighted moving averages. Int J Qual Reliab Manage 9(6):18–24
22. Mahmoud MA, Woodall WH (2010) An evaluation of the double exponentially weighted moving average control chart. Commun Stat Simul Comput 39(5):933–949
23. Khoo MBC, Teh SY, Wu Z (2010) Monitoring process mean and variability with one double EWMA chart. Commun Stat Theor Methods 39(20):3678–3694
24. Adeoti OA, Malela-Majika J-C (2020) Double exponentially weighted moving average control chart with supplementary runs-rules. Qual Technol Quant Manage 17(2):149–172
25. Raza MA, Nawaz T, Aslam M, Bhatti SH, Sherwani RAK (2020) A new nonparametric double exponentially weighted moving average control chart. Qual Reliab Eng Int 36(1):68–87
26. Zhang L, Govindaraju K, Lai CD, Bebbington MS (2003) Poisson DEWMA control chart. Commun Stat Simul Comput 32(4):1265–1283
27. Alevizakos V, Koukouvinos C (2020) Monitoring a zero-inflated Poisson process with EWMA and DEWMA control charts. Qual Reliab Eng Int 36:88–111
28. Alevizakos V, Koukouvinos C (2021) Monitoring of zero-inflated binomial processes with a DEWMA control chart. J Appl Stat 48(7):1319–1338
29. Alevizakos V, Koukouvinos C (2019) A double exponentially weighted moving average control chart for monitoring COM-Poisson attributes. Qual Reliab Eng Int 35(7):2130–2151

30. Steutel FW, van Harn K (1979) Discrete analogues of self-decomposability and stability. Ann Prob 7(5):893–899
31. Weiß CH, Kim HY (2013) Parameter estimation for binomial AR(1) models with applications in finance and industry. Stat Pap 54(3):563–590
32. Roberts SW (1959) Control chart tests based on geometric moving averages. Technometrics 1(3):239–250
33. Weiß CH (2011) The Markov chain approach for performance evaluation of control charts - a tutorial. In: Werther SP (ed) Process Control: Problems, Techniques and Applications, pp 205–228. Nova Science Publishers Inc
34. Weiß CH, Testik MC (2012) Detection of abrupt changes in count data time series: Cumulative sum derivations for INARCH(1) models. J Qual Technol 44(3):249–264
35. R Core Team R A Language and Environment for Statistical Computing. R Foundation for Statistical Computing, Vienna, Austria (2021). http://www.R-project.org
36. Weiß CH, Testik MC (2015) On the Phase I analysis for monitoring time-dependent count processes. IIE Trans 47(3):294–306

On Approaches for Monitoring Categorical Event Series

Christian H. Weiß

Abstract In many manufacturing applications, the monitoring of categorical event series is required, i. e., of processes, where the quality characteristics are measured on a qualitative scale. We survey three groups of approaches for this task. First, the categorical event series might be transformed into a count process (e. g., event counts, discrete waiting times). After having identified an appropriate model for this count process, diverse control charts are available for the monitoring of the generated counts. Second, control charts might be directly applied to the considered categorical event series, using different charts for nominal than for ordinal data. The latter distinction is also crucial for the respective possibilities of analyzing and modeling these data. Finally, also rule-based procedures from machine learning might be used for the monitoring of categorical event series, where the generated rules are used to predict the occurrence of critical events. Our comprehensive survey of methods and models for categorical event series is complemented by two real-data examples from manufacturing industry, about nominal types of defects and ordinal levels of quality.

Keywords Attributes control charts · Count time series · Episode mining · Nominal time series · Ordinal time series · Temporal association rules

1 Introduction

Methods from *statistical process control* (SPC) allow to monitor quality-related processes as they occur, for example, in manufacturing and service industries as well as in health surveillance. Here, the most well-known SPC tool is the *control chart*, where certain quality statistics are computed sequentially in time and used to decide about the actual state of the process. More precisely, we do not wish to intervene in the manufacturing process as long as it is *in control*, i. e., if the monitored statistics are stationary according to a specified time series model (e. g., independent and

C. H. Weiß (✉)
Department of Mathematics and Statistics, Helmut Schmidt University, 22043 Hamburg, Germany
e-mail: weissc@hsu-hh.de

K. P. Tran (ed.), *Control Charts and Machine Learning for Anomaly Detection in Manufacturing*, Springer Series in Reliability Engineering,
https://doi.org/10.1007/978-3-030-83819-5_5

identically distributed (i. i. d.) with a specified marginal distribution). By contrast, if deviations from this in-control model are present, such as shifts or drifts in some of the model parameters, the process is called *out of control*. In traditional control chart applications, we compare the plotted statistics against the given control limits. If a statistic is plotted beyond these limits, an alarm is triggered to indicate a possible out-of-control situation. Of course, we are interested in generating a true alarm as soon as possible, whereas a false alarm should be avoided for as long as possible. Here, the waiting time until the first alarm (already the first alarm requires action) is commonly referred to as the *run length* of the control chart, and this should be large (low) if the process is in control (out of control). For these and further basics on SPC and control charts, see the textbook by Montgomery [33].

Most of the SPC literature discusses the case where the monitored quality characteristics are measured on a continuous quantitative scale, such as real numbers or real vectors; the corresponding control charts are then referred to as variables charts. But in more and more applications, we are concerned with discrete-valued quality characteristics, which have to be monitored using so-called *attributes charts*. In this chapter, we focus on a particular type of discrete-valued process, namely on *qualitative data* monitored sequentially in time, thus leading to a *categorical event series* $(X_t)_{t\in\mathbb{N}=\{1,2,\ldots\}}$. More precisely, we consider quality features X_t having a finite range consisting of either unordered but distinguishable categories (*nominal data*), or categories exhibiting a natural order (*ordinal data*). We uniquely denote the range of X_t as $\mathcal{S} = \{s_0, s_1, \ldots, s_d\}$ with some $d \in \mathbb{N}$, where the possible outcomes are arranged in either a lexicographical order (nominal case) or their natural order (ordinal case). In the special case $d = 1$, we refer to X_t as being *binary*, and it is then common to use the 0–1 coding $s_0 := 0$ and $s_1 := 1$. See Weiß [59] for further details on categorical time series.

The monitoring of categorical event series is an important task in many manufacturing applications. For example, Mukhopadhyay [36] monitors a nominal quality characteristic referring to six possible types of paint defect on manufactured ceiling fan covers, while Marcucci [31] distinguishes between three ordinal quality levels for bricks. Similar applications are reported by Spanos and Chen [45] regarding four levels for quality features of the photoresist line profile in a plasma etching process, Li et al. [28] on four categories of flash on the head of electric toothbrushes, or Li et al. [29] on three levels of injected glue on the base plates of manufactured mopheads. Non-manufacturing examples are reported by, e. g., Bashkansky and Gadrich [4], who monitor the condition of incoming patients as well as the severity of traffic accidents (three ordinal levels in both cases), and by Perry [39], who monitors a nominal process as part of a network monitoring problem. Generally, many different types of categorical event series occur in modern manufacturing systems, where the components of the production environment permanently emit status messages, signals on machine malfunctions, reports on quality deviations, etc. [15]. Similar challenges exist for computer systems [64], where the computer audit data is monitored for anomaly detection (especially intrusion detection), and for telecommunication networks [21], where sequences of alarm messages are analyzed for fault identification.

Thus, obviously, there is a great need for procedures to monitor the categorical event series in manufacturing applications (and beyond).

In this chapter, three general approaches (and corresponding methods) for monitoring categorical event series are surveyed. First, instead of monitoring the categorical events directly, one may consider counts of events for successive time intervals. So the original categorical data are transformed into a count time series, which is then monitored by, e. g., control charts for count data. Setting-up such control charts also requires to model the event counts series in an appropriate way; these aspects are discussed in Sect. 2. Second, one develops control charts directly for the actual categorical event series, see Sect. 3. Here, approaches for analyzing and modeling the categorical time series are relevant, where different solutions are required for nominal vs. ordinal data. But as criticized by Göb [15, p. 300], stochastic models and statistical approaches for discrete-valued time series are often "insufficiently communicated and not suitably tailored for application". Therefore, in practice, categorical event series are commonly analyzed by machine learning procedures, which do not suffer from narrowing assumptions and offer scalability with respect to the amount and complexity of data. Relevant rule-based machine learning approaches for event sequence monitoring are discussed in Sect. 4. Some of the presented methods are illustrated by real-data examples in Sect. 5. Finally, Sect. 6 concludes the article and outlines directions for future research.

2 Monitoring Time Series of Event Counts

There are several ways of transforming a categorical event series $(X_t)_{\mathbb{N}}$ into a count process, say $(Y_t)_{\mathbb{N}}$, which is then monitored instead of the original qualitative data. First, one can count the categorical events within fixed time intervals, as it was done, for example, by Ye et al. [65] for detecting intrusions into a computer system, and by Lambert and Liu [25] for detecting possible malfunctions in a communication network. This is related to the traditional sampling approach, where samples or segments are taken from the quality process and used to compute an event count, such as the number of defective or non-conforming items in the sample [33]. While we are usually concerned with bounded counts in the second case (with the upper bound being given by the sample size), counts might become arbitrarily large in the first scenario. In both cases, the dependence structure of the original categorical event series affects the distribution of the monitored event counts [60]. Finally, it is also common to determine the discrete waiting time until a certain event happens, such as the number of manufactured items until the occurrence of the next defective one [8]. But also more sophisticated types of runs charts have been proposed, where one waits for the occurrence of certain patterns in the categorical event series [56]. In any case, one ends up with a process $(Y_t)_{\mathbb{N}}$ consisting of *counts*, i. e., of non-negative integers from either the full set $\mathbb{N}_0 = \{0, 1, \ldots\}$ or a bounded subset thereof, $\{0, \ldots, n\}$ with some $n \in \mathbb{N}$ (in some applications, see Sect. 5.1, we might even be concerned with sample size varying in time). In Sect. 2.1, basic concepts regarding the analysis and modeling of count time series are discussed, and references for further information are provided. Afterwards in Sect. 2.2, some control charts are presented for the different types of event counts outlined before.

2.1 Analysis and Modeling of Count Time Series

Let $(Y_t)_{\mathbb{N}}$ be the derived *count process* that is to be monitored ("Phase II"). When developing an in-control model for setting up the control chart for $(Y_t)_{\mathbb{N}}$, we first need a data sample (i. e., a *count time series* $y_1, \ldots, y_T$) collected under in-control conditions ("Phase-I data") that is used for model fitting. While standard tools from time series analysis such as the time series plot or the (partial) autocorrelation function ((P)ACF) can be applied to $y_1, \ldots, y_T$ in the usual way, the well-known autoregressive moving-average (ARMA) or generalized AR conditional heteroscedasticity (GARCH) models cannot be used for describing these data, as these models are not able to ensure the count-data range, i. e., to generate only non-negative (and possibly bounded) integer values. Therefore, tailor-made models for count time series have to be used, see Weiß [59, 62] for detailed surveys. To give the reader an impression how such count time series models look like, let us briefly discuss some popular examples.

Probably the most well-known model for unbounded counts with a Markovian dependence structure (i. e., the upcoming count Y_{t+1} only depends on the present count Y_t, but not on further past counts $Y_{t-1}, Y_{t-2}, \ldots$) is the INAR(1) model dating back to McKenzie [32], the integer-valued counterpart to the ordinary first-order AR model. To preserve the discreteness of the range, it substitutes the AR(1)'s multiplication by a random operator called binomial thinning, defined as $\alpha \circ Y | Y \sim \text{Bin}(Y, \alpha)$ for the thinning probability $\alpha \in (0, 1)$. Due to this conditional binomial distribution, $\alpha \circ Y$ generates integer values from $\{0, \ldots, Y\}$ but having the same mean as if we would multiply Y by α instead, i. e., $E[\alpha \circ Y] = \alpha \cdot E[Y] = E[\alpha \cdot Y]$. Altogether, the INAR(1) model recursion is given by $Y_t = \alpha \circ Y_{t-1} + \epsilon_t$, where the thinnings are executed independently, and where the innovations $(\epsilon_t)_{\mathbb{N}}$ are an i. i. d. count process. As a result, the INAR(1)'s ACF takes the same form as in the AR(1) case, i. e., $\rho(h) = Corr[Y_t, Y_{t-h}] = \alpha^h$ for time lags $h \in \mathbb{N}$, and its PACF vanishes for lags $h \geq 2$. Furthermore, like the AR(1) model with normally distributed innovations also has normal observations (Gaussian AR(1) model), the INAR(1) model with Poisson-distributed innovations ϵ_t also has Poisson observations Y_t (Poisson INAR(1) model). But also different types of marginal distributions can be achieved, such as geometric or negative binomial observations, see Weiß [59] for details (there, also higher-order AR and MA counterparts are discussed). However, as a limitation, the INAR(1) model is suitable only for unbounded counts.

A modification of the INAR(1) model for bounded counts with range $\{0, \ldots, n\}$ has also been proposed by McKenzie [32]. The binomial AR(1) model substitutes the innovation ϵ_t in the INAR(1) recursion by a further thinning operator, namely $Y_t = \alpha \circ Y_{t-1} + \beta \circ (n - Y_{t-1})$. Note that the first summand is $\leq Y_{t-1}$, the second one is

$\leq n - Y_{t-1}$, so altogether, a count being $\leq n$ is generated. The stationary marginal distribution is binomial this time, while the ACF is still exponentially decaying, $\rho(h) = (\alpha - \beta)^h$. Again, several extensions and modifications of the basic binomial AR(1) model exist in the literature, see Weiß [59, 62] for references.

Finally, also several regression-type models for count time series have been developed. If mimicking the ordinary AR(1) model, these regression approaches assume a linear conditional mean $M_t = E[Y_t \mid Y_{t-1}, \ldots]$ of the form $M_t = a + b \cdot Y_{t-1}$ and use, for example, a conditional Poisson or binomial distribution for generating the unbounded or bounded counts, respectively. Such models are commonly referred to as INGARCH models [13], although this name is a bit misleading as the ACF is of ARMA-type, satisfying a set of Yule–Walker equations. Besides these conditionally linear INGARCH models, also several non-linear regression models have been proposed in the literature, e. g., models with a log-link, where $\ln M_t$ is a linear expression in past observations [59], or models using the nearly linear softplus function [63]. Such generalized linear models (GLMs) are also commonly used if deterministic patterns such as seasonality or trend have to be considered [18].

2.2 *Control Charts for Count Time Series*

During the last decade, control charts for count processes received a lot of research interest, see Weiß [57] for a survey. Since this chapter is mainly concerned with categorical event series, we limit the subsequent discussion to such contributions where we have a clear connection between the counts and the categorical events. For simplicity, let us assume for the moment that there are just two possible outcomes for the monitored event, such as "defect—yes or no" (later in Sect. 3.2, we consider the general scenario of multiple event categories). So the underlying categorical event series is in fact a binary process $(X_t)_{\mathbb{N}}$ with a certain serial dependence structure. Then, a possible approach for process monitoring is to take segments of length $n \in \mathbb{N}$ from $(X_t)_{\mathbb{N}}$ at the inspection times $t_1, t_2, \ldots \in \mathbb{N}$ (we focus on the constant sample size n here, while modifications for varying sample sizes are discussed later in Sect. 5.1) and to count the number of events in each segment, i. e., to compute $Y_r = X_{t_r} + \ldots + X_{t_r+n-1}$ for $r = 1, 2, \ldots$ The stochastic properties of $(Y_r)_{\mathbb{N}}$ depend on those of $(X_t)_{\mathbb{N}}$ as well as on the exact sampling strategy. For example, if $(X_t)_{\mathbb{N}}$ is i. i. d., then so is $(Y_r)_{\mathbb{N}}$—independent of the sampling strategy. Furthermore, the counts Y_r follow an ordinary binomial distribution [14]. If $(X_t)_{\mathbb{N}}$ is a Markov chain, by contrast, then the counts Y_r exhibit extra-binomial variation and follow the Markov-binomial distribution [54]. But if the inspection times $t_1, t_2, \ldots$ are sufficiently distant, then $(Y_r)_{\mathbb{N}}$ is still approximately independent. In other cases, the counts $(Y_r)_{\mathbb{N}}$ might be binomial but serially dependent, such as for the aforementioned binomial AR(1) model [41]. In any case, we are concerned with bounded counts with range $\{0, \ldots, n\}$, while unbounded counts would happen for the time-interval approach as in Lambert and Liu [25], Ye et al. [65]. There are several well-established types of control charts for such bounded counts, the exact chart design of which differs for the different scenarios outlined before.

The most basic chart for bounded counts is the *np*-chart (for unbounded counts, it is called *c*-chart), where the counts $Y_1, Y_2, \ldots$ are directly plotted on the chart and compared to given control limits $0 \leq l < u$ (equivalently, we could plot the sample fractions Y_r/n instead, leading to the *p*-chart) [33]. More precisely, an alarm is triggered for the rth count if $Y_r > u$ or $Y_r < l$ happens (the latter is only possible if $l > 0$). The limits l, u are commonly chosen such that the resulting chart shows a certain *average run length* (ARL) performance. Here, ARL refers to the *mean* waiting time until the first alarm, and this should be sufficiently large (low) if the process is in control (out of control), because we are concerned with a false (true) alarm in this case, recall Sect. 1. The crucial point is how to compute the ARL. If the $(Y_r)_{\mathbb{N}}$ are i. i. d., then the run length distribution is geometric with mean $\text{ARL} = 1/P\big(Y_r \notin [l; u]\big)$, where the latter is calculated from, e. g., the binomial or the Markov-binomial distribution, see the above discussion. If $(Y_r)_{\mathbb{N}}$ constitutes a Markov chain, then the ARL can be determined by the Markov chain approach of Brook and Evans [9]. For more complex serial dependence structures, ARLs are typically approximated based on simulations, see Weiß [59] for a further discussion.

The *np*-chart, as any *Shewhart chart*, has the disadvantage of being rather insensitive towards small shifts in the process, because the decision at time r solely relies on the rth sample count Y_r. To achieve an improved ARL performance, control charts with an inherent memory have been proposed, such as the *cumulative sum* (CUSUM) chart dating back to Page [38], or the *exponentially weighted moving-average* (EWMA) chart dating back to Roberts [43]. The basic (upper) CUSUM chart is defined by the recursive scheme

$$C_0 = c_0, \qquad C_r = \max\big\{0,\ Y_r - k + C_{r-1}\big\} \quad \text{for } r = 1, 2, \ldots \tag{1}$$

It accumulates positive deviations from the reference value $k > 0$ and triggers an alarm if the (upper) control limit $h > 0$ is exceeded. It is thus able to detect increases in the counts' mean (a CUSUM chart for decreases is defined by $C_r = \max\big\{0,\ k - Y_r + C_{r-1}\big\}$ instead). If the reference value k is integer-valued (rational), then also C_r can only take integer (rational) numbers, which allows for an exact ARL computation in some cases. Namely, if $(Y_r)_{\mathbb{N}}$ is i. i. d., then the Markov chain approach as described in Brook and Evans [9] can be applied, while for $(Y_r)_{\mathbb{N}}$ being itself a Markov chain (such as a binomial AR(1) process), the modifications as in Rakitzis et al. [41] have to be used. More sophisticated types of CUSUM chart are obtained if the statistics are derived from the process model's log-likelihood ratio [59]; this log-LR CUSUM approach can also be extended to GLMs having seasonality or trend [18].

The standard EWMA chart of Roberts [43] is defined by the recursion $Z_r = \lambda \cdot Y_r + (1 - \lambda) \cdot Z_{r-1}$ with smoothing parameter $\lambda \in (0; 1]$ and control limits $0 \leq l < u$. The choice $\lambda = 1$ leads to the *np*-chart, i. e., it corresponds to no smoothing at all. Because of the multiplications for computing Z_r, however, the discrete nature of the counts Y_r evaporates as time progresses, thus making an exact ARL computation impossible. For this reason, it might be advantageous to consider a discretized version such as the rounded EWMA chart proposed by Gan [14] and used by Weiß [54] for Markov-binomial counts, which is defined by

$$Q_r = \text{round}\big(\lambda \cdot X_r + (1-\lambda) \cdot Q_{r-1}\big) \quad \text{for } r = 1, 2, \ldots \tag{2}$$

Here, ARLs can be computed by adapting the Markov chain approach of Brook and Evans [9]. Another option is to use the EWMA-type recursion developed by Morais et al. [34], where the multiplications in the EWMA recursion are substituted by binomial thinnings, in analogy to the thinning-based count time series models surveyed in Sect. 2.1.

Finally, let us discuss a completely different monitoring approach for the categorical event series $(X_t)_{\mathbb{N}}$, which also leads to the monitoring of counts. Let us start with the binary case (defect—yes or no), and assume that the probability for getting a defect is rather low (high-quality process). Then, $(X_t)_{\mathbb{N}}$ can be monitored by waiting for the respective next defect and by applying a control chart to the obtained waiting times R_i, $i = 1, 2, \ldots$ (run lengths), see Szarka and Woodall [48] for a review. It is clear that the waiting times decrease if defects happen more frequently, i. e., control charts mainly focus on possible decreases in the mean of $(R_i)_{\mathbb{N}}$. The distribution of $(R_i)_{\mathbb{N}}$ depends on the serial dependence structure of $(X_t)_{\mathbb{N}}$. If $(X_t)_{\mathbb{N}}$ is i. i. d., then $(R_i)_{\mathbb{N}}$ consists of independent geometric counts [8], while $(X_t)_{\mathbb{N}}$ being a Markov chain still leads to independent run lengths $(R_i)_{\mathbb{N}}$, but having additional dispersion compared to a geometric distribution [7]. In both cases, ARLs are computed via $\text{ARL} = 1/P\big(R_i \notin [l; u]\big)$, in analogy to the *np*-chart. Besides monitoring $(R_i)_{\mathbb{N}}$ with a Shewhart chart, Bourke [8] also suggests to use a moving-sum chart (i. e., for fixed window length $w \in \mathbb{N}$, the sum $R_{i-w+1} + \ldots + R_i$ is plotted on the chart after having observed the ith run) or a CUSUM chart. Furthermore, such waiting-time charts can be applied to much more complex patterns then just a single defect. In Weiß [56], for example, a truly categorical event series $(X_t)_{\mathbb{N}}$ (i. e., with more than just two categories) is considered and one waits for constant segments of specified categories.

3 Monitoring Categorical Event Series

Although the dividing line between Sects. 2 and 3 is sometimes not sharp, here, we focus on such monitoring procedures that explicitly account for the categorical nature of the data $(X_t)_{\mathbb{N}}$. Following Weiß [60], we distinguish between control charts for statistics relying on samples or segments taken from the categorical event series, such as the basic χ^2-chart dating back to Duncan [11], and charts for continuously monitoring the process, such as the EWMA chart used by Ye et al. [66] for intrusion detection, or the categorical CUSUM chart proposed by Ryan et al. [44]. These and many further control chart proposals for nominal or ordinal time series data are surveyed in Sect. 3.2, where the actual chart design again relies on appropriate model assumptions. Basic modeling approaches are discussed in Sect. 3.1 and, in particular, ways of analyzing such categorical time series. As the data are qualitative, standard tools from time series analysis cannot be used for model identification. Instead, tailor-made solutions are required, where the ordinal case has to be distinguished from the nominal one.

3.1 Analysis and Modeling of Categorical Time Series

For a categorical event series $(X_t)_{\mathbb{N}}$, the range consists of qualitative categories, denoted by $\mathcal{S} = \{s_0, s_1, \ldots, s_d\}$ with $d \in \mathbb{N}$, such that arithmetic operations cannot be applied to $\mathcal{S}$. Consequently, moments are not defined for $(X_t)_{\mathbb{N}}$, i. e., we cannot compute the mean, variance, ACF, etc. if analyzing $(X_t)_{\mathbb{N}}$. If the range of $(X_t)_{\mathbb{N}}$ is ordinal, at least quantiles can be defined and a time series plot is possible by arranging the possible outcomes in their natural order along the Y axis. In the nominal case, by contrast, we can compute the mode(s) to infer the location of X_t, and a rate evolution graph may serve as a substitute of the time series plot [59, Section 6]. This brief discussion already shows that the analysis of categorical event series cannot be done with standard tools, but tailor-made solutions are required and need to be implemented.

Analytic tools for nominal time series commonly rely on the probability mass function (PMF), whereas for ordinal time series, one uses the cumulative distribution function (CDF) to account for the natural order among the categories. An overview of some popular statistics is provided by Table 1, see Klein and Doll [20], Kvålseth [24], Weiß [59, 61] for further details. Both for nominal and ordinal random variables, minimal dispersion is attained in the case of a one-point distribution, expressed as

$$\boldsymbol{p} \in \left\{ \begin{pmatrix} 1 \\ 0 \\ \vdots \\ 0 \end{pmatrix}, \begin{pmatrix} 0 \\ 1 \\ \vdots \\ 0 \end{pmatrix}, \ldots, \begin{pmatrix} 0 \\ \vdots \\ 1 \\ 0 \end{pmatrix}, \begin{pmatrix} 0 \\ \vdots \\ 0 \\ 1 \end{pmatrix} \right\} =: \{\boldsymbol{e}_0, \ldots, \boldsymbol{e}_d\} \subset [0;1]^{d+1} \tag{1}$$

and

$$\boldsymbol{f} \in \left\{ \begin{pmatrix} 1 \\ \vdots \\ 1 \end{pmatrix}, \begin{pmatrix} 0 \\ 1 \\ \vdots \\ 1 \end{pmatrix}, \ldots, \begin{pmatrix} 0 \\ \vdots \\ 0 \\ 1 \end{pmatrix}, \begin{pmatrix} 0 \\ \vdots \\ 0 \\ 0 \end{pmatrix} \right\} =: \{\boldsymbol{c}_0, \ldots, \boldsymbol{c}_d\} \subset [0;1]^{d}, \tag{2}$$

respectively. By contrast, the concepts of maximal dispersion differ: maximal dispersion in the nominal sense happens for a uniform distribution, $\boldsymbol{p} = (\frac{1}{d+1}, \ldots, \frac{1}{d+1})^\top$, while maximal dispersion in the ordinal sense is given by the extreme two-point distribution, $\boldsymbol{f} = (\frac{1}{2}, \ldots, \frac{1}{2})^\top$, i. e., with probability mass 1/2 in the outer-most categories. These different dispersion concepts are taken into account by the (normalized) dispersion measures IQV and IOV in Table 1 [see [24]]. In the ordinal case, it is also possible to define a (normalized) skewness measure, see Table 1 as well as Klein and Doll [20], while there is no meaningful concept of (a)symmetry for a nominal distribution. Finally, signed serial dependence at lag $h \in \mathbb{N}$ can be measured in terms

Table 1 Statistics for analyzing categorical event series

Nominal range	Ordinal range
Marginal PMF: $\boldsymbol{p} = (p_0, \ldots, p_d)^\top \in [0; 1]^{d+1}$ with $p_i = P(X = s_i)$	Marginal CDF: $\boldsymbol{f} = (f_0, \ldots, f_{d-1})^\top \in [0; 1]^d$ with $f_i = P(X \leq s_i)$
Bivariate lag-h PMF: $p_{ij}(h) = P(X_t = s_i, X_{t-h} = s_j)$	Bivariate lag-h CDF: $f_{ij}(h) = P(X_t \leq s_i, X_{t-h} \leq s_j)$
Index of qualitative variation: $\mathrm{IQV} = \frac{d+1}{d}\left(1 - \sum_{i=0}^{d} p_i^2\right)$	Index of ordinal variation: $\mathrm{IOV} = \frac{4}{d} \sum_{i=0}^{d-1} f_i(1 - f_i)$
	Ordinal skewness: $\mathrm{skew} = \frac{2}{d} \sum_{i=0}^{d-1} f_i - 1$
Nominal Cohen's κ: $\kappa_{\mathrm{nom}}(h) = \dfrac{\sum_{j=0}^{d}\left(p_{jj}(h) - p_j^2\right)}{1 - \sum_{i=0}^{d} p_i^2}$	Ordinal Cohen's κ: $\kappa_{\mathrm{ord}}(h) = \dfrac{\sum_{j=0}^{d-1}\left(f_{jj}(h) - f_j^2\right)}{\sum_{i=0}^{d-1} f_i(1 - f_i)}$

of Cohen's κ, where positive (negative) values express the extend of (dis)agreement between X_t and X_{t-h} [59, 61].

The sample counterparts to the measures in Table 1 are defined by replacing all (cumulative) probabilities by (cumulative) relative frequencies. The latter, in turn, can be computed as sample means about appropriately defined *binarizations*. A nominal event series $(X_t)_{\mathbb{N}}$ is equivalently expressed as $(\boldsymbol{Y}_t)_{\mathbb{N}}$ with $Y_{t,i} = \mathbb{1}_{\{X_t = s_i\}}$ for $i \in \{0, \ldots, d\}$, and an ordinal event series $(X_t)_{\mathbb{N}}$ by $(\boldsymbol{Z}_t)_{\mathbb{N}}$ with $Z_{t,i} = \mathbb{1}_{\{X_t \leq s_i\}}$ for $i \in \{0, \ldots, d-1\}$. Here, $\mathbb{1}_A$ denotes the indicator function, which takes the value 1 (0) if A is true (false). So the range of the nominal binarization $(\boldsymbol{Y}_t)_{\mathbb{N}}$ is given by (1), the one of the ordinal binarization $(\boldsymbol{Z}_t)_{\mathbb{N}}$ by (2). Then, the sample PMF equals $\hat{\boldsymbol{p}} = \frac{1}{T} \sum_{t=1}^{T} \boldsymbol{Y}_t$, the sample CDF $\hat{\boldsymbol{f}} = \frac{1}{T} \sum_{t=1}^{T} \boldsymbol{Z}_t$, and the bivariate (cumulative) relative frequencies are $\hat{p}_{ij}(h) = \frac{1}{T-h} \sum_{t=h+1}^{T} Y_{t,i}\, Y_{t-h,j}$ and $\hat{f}_{ij}(h) = \frac{1}{T-h} \sum_{t=h+1}^{T} Z_{t,i}\, Z_{t-h,j}$, respectively. The sample counterparts to the measures of Table 1 can then be used for identifying an appropriate in-control model for the categorical event series $(X_t)_{\mathbb{N}}$.

Models for categorical event series, see Weiß [59] for a survey, often suffer from a large number of model parameters. For higher-order Markov models, this number increases exponentially in the model order such that solutions for reducing the model complexity are required. These might be the variable-length Markov models by Bühlmann and Wyner [10], where the model order depends on the actual past, or the mixture transition distribution models by Raftery [40], where parametric relations between the transition probabilities are introduced. Markov models with a reduced number of model parameters can also be achieved by GLM approaches such as in

Höhle [17], whereas the Hidden-Markov models dating back to Baum and Petrie [5] lead to a non-Markovian categorical process that is generated by a latent Markov chain. However, the most parsimonious class of models for categorical event series $(X_t)_{\mathbb{N}}$ appear to be the discrete ARMA(p, q) models proposed by Jacobs and Lewis [19], where $\mathrm{p}, \mathrm{q} \in \mathbb{N}_0$. These rely on a random choice mechanism, implemented by the i. i. d. multinomial random vectors

$$\boldsymbol{D}_t \;=\; (\alpha_{t,1}, \ldots, \alpha_{t,\mathrm{p}}, \beta_{t,0}, \ldots, \beta_{t,\mathrm{q}}) \quad\sim\; \mathrm{Mult}(1;\; \phi_1, \ldots, \phi_{\mathrm{p}}, \varphi_0, \ldots, \varphi_{\mathrm{q}}),$$

which are also independent of the i. i. d. categorical innovations $(\epsilon_t)_{\mathbb{Z}}$ with range $\mathcal{S}$. Then, $(X_t)_{\mathbb{N}}$ is defined by

$$X_t \;=\; \alpha_{t,1} \cdot X_{t-1} + \ldots + \alpha_{t,\mathrm{p}} \cdot X_{t-\mathrm{p}} \;+\; \beta_{t,0} \cdot \epsilon_t + \ldots + \beta_{t,\mathrm{q}} \cdot \epsilon_{t-\mathrm{q}}, \tag{3}$$

where we assume $0 \cdot s = 0$, $1 \cdot s = s$, and $s + 0 = s$ for each $s \in \mathcal{S}$. So X_t is generated by choosing the outcome of either one of the past observations $X_{t-1}, \ldots, X_{t-\mathrm{p}}$, or one of the available innovations $\epsilon_t, \ldots, \epsilon_{t-\mathrm{q}}$. The stationary marginal distribution of X_t is the one of ϵ_t, and both κ-measures from Table 1 satisfy the Yule–Walker equations

$$\kappa(h) \;=\; \textstyle\sum_{j=1}^{\mathrm{p}} \phi_j\, \kappa(|h-j|) \;+\; \sum_{i=0}^{\mathrm{q}-h} \varphi_{i+h}\, r(i) \qquad \text{for } h \geq 1, \tag{4}$$

where $r(i) = \sum_{j=\max\{0,i-\mathrm{p}\}}^{i-1} \phi_{i-j}\, r(j) + \varphi_i\, \mathbb{1}(0 \leq i \leq \mathrm{q})$ [59]. For example, for $(\mathrm{p}, \mathrm{q}) = (1, 0)$, we get a parsimonious Markov chain with $\kappa(h) = \phi_1^h$, in analogy to the AR(1)-like models discussed in Sect. 2.1.

For the particular case of ordinal event series $(X_t)_{\mathbb{N}}$, one may also use models for time series of bounded counts (recall Sect. 2.1) in view of the rank-count approach discussed by Weiß [61], i.e., we write $X_t = s_{I_t}$ with the rank count I_t having the bounded range $\{0, \ldots, d\}$. Then, the rank-count process $(I_t)_{\mathbb{N}}$ is modeled instead of the original ordinal event series $(X_t)_{\mathbb{N}}$. In a similar spirit, one may assume that X_t is generated by a latent, continuously distributed random variable, say L_t with range $\mathbb{R}$. For the latent-variable approach, see Agresti [2] for details, we have to specify the threshold parameters $-\infty = \eta_{-1} < \eta_0 < \ldots < \eta_{d-1} < \eta_d = +\infty$. Then, X_t falls into the jth category iff L_t falls into the jth interval, $j \in \{0, \ldots, d\}$, i.e., $X_t = s_j$ iff $L_t \in [\eta_{j-1}; \eta_j)$. Thus, if F_L denotes the CDF of L_t, we have $f_j = F_L(\eta_j)$, where the choice of the standard logistic (normal) distribution for L_t leads to the cumulative logit (probit) model. These models are commonly combined with a regression approach, such as the GLMs considered by Höhle [17], Li et al. [29]. The latent-variable approach is also related to the step gauge discussed by Steiner [47], Steiner et al. [46], where a real-valued quality characteristic is not measured exactly but only classified into one of finitely many successive groups.

3.2 *Control Charts for Categorical Time Series*

In what follows, we survey a couple of approaches for monitoring categorical event series $(X_t)_{\mathbb{N}}$. To impose some structure, we distinguish between control charts mainly designed for a continuous process monitoring (i. e., the monitored statistic is updated with each incoming observation), and those for a sample-based monitoring (i. e., samples or segments of specified size are taken from $(X_t)_{\mathbb{N}}$ and used to compute the plotted statistics). But similar to the dividing line between Sects. 2 and 3, this distinction does not lead to two disjoint groups: The sample-based procedures in Sect. 3.2.2 can be used for continuous process monitoring by using a moving-window approach [56], while some of the procedures from Sect. 3.2.1 can also be adapted to a sample-based monitoring.

3.2.1 Continuous Process Monitoring

A common approach for continuously monitoring a categorical event series $(X_t)_{\mathbb{N}}$ is to use a type of CUSUM chart, which is derived from the log-likelihood ratio (log-LR) corresponding to $(X_t)_{\mathbb{N}}$. If $(X_t)_{\mathbb{N}}$ is i. i. d. with marginal PMF $\boldsymbol{p}_0$ in the in-control state, and if we anticipate $(X_t)_{\mathbb{N}}$ to switch to $\boldsymbol{p}_1$ in the out-of-control case, then Ryan et al. [44] propose to monitor

$$\mathrm{C}_t = \max\{0,\ \ell R_t + \mathrm{C}_{t-1}\} \qquad \text{with } \ell R_t = \sum_{j=0}^{d} Y_{t,j}\, \ln\left(\frac{p_{1,j}}{p_{0,j}}\right), \tag{5}$$

where $Y_{t,j}$ refers to the nominal binarization of X_t discussed in Sect. 3.1. An alarm is triggered if the upper control limit $h > 0$ is violated, where h is again chosen based on ARL considerations. Weiß [60] extends this approach to the monitoring of Markov-dependent processes $(X_t)_{\mathbb{N}}$, whereas Höhle [17] presents a log-LR CUSUM chart with respect to a categorical logit regression model. For $d = 1$, (5) reduces to the Bernoulli CUSUM discussed by Reynolds and Stoumbos [42] (and by Mousavi and Reynolds [35] in the binary Markov case).

At this point, a relation to Sect. 2.2 should be noted. There, we discussed several waiting-time charts, which also allow for a continuous monitoring of $(X_t)_{\mathbb{N}}$ (but we discussed these charts already in Sect. 2.2 due to the count nature of the waiting times). As argued by Reynolds and Stoumbos [42], their Bernoulli CUSUM chart is essentially equivalent to the geometric CUSUM chart proposed by Bourke [8] for monitoring runs in an i. i. d. binary process.

Two further points are worth mentioning. First, Ryan et al. [44] also proposed a modification of (5) that can be used for a sample-based monitoring of $(X_t)_{\mathbb{N}}$, see (8) in Sect. 3.2.2 below. Second, the CUSUM approach (5) is essentially the same as that discussed by Steiner et al. [47], although there, it was formulated for grouped data resulting from gauging. As the main difference, the anticipated out-of-control PMF $\boldsymbol{p}_1$ is derived by assuming, e. g., a mean shift in the underlying real-valued quality characteristic. In this way, ordinal information is incorporated into the monitored statistic (5).

Besides the waiting-time and CUSUM approaches for continuously monitoring a categorical event series $(X_t)_{\mathbb{N}}$, also the EWMA chart proposed by Ye et al. [66] might be used. They suggest to apply the EWMA approach to the nominal binarizations $(\boldsymbol{Y}_t)_{\mathbb{N}}$, i. e., to compute the smoothed estimators $\boldsymbol{P}_t^{(\lambda)} = \lambda\, \boldsymbol{Y}_t + (1-\lambda)\, \boldsymbol{P}_{t-1}^{(\lambda)}$ of the true marginal distribution $\boldsymbol{p}$. For chart design, it is necessary to estimate the mean and covariance of $\boldsymbol{P}_t^{(\lambda)}$ from given Phase-I data, leading to the sample estimates $\overline{\boldsymbol{P}}$ and $\mathbf{S}$, respectively. Then, Ye et al. [66] suggest to either plot Hotelling's T^2-statistic, $\mathrm{T}_t^2 = \big(\boldsymbol{P}_t^{(\lambda)} - \overline{\boldsymbol{P}}\big)^\top\, \mathbf{S}^{-1} \big(\boldsymbol{P}_t^{(\lambda)} - \overline{\boldsymbol{P}}\big)$, on a control chart, or Pearson's χ^2-statistic given by $\mathrm{X}_t^2 = \sum_{j=0}^{d} \big(P_{t,j}^{(\lambda)} - \overline{P}_j\big)^2 / \overline{P}_j$.

Finally, let us refer to the Shewhart-type charts proposed by Bersimis et al. [6], Koutras et al. [23] for monitoring a bivariate ordinal process. These charts define a set of patterns among the bivariate ordinal outcomes, where an alarm is triggered if one of the patterns occurs in the process.

3.2.2 Sample-Based Process Monitoring

The large majority of proposals for monitoring a categorical event series $(X_t)_{\mathbb{N}}$ fall into the class of sample-based approaches. For ease of presentation, we again restrict the subsequent presentation to the constant sample size n, while necessary modifications for varying sample sizes are later considered in Sect. 5.1. Using the nominal binarizations $(\boldsymbol{Y}_t)_{\mathbb{N}}$, one first computes the vectors of (absolute) sample frequencies $(\boldsymbol{N}_r)_{\mathbb{N}}$, i. e., $\boldsymbol{N}_r = \boldsymbol{Y}_{t_r} + \ldots + \boldsymbol{Y}_{t_r+n-1}$, which are multinomially distributed according to $\mathrm{Mult}(n,\, \boldsymbol{p})$ if $(\boldsymbol{Y}_t)_{\mathbb{N}}$ is i. i. d. If the sampled $\boldsymbol{Y}_{t_r}, \ldots, \boldsymbol{Y}_{t_r+n-1}$ are serially dependent instead, the distribution of $\boldsymbol{N}_r$ differs from a multinomial one. For example, if $(X_t)_{\mathbb{N}}$ is a discrete AR(1) process, recall (3), then $\boldsymbol{N}_r$ follows the Markov-multinomial distribution, see Weiß [60] for details. Note that the special case of a sample-based monitoring of a binary process $(X_t)_{\mathbb{N}}$, so relying on (Markov-)binomial counts rather the (Markov-)multinomial vectors, was already discussed in Sect. 2.2. Thus, here, we concentrate on the truly categorical case ($d > 1$), and if not stated otherwise, we assume that $(\boldsymbol{N}_r)_{\mathbb{N}}$ are i. i. d. multinomial vectors under in-control conditions. Furthermore, in the case of an ordinal event series $(X_t)_{\mathbb{N}}$, we shall also consider the cumulative frequencies $\boldsymbol{C}_r = \boldsymbol{Z}_{t_r} + \ldots + \boldsymbol{Z}_{t_r+n-1}$ derived from the ordinal binarizations $(\boldsymbol{Z}_t)_{\mathbb{N}}$, recall Sect. 3.1.

The most well-known approach for a sample-based monitoring of $(X_t)_{\mathbb{N}}$ is the χ^2-chart proposed by Duncan [11], see Koutras et al. [31], Marcucci [36], Mukhopadhyay [22] for further discussions and extensions. With $\boldsymbol{p}_0$ being the marginal PMF of $(X_t)_{\mathbb{N}}$ under in-control conditions, it plots the Pearson statistics

$$\mathrm{X}_r^2 \;=\; \sum_{j=0}^{d} \frac{\big(N_{r,j} - n\, p_{0,j}\big)^2}{n\, p_{0,j}} \tag{6}$$

derived from the sample frequencies $(\boldsymbol{N}_r)_{\mathbb{N}}$. Pearson's χ^2-statistic, which is commonly used for goodness-of-fit testing, measures any kind of deviation from the hypothetical in-control distribution $\boldsymbol{p}_0$. In quality-related applications, however, one is commonly concerned with the situation that one of the categories, say the 0th category s_0, expresses conforming items and is thus much more frequently observed than the remaining defect categories $s_1, \ldots, s_d$. So $\boldsymbol{p}_0$ is often close to the one-point distribution $\boldsymbol{e}_0$ in (1) (low dispersion), while deteriorations of quality go along with deviations towards, e. g., a uniform distribution (maximal nominal dispersion) or an extreme two-point distribution (maximal ordinal dispersion). Thus, a reasonable alternative to monitoring (6) is to monitor a categorical dispersion measure instead, such as the IQV from Table 1 in the nominal case (this was done by Weiß [56, 60] for i. i. d. and serially dependent data, respectively), or the IOV in the ordinal case [see [4]]. So the corresponding sample statistics compute as

$$\mathrm{IQV}_r = \tfrac{d+1}{d}\left(1-\sum_{j=0}^{d}\frac{N_{r,j}^2}{n^2}\right) \quad \text{and} \quad \mathrm{IOV}_r = \tfrac{4}{d}\sum_{j=0}^{d-1}\frac{C_{r,j}}{n}\left(1-\frac{C_{r,j}}{n}\right), \tag{7}$$

respectively. Also the "p-tree method" of Duran and Albin [12] should be mentioned here, where conditional (ordinal) events of the form $X = s_j \mid X \geq s_j$ are monitored by plotting the statistics $N_{r,j}/(n - C_{r,j-1})$ for $j = 1, \ldots, d-1$ as well as $N_{r,0}/n$ on multiple charts simultaneously.

At this point, let us have a more detailed look on control charts for ordinal event series $(X_t)_{\mathbb{N}}$. Sometimes, people analyze ordinal data by assigning numerical values ("scores") to the possible outcomes and by treating the transformed data like quantitative data [2, Section 2.1.1]. In a quality context, this happens for the demerit control charts [33, Section 7.3.3], where the chosen demerit weights $0 = w_0 < w_1 < \ldots < w_d$ try to reflect the severity of the defect categories; the computed demerit statistics are $D_r = w_1\, N_{r,1} + \ldots + w_d\, N_{r,d}$. In Nembhard and Nembhard [37], for example, who also allowed for possible serial dependence between the demerit statistics, a manufacturing process for plastic containers with three seels is considered, where a leak in the inner seal is less critical than one in the middle seal, and a middle-seal leak is less critical than an outer-seal one. This is accounted for by assigning the weights 1, 3, and 10, respectively, to these events, but there is the danger of arbitrariness in choosing the weights. In the step gauge scenario considered by Steiner [46], the weights are derived from the gauge limits by a midpoint scheme, while Perry [39], who considers a nominal event series $(X_t)_{\mathbb{N}}$, defines the weights by statistical reasoning as $1/(n\, p_{0,j})$ for $j = 0, \ldots, d$. Both Steiner [46] and Perry [39] apply an EWMA approach to their weighted class counts. In a sense, also the sample version of the log-LR CUSUM scheme (5) proposed by Ryan et al. [44], Steiner et al. [47],

$$\mathrm{C}_r = \max\{0,\ \ell R_r + \mathrm{C}_{r-1}\} \qquad \text{with } \ell R_t = \sum_{j=0}^{d} N_{r,j}\, \ln\!\left(\frac{p_{1,j}}{p_{0,j}}\right), \tag{8}$$

relies on a weighted class count, now with the weights being $\ln\big(p_{1,j}/p_{0,j}\big)$. A similar motivation applies to the so-called "simple ordinal categorical" (SOC) chart proposed by Li et al. [28], which relies on likelihoods computed from a latent-variable approach. In case of a logit model, the plotted statistics finally take a simple form,

$$\mathrm{SOC}_r = \left|\sum_{j=0}^{d}(f_{0,j-1}+f_{0,j}-1)\,N_{r,j}\right| \quad \text{with } f_{0,-1}:=0, \tag{9}$$

where the average cumulative proportions $\frac{1}{2}\,(f_{0,j-1}+f_{0,j})$ are known as "ridits" in the literature [2, p. 10]. In addition, Li et al. [28] suggest to substitute the raw frequencies $\boldsymbol{N}_r$ in (9) by smoothed frequencies resulting from an EWMA approach, i. e., $\boldsymbol{N}_r^{(\lambda)} = \lambda\,\boldsymbol{N}_r + (1-\lambda)\,\boldsymbol{N}_{r-1}^{(\lambda)}$. Such EWMA-smoothed frequencies are also used by Wang et al. [52], whose average cumulative data (ACD) chart plots the quadratic-form statistics

$$\mathrm{ACD}_r = n^{-1}\,\big(\boldsymbol{N}_r - n\,\boldsymbol{p}_0\big)^{\top}\,\mathbf{V}\,\big(\boldsymbol{N}_r - n\,\boldsymbol{p}_0\big), \tag{10}$$

where $\mathbf{V}$ denotes a weight matrix that needs to be specified prior to monitoring. Note that the particular choice $\mathbf{V}^{-1} = \mathrm{diag}(\boldsymbol{p}_0)$ just leads to the Pearson statistic (6). But for ordinal data, Wang et al. [52] recommend to choose $\mathbf{V} = \mathbf{L}^{\top}\,\mathrm{diag}(\boldsymbol{w})\,\mathbf{L}$ with weight vector $\boldsymbol{w}$, e. g., $\boldsymbol{w} = \mathbf{1} = (1,\ldots,1)^{\top}$, and $\mathbf{L}$ being of the following triangular structure: the lower (upper) triangle is filled with 2 (0), and the main diagonal with 1. Then, ACD_r can be rewritten as

$$\mathrm{ACD}_r = n^{-1}\sum_{j=0}^{d} w_j\,\Big(C_{r,j-1}+C_{r,j}-n\,(f_{0,j-1}+f_{0,j})\Big)^2, \tag{11}$$

i. e., the statistic ACD_r again relies on ridits, in analogy to (9). Bai and Li [3] extend the SOC chart (9) to the (univariate) location-scale ordinal (LSO) chart, which is also sensitive to changes in the dispersion structure (i. e., the scale of the latent variable). The monitored statistic is of quadratic form like in (10). For multivariate extensions of the quadratic-form approach (10), see the multivariate ordinal categorical (MOC) chart of Wang et al. [51] as well as the multivariate LSO chart of Bai and Li [3].

Similar to Steiner et al. [47] and Li et al. [28], also Tucker et al. [49] takes the likelihood derived for a latent-variable model for the ordinal event series $(X_t)_{\mathbb{N}}$ as the starting point. But this time, the location parameter's maximum likelihood estimate is computed for each sample. It is then plotted on the chart after an appropriate standardization. Finally, Li et al. [29] account for first-order serial dependence in the ordinal event series $(X_t)_{\mathbb{N}}$ by determining bivariate frequencies $N_{r,ij}$ in the rth sample, referring to the pairwise events $(X_{t-1}, X_t) = (s_i, s_j)$. These bivariate counts are subjected to an EWMA smoothing. Then, quadratic-form statistics similar to (10) are computed for each sample, where the weight matrix is derived from an ordinal log-linear model for $(X_t)_{\mathbb{N}}$, and which are then plotted on the serially dependent ordinal categorical (SDOC) control chart.

4 Machine Learning Approaches for Event Sequence Monitoring

Besides the statistically sound approach of control charts for monitoring categorical event sequences, also machine learning approaches have been established for this purpose. These do without explicit stochastic model assumptions[1] while having the advantage of being scalable, see Göb [15] for a critical discussion. The procedures described in the sequel can be subsumed under the discipline of temporal data mining, see Laxman and Sastry [26] for a survey, and they also belong to the class of rule-based procedures, see Weiß [58] for a brief overview. More precisely, we focus on procedures of *episode mining*, where rules are generated based on available data from a single categorical event sequence ("Phase-I data"), and which are then applied to forecasting events in the ongoing process ("Phase-II application"). While control charts trigger an alarm once the control limits are violated, such rule-based procedures require action once a critical event is predicted.

The task of episode mining was first considered by Mannila et al. [30] and further developed by several authors, see the references in Zimmermann [68]. The original motivation of Mannila et al. [30] was telecommunication alarm management, where an online analysis of the incoming alarm stream is required, e. g., to predict severe faults of the system. But many further applications have been reported meanwhile. Laxman et al. [27], for example, applied episode mining to sequences of status codes stemming from the assembly lines of a car manufacturing plant, while Yuan et al. [67] used episode mining for failure prediction in mechatronic systems (with a case study referring to a medical imaging system). A further example is provided by Vasquez Capacho et al. [50], who consider alarm management systems for industrial plant safety, which support the operators to adequately react on an "alarm flood", e. g., to distinguish between normal and dangerous conditions of a vacuum oven used in a refinery.

For the illustration of episode mining, let us discuss the approaches proposed by Mannila et al. [30] in some detail. The aim is to derive (and later to apply) rules such as, e. g., if the segment $(x_{t-2}, x_{t-1}, x_t) = (s_0, s_1, s_0)$ is observed ("episode"), then we expect $X_{t+1} = s_2$ with some given "confidence". Let us use the notation "$\boldsymbol{a} \Rightarrow \boldsymbol{b}$" with $\boldsymbol{a} = (s_0, s_1, s_0)$ and $\boldsymbol{b} = (s_0, s_1, s_0, s_2)$ for such a rule. Here, to prevent spurious correlation, only episodes satisfying a certain "support" requirement are considered, i. e., the frequency (in some sense) of the episodes has to exceed a given threshold value $\text{supp}_{\min}$. The permitted episodes are not limited to single tuples from $\mathcal{S} \times \mathcal{S}^2 \times \ldots$, but also sets of tuples are possible, which result from using operators such as the wildcard "*" for an arbitrary symbol from $\mathcal{S}$ or the parallel episode "[...]" allowing for an arbitrary ordering of the specified symbols. For example, $\big(s_0, [s_1, s_2]\big)$ corresponds to the set $\big\{(s_0, s_1, s_2), (s_0, s_2, s_1)\big\}$, and $\big(s_0, *\big)$ to $\big\{(s_0, s_0), \ldots, (s_0, s_d)\big\}$.

[1] Although the presented approaches for episode mining do not explicitly use stochastic assumptions, several connections to models from Sect. 3.1 have been established in the literature, namely to Markov models by Gwadera et al. [16], to Hidden-Markov models by Laxman et al. [27], and to variable-length Markov models by Weiß [55].

Since episode mining targets at extremely long categorical sequences, it is crucial to develop highly efficient algorithms for determining the "frequent episodes" as well as the corresponding rules. In Mannila et al. [30], this is achieved by applying (among others) the famous Apriori principle, which dates back to Agrawal and Srikant [1] and relies on the fact that an episode $\boldsymbol{a}$ can only be frequent (i. e., $\text{supp}(\boldsymbol{a}) > \text{supp}_{\text{min}}$ for fixed threshold value supp_{min}) if all its sub-episodes are frequent as well. So the frequent episodes are detected in a bottom-up manner: given the set $\mathcal{F}_{k-1}$ of frequent episodes of length $k-1$, the set $\mathcal{C}_k$ with candidate episodes of length k is constructed by combining the episodes from $\mathcal{F}_{k-1}$, and then their support is determined to filter out the frequent episodes for $\mathcal{F}_k \subseteq \mathcal{C}_k$. Here, Mannila et al. [30] consider two types of support measure: In their "Winepi" algorithm, supports are computed by passing a moving window of fixed length w (chosen sufficiently large) along the categorical event sequence, and $\text{supp}(\boldsymbol{a})$ is defined as the number of windows covering $\boldsymbol{a}$, whereas their "Minepi" algorithm defines $\text{supp}(\boldsymbol{a})$ as the actual number of occurrences of $\boldsymbol{a}$ in the given event sequence (x_t). In any case, for frequent episodes $\boldsymbol{a}, \boldsymbol{b}$ with $\boldsymbol{a}$ being a sub-episode of $\boldsymbol{b}$, the rule $\boldsymbol{a} \Rightarrow \boldsymbol{b}$ is evaluated by computing its confidence $\text{conf}(\boldsymbol{a} \Rightarrow \boldsymbol{b}) = \text{supp}(\boldsymbol{b})/\text{supp}(\boldsymbol{a})$, where this kind of conditional frequency is interpreted as expressing the predictive power of $\boldsymbol{a} \Rightarrow \boldsymbol{b}$. Only those rules are used in Phase II, the confidence of which exceeds a specified threshold value conf_{min} ("interesting rules"). Note that the final set of rules is only limited by the given support and confidence requirements, i. e., we are concerned with unsupervised learning here. This differs from the supervised approach by Weiss [53], where only rules regarding pre-specified target events are constructed.

Example 1. Let us illustrate the Minepi algorithm with a toy example. The range $\mathcal{S}$ with $d = 2$ is given by the symbols $s_0 = \texttt{a}$, $s_1 = \texttt{b}$, $s_2 = \texttt{c}$, and the available event series is of length $T = 11$:

$$\texttt{a, b, a, c, b, a, b, c, a, b, a.}$$

For episodes, we require more than 10% frequency, i. e., we set $\text{supp}_{\text{min}} = 1$. Furthermore, we declare rules as interesting if $\text{conf}_{\text{min}} = 0.5$ is exceeded. To keep it simple, we restrict to single tuples as the possible episodes, i. e., we do not consider sets of tuples as originating from wildcards etc.

- The algorithm starts with episodes of length $k = 1$, we get (a) with support $5 > \text{supp}_{\text{min}}$, (b) with support $4 > \text{supp}_{\text{min}}$, and (c) with support $2 > \text{supp}_{\text{min}}$. So the frequent episodes of length 1 are $\mathcal{F}_1 = \{(\texttt{a}), (\texttt{b}), (\texttt{c})\} = \mathcal{S}^1$.
- For $k = 2$, we first construct the candidate set $\mathcal{C}_2$ from $\mathcal{F}_1$, which simply equals $\mathcal{C}_2 = \mathcal{S}^2$, i. e., it covers all possible tuples of length 2. With the next data pass, we compute the supports $3 > \text{supp}_{\text{min}}$ for (a, b), (b, a), $1 \leq \text{supp}_{\text{min}}$ for (a, c), (b, c), (c, a), (c, b), and 0 otherwise. So $\mathcal{F}_2 = \{(\texttt{a}, \texttt{b}), (\texttt{b}, \texttt{a})\}$.

Before passing to $k = 3$, let us first discuss some possible variations. If we would also allow for sets of tuples, then further episodes would be frequent, such as $[\texttt{a}, \texttt{c}] = \{(\texttt{a}, \texttt{c}), (\texttt{c}, \texttt{a})\}$ and $(\texttt{c}, *) = \{(\texttt{c}, \texttt{a}), (\texttt{c}, \texttt{b}), (\texttt{c}, \texttt{c})\}$, both having support 2. If we would use Winepi instead of Minepi, say with window length $w = 5$, then supports would be computed differently, by considering the window contents

$$(\underline{\mathtt{a},\mathtt{b}},\mathtt{a},\mathtt{c},\mathtt{b}),\ (\mathtt{b},\mathtt{a},\mathtt{c},\mathtt{b},\mathtt{a}),\ (\mathtt{a},\mathtt{c},\mathtt{b},\underline{\mathtt{a},\mathtt{b}}),\ (\mathtt{c},\mathtt{b},\underline{\mathtt{a},\mathtt{b}},\mathtt{c}),$$
$$(\mathtt{b},\underline{\mathtt{a},\mathtt{b}},\mathtt{c},\mathtt{a}),\ (\underline{\mathtt{a},\mathtt{b}},\mathtt{c},\underline{\mathtt{a},\mathtt{b}}),\ (\mathtt{b},\mathtt{c},\underline{\mathtt{a},\mathtt{b}},\mathtt{a}).$$

For example, the episode $(\mathtt{a},\mathtt{b})$ would be contained in six out of seven windows (highlighted by underlining). But let's return to Minepi with simple episodes.

- For $k = 3$, we first construct the candidate set $\mathcal{C}_3$ from $\mathcal{F}_2 = \{(\mathtt{a},\mathtt{b}),(\mathtt{b},\mathtt{a})\}$, leading to $\mathcal{C}_3 = \{(\mathtt{a},\mathtt{b},\mathtt{a}),(\mathtt{b},\mathtt{a},\mathtt{b})\}$. Any other 3-tuple contains at least one none-frequent sub-episode, contradicting the Apriori principle. The candidate episodes have the supports $2 > \text{supp}_{\text{min}}$ for $(\mathtt{a},\mathtt{b},\mathtt{a})$ and $1 \leq \text{supp}_{\text{min}}$ for $(\mathtt{b},\mathtt{a},\mathtt{b})$, so $\mathcal{F}_3 = \{(\mathtt{a},\mathtt{b},\mathtt{a})\}$.

Since $\mathcal{C}_4 = \emptyset$, the algorithm stops with the frequent episodes in $\mathcal{F}_1 \cup \mathcal{F}_2 \cup \mathcal{F}_3$. From these, we get the interesting rules $(\mathtt{a}) \Rightarrow (\mathtt{a},\mathtt{b})$ with confidence $3/5$, $(\mathtt{b}) \Rightarrow (\mathtt{b},\mathtt{a})$ with confidence $3/4$, and $(\mathtt{a},\mathtt{b}) \Rightarrow (\mathtt{a},\mathtt{b},\mathtt{a})$ with confidence $2/3$, whereas $(\mathtt{a}) \Rightarrow (\mathtt{a},\mathtt{b},\mathtt{a})$ has confidence $2/5 \leq \text{conf}_{\text{min}} = 0.5$. Among the interesting rules, $(\mathtt{a},\mathtt{b}) \Rightarrow (\mathtt{a},\mathtt{b},\mathtt{a})$ might be judged as being redundant, because the less specific rule $(\mathtt{b}) \Rightarrow (\mathtt{b},\mathtt{a})$ leads to the same event prediction.

The "interesting" episode rules being generated such as in Example 1 can then be applied for prospective process monitoring. In an SPC context, such an application might look as follows. Let $\mathcal{A} \subset \mathcal{S}$ denote those categorical events that are classified as being critical, in the sense that immediate action is required once an event from $\mathcal{A}$ is observed. Then, from the set $\mathcal{R}$ of all rules $\boldsymbol{a} \Rightarrow \boldsymbol{b}$ generated during Phase I, we select those rules where the precursor $\boldsymbol{a}$ does not contain an event from $\mathcal{A}$, but where the successor $\boldsymbol{b}$ does; let us denote this subset as $\mathcal{R}_{\mathcal{A}}$. Then, having observed a new event x_t at time t, we take the available past $\ldots, x_{t-1}, x_t$ and check if rules from $\mathcal{R}_{\mathcal{A}}$ are applicable, i.e., if there are precursors $\boldsymbol{a}$ in $\mathcal{R}_{\mathcal{A}}$ that match $\ldots, x_{t-1}, x_t$ and end up in x_t. In this case, an alarm is triggered and the relevant critical rules are presented. The operator then decides on appropriate countermeasures to prevent the occurrence of the critical events.

5 Real Applications: Sample-Based Monitoring of Categorical Event Series

Due to page limitations, it is not possible to provide real-data examples for each approach being discussed in this chapter. Instead, we focus on the most common situation of categorical event series monitoring in practice, where independent samples are taken from time to time, recall Sect. 3.2.2. We discuss two quite different real applications: In Sect. 5.1, a nominal event sequence regarding paint defects on ceiling fan covers [36] is discussed, where we are concerned with the additional difficulty of varying sample sizes. Section 5.2, by contrast, is concerned with an ordinal event sequence regarding so-called "flash" on toothbrush heads [28], where samples of equal size are drawn.

5.1 A Nominal Event Sequence on Paint Defects

In Table 1 of Mukhopadhyay [36], data regarding 24 samples taken from a manufacturing process of ceiling fan covers are presented. The manufactured items are checked for possible paint defects (among $d = 6$ defect categories), and each item is either classified as being conforming (so state s_0 = "no defect"), or it is classified according to the most predominant defect, with the non-conforming categories being s_1 = "poor covering", s_2 = "overflow", s_3 = "patty defect", s_4 = "bubbles", s_5 = "paint defect", and s_6 = "buffing". Note that there is no ordering among the defect categories, so we are concerned with a nominal event sequence here. For illustration, like in Weiß [60], let us assume that the in-control probabilities are given by $\boldsymbol{p}_0 = (0.769, 0.081, 0.059, 0.021, 0.023, 0.022, 0.025)^\top$, i. e., if the rth sample has sample size n_r, then the corresponding vector $\boldsymbol{N}_r$ of sample frequencies is assumed to follow the multinomial distribution $\text{Mult}(n_r, \boldsymbol{p}_0)$ under in-control conditions. Note that by contrast with the discussion in Sect. 3.2.2, we are concerned with (heavily) deviating sample sizes here, where $n_1, \ldots, n_{24}$ vary between 20 and 404. This also affects the situation discussed in Sect. 2.2, i. e., if we aggregate the different types of defect into the binary situation "defect—yes or no". Then, the rth defect count Y_r follows the binomial distribution $\text{Bin}(n_r, \pi_0)$ with $\pi_0 = p_{0,1} + \ldots + p_{0,d} = 0.231$ in the in-control case.

To monitor independent samples $(\boldsymbol{N}_r)$ from a nominal event sequence, the Pearson χ^2-chart (6) and the IQV chart (7) appear to be most relevant, while for the aggregated defect counts (Y_r), the np-chart presented in Sect. 2.2 constitutes a further solution. The chart design should be based on ARL considerations, where we use the popular textbook choice $\text{ARL}_0 = 370.4$ as the target in-control ARL. Let us start with the most simple case, the np-chart. Using the formula $\text{ARL} = 1/P\big(Y \notin [l; u]\big)$ for Shewhart charts, we can compute the control limits as the $\alpha/2$- and $(1-\alpha/2)$-quantile of the in-control binomial distribution, where $\alpha = 1/370.4 \approx 0.00270$. But since the sample sizes n_r vary, also these "probability limits" $[l_r; u_r]$ vary and, in addition, we do not get a unique ARL performance (the same happens for the equivalent representation as a p-chart, where Y_r/n_r is plotted against $[l_r/n_r; u_r/n_r]$). For discrete attributes-data processes, one generally does not meet the target ARL exactly, and here, the extend of deviation from ARL_0 changes with r. To illustrate this issue, the first few sample sizes are $n_1 = 176$, $n_2 = 160$, ..., leading to the limits $[l_1; u_1] = [25; 58]$, $[l_2; u_2] = [22; 54]$, ... with individual in-control ARL values 444.4, 526.6, ... But since we do not know the system for choosing the samples sizes, we cannot conclude on the overall in-control ARL. The situation gets even worse for the χ^2- and IQV chart, because then,

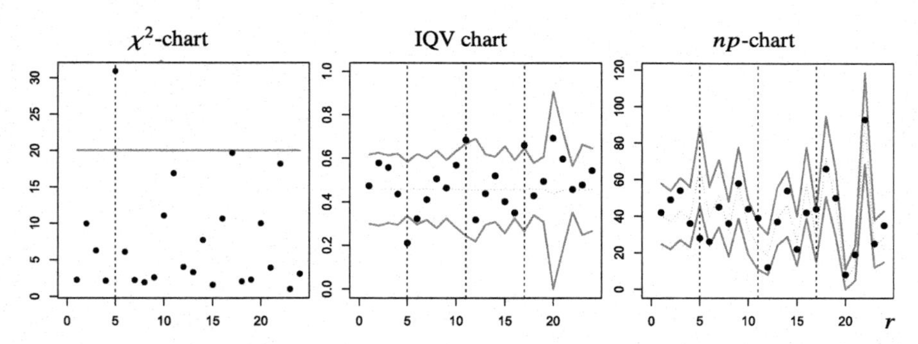

Fig. 1 Control charts for paint defects data, where times of alarms are highlighted by dashed lines

we do not even know the exact sample distribution of X_r^2 and IQV_r, respectively, but only asymptotic approximations [60]:

- X_r^2 is asymptotically χ_d^2-distributed;
- IQV_r is asymptotically normally distributed with mean $\mu_r = (1 - 1/n_r)\,\frac{d+1}{d}\,(1 - \sum_{j=0}^{d} p_{0,j}^2)$ and variance $\sigma_r^2 = 4/n_r\,\left(\frac{d+1}{d}\right)^2 \left(\sum_{j=0}^{d} p_{0,j}^3 - (\sum_{j=0}^{d} p_{0,j}^2)^2\right)$.

In Section 3.4 of Weiß [60], it was shown that these asymptotics only lead to rough approximations of the actually intended chart design. But since we cannot improve the chart design based on simulations (as done in Weiß [60]) due to the lack of a systematic choice of n_r, we continue with the asymptotic chart design here. More precisely, for the upper-sided χ^2-chart, the upper-limit is (uniquely) computed as the $(1-\alpha)$-quantile of the χ_d^2-distribution, leading to the value 20.062, and for the two-sided IQV chart, we compute the rth limits as $\mu_r \mp z\,\sigma_r$, where $z \approx 3.000$ ("$3\,\sigma$-limits") is the $(1-\alpha/2)$-quantile of the standard normal distribution, N(0, 1).

The obtained control charts are plotted in Fig. 1. The χ^2-chart triggers an alarm for the 5th sample (highlighted by a dashed line), but does not give further signals. So it appears that the 5th sample was drawn under out-of-control conditions – but which kind of out-of-control scenario exactly? It is a common problem with the χ^2-chart that more detailed insights cannot be gained from the plotted statistics such that further post-hoc analyses are required here. The IQV chart in Fig. 1 (where we also added a graph of μ_r as the center line) is more informative, and it also triggers two further alarms for the 11th and 17th sample. The alarm at the 5th sample is caused by a violation of the lower limit, so the dispersion of $\boldsymbol{N}_5/n_5$ decreased compared to $\boldsymbol{p}_0$. This mean that $\boldsymbol{N}_5/n_5$ moved towards a one-point distribution in $s_0 =$ "no defect", i. e., we have a quality improvement compared to the in-control state (another explanation might be problems in quality evaluation, i. e., defects might have been overlooked at $r = 5$). The alarms at $r = 11, 17$, by contrast, result from a (slight) violation of the upper limit, so indicating a quality deterioration this time. For the present example, the same conclusions can also be drawn from the np-chart in Fig. 1 (with additional center line $n_r\,\pi_0$), where we again have a violation of the lower (upper) limit at $r = 5$ $(r = 11, 17)$. Note that we can easily recognize the

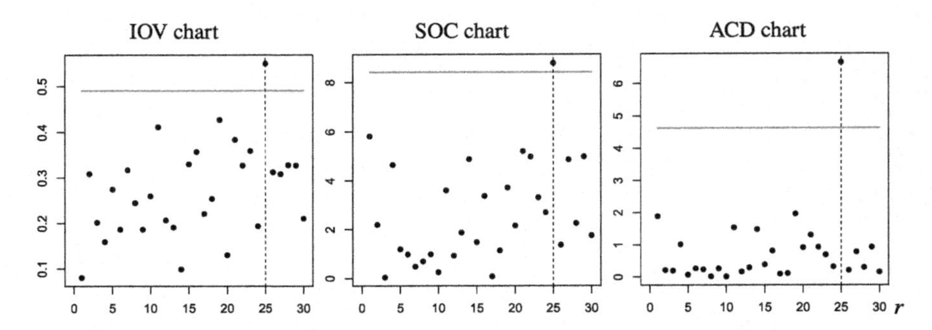

Fig. 2 Control charts (without EWMA smoothing) for flash data, where times of alarms are highlighted by dashed lines

sample size from the width of the limits of the IQV chart as well as from the center line of the np-chart, with the smallest (largest) sample at $r = 20$ ($r = 22$), namely $n_{20} = 20$ and $n_{22} = 404$. The χ^2-chart is a black box in this sense.

5.2 An Ordinal Event Sequence Regarding Flash on Toothbrush Heads

In Li et al. [28], the manufacturing of electric toothbrushes is considered, where in one of the production steps, the two parts of the brush head are welded together. In this step, excess plastic (referred to as "flash") might be generated, which could injure users and should thus be avoided. Therefore, an important quality characteristic of the manufactured toothbrush heads is given by the extent of flash, with the $d + 1 = 4$ ordinal levels s_0 = "slight", s_1 = "small", s_2 = "medium", and s_3 = "large". In their Phase-I analysis, Li et al. [28] identified the in-control PMF $\boldsymbol{p}_0 = (0.8631, 0.0804, 0.0357, 0.0208)^\top$, which shall now be used for control chart design. For Phase-II application, 30 samples of the unique sample size $n = 64$ are available.

To account for the ordinal nature of the data, we do not use the nominal χ^2- and IQV chart this time, although these could be applied to ordinal data as well. Instead, we use the IOV chart (7) for monitoring the ordinal dispersion of the samples, as well as the SOC chart (9) and the ACD chart (11) with $\boldsymbol{w} = \mathbf{1}$, both relying on ridits. In a first step, we apply these charts without an EWMA smoothing of the frequencies $\boldsymbol{N}_r$ (corresponding to smoothing parameter $\lambda = 1$), i.e., we use them as Shewhart charts. In this case, we could determine the control limits of the IOV chart based on asymptotic considerations, as IOV_r is asymptotically normally distributed with mean $\mu = (1 - 1/n)\,\frac{4}{d}\,\sum_{j=0}^{d-1} f_{0,j}\,(1 - f_{0,j})$ and variance $\sigma^2 = 16/n/d^2\,\sum_{i,j=0}^{d-1}(1 - f_{0,i})(1 - f_{0,j})\,(f_{\min\{i,j\}} - f_{0,i}\,f_{0,j})$, see Weiß [61]. But this leads to an only rough approximation of the intended chart design (again with

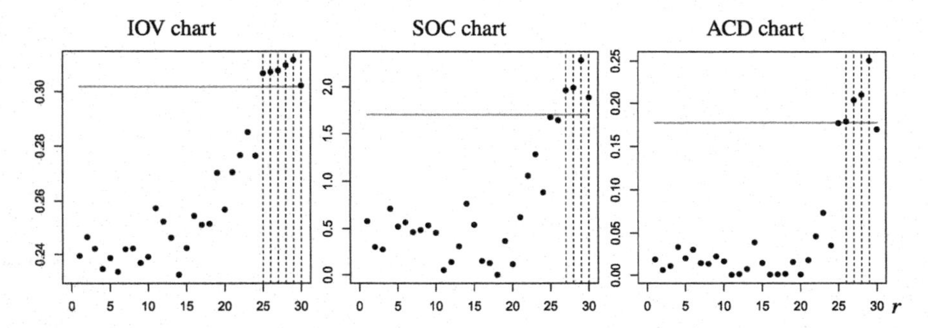

Fig. 3 Control charts (with EWMA smoothing) for flash data, where times of alarms are highlighted by dashed lines

target $\text{ARL}_0 = 370.4$), which can be checked with simulations. Generally, since we have a unique sample size n, the ARLs of all charts (also those with EWMA smoothing) can be computed by simulations this time. We use 10^4 replications, and we approximate the ARL by the sample mean across the 10^4 simulated run lengths of the considered chart. In this way, we obtain the upper control limits of the IOV chart (we concentrate on the upper-sided chart this time, because increased IOV values correspond to a deterioration of quality) as 0.4915, of the SOC chart as 8.432, and of the ACD chart as 4.638. The resulting Shewhart control charts are plotted in Fig. 2. It can be seen that all charts lead to the same results: an alarm is triggered for the 25th sample.

Before further analyzing this result, let us also look at the control charts in Fig. 3. Here, we used the same chart types as before, but with an additional EWMA smoothing for the sample frequencies: $\boldsymbol{N}_r^{(\lambda)} = \lambda\, \boldsymbol{N}_r + (1-\lambda)\, \boldsymbol{N}_{r-1}^{(\lambda)}$ with $\boldsymbol{N}_0^{(\lambda)} = n\, \boldsymbol{p}_0$, where the smoothing parameter is chosen as $\lambda = 0.1$ (the same value was also used by Li et al. [28], Wang et al. [52]). Then, the upper control limits become 0.3021 for the IOV chart, 1.707 for the SOC chart, and 0.1777 for the ACD chart. Because of the smoothing, we now have more narrow control limits than in Fig. 2. The charts of Fig. 3 show a slightly different behaviour: While the IOV chart again triggers its first alarm for $r = 25$, it takes until $r = 27$ for the SOC chart and $r = 26$ for the ACD chart. Taking all results together, there seems to be a problem with the 25th sample. This sudden change was rather strong such that it was immediately detected by the Shewhart charts of Fig. 2, whereas some of the EWMA charts in Fig. 3 reacted with a delay (EWMA charts are well-suited for a persistent change because of their inherent memory). In fact, we have $\boldsymbol{N}_{25}/n \approx (0.7344, 0.1094, 0.06250.0938)^\top$; so compared to $\boldsymbol{p}_0 = (0.8631, 0.0804, 0.0357, 0.0208)^\top$, the conforming probability for $s_0 =$ "slight" is notably reduced, while especially the probability for the worst state $s_3 =$ "large" was increased. So $\boldsymbol{N}_{25}/n$ moved towards an extreme two-point distribution, which explains the good performance of the IOV charts.

6 Conclusions

We surveyed three approaches for the monitoring of categorical event series. First, control charts might be applied to counts generated from the categorical event series, where the stochastic properties of the resulting count time series depend on those of the original categorical process. Second, the control charts might be applied to the categorical process itself, carefully distinguishing the cases of nominal and ordinal data. Third, if huge amounts of data need to be managed, rule-based procedures from machine learning (episode mining) might be an adequate solution.

The monitoring of categorical event series is generally a quite demanding task. During Phase I, appropriate methods for time series analysis and feasible stochastic models need to be identified, while chart design for Phase-II application suffers from discreteness problems. Although various control charts for i. i. d. categorical data are meanwhile available, only few contributions exist concerning the monitoring of serially dependent categorical event series (or the count processes derived thereof). In particular, the development of tailored methods for ordinal time-series data seems to be a promising direction for future research, as this type of categorical event series is particularly relevant for real applications.

Acknowledgments The author thanks the referee for useful comments on an earlier draft of this article. The author is grateful to Professor Jian Li (Xi'an Jiaotong University, China) for providing the flash data discussed in Sect. 5.2.

References

1. Agrawal R, Srikant R (1994) Fast algorithms for mining association rules. In: Proceedings of the 20th international conference on very large databases, pp 487–499
2. Agresti A (2010) Analysis of ordinal categorical data, 2nd edn. John Wiley & Sons Inc., Hoboken
3. Bai K, Li J (2021) Location-scale monitoring of ordinal categorical processes. Naval research logistics, forthcoming
4. Bashkansky E, Gadrich T (2011) Statistical quality control for ternary ordinal quality data. Appl Stoch Models Bus Ind 27(6):586–599
5. Baum LE, Petrie T (1966) Statistical inference for probabilistic functions of finite state Markov chains. Ann Math Stat 37(6):1554–1563
6. Bersimis S, Sachlas A, Castagliola P (2017) Controlling bivariate categorical processes using scan rules. Methodol Comput Appl Probab 19(4):1135–1149
7. Blatterman DK, Champ CW (1992) A Shewhart control chart under 100% inspection for Markov dependent attribute data. In: Proceedings of the 23rd annual modeling and simulation conference, pp 1769–1774
8. Bourke PD (1991) Detecting a shift in fraction nonconforming using run-length control charts with 100% inspection. J Qual Technol 23(3):225–238
9. Brook D, Evans DA (1972) An approach to the probability distribution of CUSUM run length. Biometrika 59(3):539–549
10. Bühlmann P, Wyner AJ (1999) Variable length Markov chains. Ann Stat 27(2):480–513
11. Duncan AJ (1950) A chi-square chart for controlling a set of percentages. Ind Qual Control 7:11–15

12. Duran RI, Albin SL (2009) Monitoring and accurately interpreting service processes with transactions that are classified in multiple categories. IIE Trans 42(2):136–145
13. Ferland R, Latour A, Oraichi D (2006) Integer-valued GARCH processes. J Time Ser Anal 27(6):923–942
14. Gan FF (1990) Monitoring observations generated from a binomial distribution using modified exponentially weighted moving average control chart. J Stat Comput Simul 37(1–2):45–60
15. Göb R (2006) Data mining and statistical control – a review and some links. In: Lenz HJ, Wilrich PT (eds) Frontiers in statistical quality control 8, pp 285–308. Physica-Verlag, Heidelberg
16. Gwadera R, Atallah MJ, Szpankowski W (2005) Markov models for identification of significant episodes. In: Proceedings of the 2005 SIAM international conference on data mining, pp 404–414
17. Höhle M (2010) Online change-point detection in categorical time series. In: Kneib T, Tutz G (eds) Statistical modelling and regression structures. Festschrift in Honour of Ludwig Fahrmeir, pp 377–397. Physica-Verlag, Heidelberg
18. Höhle M, Paul M (2008) Count data regression charts for the monitoring of surveillance time series. Comput Stat Data Anal 52(9):4357–4368
19. Jacobs PA, Lewis PAW (1983) Stationary discrete autoregressive-moving average time series generated by mixtures. J Time Ser Anal 4(1):19–36
20. Klein I, Doll M (2021) Tests on asymmetry for ordered categorical variables. J Appl Stat 48(7):1180–1198
21. Klemettinen M, Mannila H, Toivonen H (1999) Rule discovery in telecommunication alarm data. J Netw Syst Manag 7(4):395–423
22. Koutras MV, Bersimis S, Antzoulakos DL (2006) Improving the performance of the chi-square control chart via runs rules. Methodol Comput Appl Prob 8(3):409–426
23. Koutras MV, Maravelakis PE, Bersimis S (2008) Techniques for controlling bivariate grouped observations. J Multivar Anal 99(7):1474–1488
24. Kvålseth TO (1995) Coefficients of variation for nominal and ordinal categorical data. Percept Mot Skills 80(3):843–847
25. Lambert D, Liu C (2006) Adaptive thresholds: monitoring streams of network counts. J Am Stat Assoc 101(473):78–88
26. Laxman S, Sastry PS (2006) A survey of temporal data mining. Sādhanā 31(2):173–198
27. Laxman S, Sastry PS, Unnikrishnan KP (2005) Discovering frequent episodes and learning hidden Markov models: a formal connection. IEEE Trans Knowl Data Eng 17(11):1505–1517
28. Li J, Tsung F, Zou C (2014) A simple categorical chart for detecting location shifts with ordinal information. Int J Prod Res 52(2):550–562
29. Li J, Xu J, Zhou Q (2018) Monitoring serially dependent categorical processes with ordinal information. IISE Trans 50(7):596–605
30. Mannila H, Toivonen H, Verkamo AI (1997) Discovery of frequent episodes in event sequences. Data Min Knowl Disc 1(3):259–289
31. Marcucci M (1985) Monitoring multinomial processes. J Qual Technol 17(2):86–91
32. McKenzie E (1985) Some simple models for discrete variate time series. Water Resour Bull 21(4):645–650
33. Montgomery DC (2009) Introduction to statistical quality control, 6th edn. John Wiley & Sons Inc., New York
34. Morais MC, Knoth S, Weiß CH (2018) An ARL-unbiased thinning-based EWMA chart to monitor counts. Seq Anal 37(4):487–510
35. Mousavi S, Reynolds MR Jr (2009) A CUSUM chart for monitoring a proportion with autocorrelated binary observations. J Qual Technol 41(4):401–414
36. Mukhopadhyay AR (2008) Multivariate attribute control chart using Mahalanobis D^2 statistic. J Appl Stat 35(4):421–429
37. Nembhard DA, Nembhard HB (2000) A demerits control chart for autocorrelated data. Qual Eng 13(2):179–190
38. Page E (1954) Continuous inspection schemes. Biometrika 41(1):100–115

39. Perry MB (2020) An EWMA control chart for categorical processes with applications to social network monitoring. J Qual Technol 52(2):182–197
40. Raftery AE (1985) A model for high-order Markov chains. J Roy Stat Soc B 47(3):528–539
41. Rakitzis AC, Weiß CH, Castagliola P (2017) Control charts for monitoring correlated counts with a finite range. Appl Stoch Models Bus Ind 33(6):733–749
42. Reynolds MR Jr, Stoumbos ZG (1999) A CUSUM chart for monitoring a proportion when inspecting continuously. J Qual Technol 31(1):87–108
43. Roberts SW (1959) Control chart tests based on geometric moving averages. Technometrics 1(3):239–250
44. Ryan AG, Wells LJ, Woodall WH (2011) Methods for monitoring multiple proportions when inspecting continuously. J Qual Technol 43(3):237–248
45. Spanos CJ, Chen RL (1997) Using qualitative observations for process tuning and control. IEEE Trans Semicond Manuf 10(2):307–316
46. Steiner SH (1998) Grouped data exponentially weighted moving average control charts. J Roy Stat Soc C 47(2):203–216
47. Steiner SH, Geyer PL, Wesolowsky GO (1996) Grouped data-sequential probability ratio tests and cumulative sum control charts. Technometrics 38(3):230–237
48. Szarka JL III, Woodall WH (2011) A review and perspective on surveillance of Bernoulli processes. Qual Reliab Eng Int 27(6):735–752
49. Tucker GR, Woodall WH, Tsui K-L (2002) A control chart method for ordinal data. Am J Math Manag Sci 22(1–2):31–48
50. Vasquez Capacho JW, Subias A, Travé-Massuyès L, Jimenez F (2017) Alarm management via temporal pattern learning. Eng Appl Artif Intell 65:506–516
51. Wang J, Li J, Su Q (2017) Multivariate ordinal categorical process control based on log-linear modeling. J Qual Technol 49(2):108–122
52. Wang J, Su Q, Xie M (2018) A univariate procedure for monitoring location and dispersion with ordered categorical data. Commun Stat Simul Comput 47(1):115–128
53. Weiss GM (1999) Timeweaver: a genetic algorithm for identifying predictive patterns in sequences of events. In: Banzhaf et al (eds) Proceedings of the genetic and evolutionary computation conference, pp 718–725. Morgan Kaufmann, San Francisco
54. Weiß CH (2009) Group inspection of dependent binary processes. Qual Reliab Eng Int 25(2):151–165
55. Weiß CH (2011) Rule generation for categorical time series with Markov assumptions. Stat Comput 21(1):1–16
56. Weiß CH (2012) Continuously monitoring categorical processes. Qual Technol Quant Manag 9(2):171–188
57. Weiß CH (2015) SPC methods for time-dependent processes of counts – a literature review. Cogent Math 2(1):1111116
58. Weiß CH (2017) Association rule mining. In: Balakrishnan et al (eds) Wiley StatsRef: statistics reference online. John Wiley & Sons Ltd., Hoboken
59. Weiß CH (2018a) An introduction to discrete-valued time series. John Wiley & Sons Inc., Chichester
60. Weiß CH (2018b) Control charts for time-dependent categorical processes. In: Knoth S, Schmid W (eds) Frontiers in statistical quality control 12. Physica-Verlag, Heidelberg, pp 211–231
61. Weiß CH (2020) Distance-based analysis of ordinal data and ordinal time series. J Am Stat Assoc 115(531):1189–1200
62. Weiß CH (2021) Stationary count time series models. WIREs Comput Stat 13(1):e1502
63. Weiß CH, Zhu F, Hoshiyar A (2022) Softplus INGARCH models. Statistica Sinica 32(3), forthcoming
64. Ye N (2003) Mining computer and network security data. In: Ye N (ed) The handbook of data mining. Lawrence Erlbaum Associations Inc., New Jersey, pp 617–636
65. Ye N, Borror C, Zhang Y (2002a) EWMA techniques for computer intrusion detection through anomalous changes in event intensity. Qual Reliab Eng Int 18(6):443–451

66. Ye N, Masum S, Chen Q, Vilbert S (2002b) Multivariate statistical analysis of audit trails for host-based intrusion detection. IEEE Trans Comput 51(7):810–820
67. Yuan Y, Zhou S, Sievenpiper C, Mannar K, Zheng Y (2011) Event log modeling and analysis for system failure prediction. IIE Trans 43(9):647–660
68. Zimmermann A (2014) Understanding episode mining techniques: benchmarking on diverse, realistic, artificial Data. Intell Data Anal 18(5):761–791

Machine Learning Control Charts for Monitoring Serially Correlated Data

Xiulin Xie and Peihua Qiu

Abstract Some control charts based on machine learning approaches have been developed recently in the statistical process control (SPC) literature. These charts are usually designed for monitoring processes with independent observations at different observation times. In practice, however, serial data correlation almost always exists in the observed data of a temporal process. It has been well demonstrated in the SPC literature that control charts designed for monitoring independent data would not be reliable to use in applications with serially correlated data. In this chapter, we suggest using certain existing machine learning control charts together with a recursive data de-correlation procedure. It is shown that the performance of these charts can be substantially improved for monitoring serially correlated processes after data de-correlation.

1 Introduction

In recent years, machine learning approaches have attracted much attention in different research areas, including statistical process control (SPC) (e.g., [1, 4, 8, 11], [12]). Some control charts based on different machine learning algorithms have been developed in the SPC literature. For instance, the k-nearest neighbors (KNN), random forest (RF) and support vector machines (SVM) have been used in developing SPC control charts. Most of these existing machine learning control charts are based on the assumption that process observations at different observation times are independent of each other. In practice, however, serial data correlation almost always exists in a time series data. It has been well demonstrated in the SPC literature that control charts designed for monitoring independent data would not be reliable to

X. Xie · P. Qiu (✉)
Department of Biostatistics, University of Florida, 2004 Mowry Road,
Gainesville, FL 32610, USA
e-mail: pqiu@ufl.edu

X. Xie
e-mail: xiulin.xie@ufl.edu

K. P. Tran (ed.), *Control Charts and Machine Learning for Anomaly Detection in Manufacturing*, Springer Series in Reliability Engineering,
https://doi.org/10.1007/978-3-030-83819-5_6

use when serial data correlation exists (e.g. [3, 15–17, 20, 22, 23, 28, 30]). Thus, it's necessary to improve these machine learning control charts by overcoming that limitation. This chapter aims to address this important issue by suggesting to apply a recursive data de-correlation procedure to the observed data before an existing machine learning control chart is used.

In the SPC literature, there has been some existing discussion about process monitoring of serially correlated data (e.g., [2, 7, 19, 20]). Many such existing methods are based on parametric time series modeling of the observed process data and monitoring of the resulting residuals. For instance, [16] proposed an exponentially weighted moving average (EWMA) chart for monitoring correlated data by assuming the *in-control* (IC) process observations to follow an ARMA model. In practice, however, the assumed parametric time series models may not be valid, and consequently these control charts may be unreliable to use (e.g., [17]). Recently, [22] suggested a more flexible data de-correlation method without using a parametric time series model for univariate cases. It only requires the serial data correlation to be stationary and short-range (i.e., the correlation between two observations become weaker when the observation times get farther away). A multivariate extension of that method was discussed in [30]. Numerical studies show that such sequential data de-correlation approaches perform well in different cases. In this chapter, we suggest improving some existing machine learning control charts by applying such a data de-correlation procedure to the observed process observations in advance. The modified machine learning control charts can handle cases with multiple numerical quality variables, and the quality variables could be continuous numerical or discrete. Numerical studies show that the performance of these modified machine learning control charts is substantially better than their original versions for monitoring processes with serially correlated data in various different cases.

The remaining parts of this chapter are organized as follows. In Sect. 2, the proposed modification for some existing machine learning control charts are described in detail. Numerical studies for evaluating their performance are presented in Sect. 3. A real-data example to demonstrate the application of the modified control charts is discussed in Sect. 4. Finally, some remarks conclude the article in Sect. 5.

2 Improve Some Machine Learning Control Charts for Monitoring Serially Correlated Data

This section is organized in three parts. In Subsect. 2.1, some representative existing machine learning control charts are briefly described. In Subsect. 2.2, a recursive data de-correlation procedure for the observed sequential data is introduced in detail. Then, the modified machine learning control charts, in which the recursive data de-correlation procedure is applied to the observed data before the original machine learning control charts, are discussed in Subsect. 2.3.

2.1 Description of Some Representative Machine Learning Control Charts

Classification is one of the major purposes of supervised machine learning, and many machine learning algorithms like the artificial neural networks, RF and SVM have demonstrated a good performance in accurately classifying input data after learning the data structure from a large training data. Since an SPC problem can be regarded as a binary class classification problem, in which each process observation needs to be classified into either the IC or the *out-of-control* (OC) status during phase II process monitoring, several machine learning algorithms making use of both IC and OC historical data have been employed for process monitoring. For instance, [31] proposed an EWMA control chart based on the probabilistic outputs of a SVM classifier that needs to be built by using both IC and OC historical data. Several other classifiers like the KNN and linear discriminant analysis were also proposed for process monitoring (e.g., [18]; [24]). In many SPC applications, however, few OC process observations would be available in advance. For instance, a production process is often properly adjusted during the Phase I SPC, and a set of IC data is routinely collected afterwards for estimating the IC process distribution or some of its parameters ([21], Chap. 1). Thus, for such applications, an IC data is usually available before the Phase II SPC, but the OC process observations are often unavailable. To overcome this difficulty, some creative ideas like the *artificial contrast*, *real-time contrast*, and *one class classification* were proposed to develop control charts without assuming the availability of OC process observations during the design stage of the related charts. Several representative machine learning control charts based on these ideas are briefly introduced below.

Control Chart Based on Artificial Contrasts
[27] proposed the idea of *artificial contrast* to overcome the difficulty that only IC data are available before the Phase II process monitoring in certain SPC applications. By this idea, an artificial dataset is first generated from a given off-target distribution (e.g., Uniform) and observations in that dataset are regarded as OC observations. Then, a machine learning algorithm (e.g., RF) is applied to the training dataset that consists of the original IC dataset, denoted as $\mathcal{X}_{IC}$, and the artificial contrast dataset, denoted as $\mathcal{X}_{AC}$. The classifier obtained by the RF algorithm is then used for online process monitoring. [14] studied the performance of such machine learning control charts by using both the RF and SVM algorithms. These machine learning control charts suffer two major limitations. First, their classification error rates cannot be transferred to the traditional average run length (ARL) metric without the data independence assumption. Second, their decisions at a given time point during phase II process monitoring are made based on the observed data at that time point only, and they have not made use of history data. To overcome these limitations, Hu and Runger (2010) suggested the following modification that consisted of two major steps. i) For process observation $\mathbf{X}_n$ at a given time point n, the log likelihood ratio is first calculated as

$$l_n = \log\left[\hat{p}_1(\mathbf{X}_n)\right] - log\left[\hat{p}_0(\mathbf{X}_n)\right],$$

where $\hat{p}_1(\mathbf{X}_n)$ and $\hat{p}_0(\mathbf{X}_n)$ are the estimated probabilities of $\mathbf{X}_n$ in each class obtained by the RF classifier. ii) A modified EWMA chart is then suggested with the following charting statistic:

$$E_n = \lambda l_n + (1 - \lambda)E_{n-1},$$

where $\lambda \in (0, 1]$ is a weighting parameter. This control chart is denoted as AC, representing "artificial contrast". Obviously, like the traditional EWMA charts, the charting statistic E_n of AC is a weighted average of the log likelihood ratios of all available observations up to the time point n.

As suggested by Hu and Runger (2010), the control limit of AC can be determined by the following 10-fold cross-validation (CV) procedure. First, 90% of the IC dataset $\mathcal{X}_{IC}$ and the artificial contrast dataset $\mathcal{X}_{AC}$ is used to train the RF classifier. Then, the E_n with a control limit h is applied to the remaining 10% of the IC dataset $\mathcal{X}_{IC}$ to obtain a run length (RL) value. The above CV procedure is then repeated for $C = 1,000$ times, and the average of the C RL values is used to approximate the ARL_0 value for the given h. Finally, h can be searched by a numerical algorithm (e.g., the bisection searching algorithm) so that the assumed ARL_0 value is reached.

Control Chart Based on Real Time Contrasts

The artificial contrasts $\mathcal{X}_{AC}$ used in AC are generated from a subjectively chosen off-target distribution (e.g., Uniform), and thus may not represent the actual OC observations well. Consequently, the RF classifier trained using $\mathcal{X}_{IC}$ and $\mathcal{X}_{AC}$ may not be effective for monitoring certain processes. To improve the chart AC, [10] propose a *real time contrast (RTC)* approach, in which the most recent observations within a moving window of the current time point are used as the contrasts. In their proposed approach, the IC dataset is first divided into two parts: a randomly selected N_0 observations from $\mathcal{X}_{IC}$, denoted as $\mathcal{X}_{IC_0}$, is used for training the RF classifier, the remaining IC data, denoted as $\mathcal{X}_{IC_1}$, is used for determining the control limit. The process observations in a window of the current observation time point n are treated as OC data and denoted as $\mathcal{X}_{AC_n} = \{\mathbf{X}_{n-w+1}, \mathbf{X}_{n-w+2}, \ldots, \mathbf{X}_n\}$, where w is the window size. Then, the RF classifier can be re-trained sequentially over time using the training dataset that combines $\mathcal{X}_{IC_0}$ and $\mathcal{X}_{AC_n}$, and the decision rule can be updated accordingly once the new observation $\mathbf{X}_n$ is collected at time n.

[10] suggested using the following estimated "out-of-bag" correct classification rate for observations in $\mathcal{X}_{IC_0}$ as the charting statistic:

$$P_n = \frac{\sum P_{OOB}(\mathbf{X}_i) I(\mathbf{X}_i \in \mathcal{X}_{IC_0})}{|\mathcal{X}_{IC_0}|},$$

where $|\mathcal{X}_{IC_0}|$ denotes the number of observations in the set $\mathcal{X}_{IC_0}$, and $P_{OOB}(\mathbf{X}_i)$ is the estimated "out-of-bag" correct classification probability for the IC observation $\mathbf{X}_i$ that is obtained from the RF classification. As discussed in [10], there could be several alternative charting statistics, such as the estimated "out-of-bag" correct classification rate for observations in $\mathcal{X}_{AC_n}$. But, they found that the chart based on the above P_n, denoted as RTC, had some favorable properties.

The control limit of the chart RTC can be determined by the following bootstrap procedure suggested by [10]. First, we draw with replacement a sample from the dataset $\mathcal{X}_{IC_1}$. Then, the chart RTC with control limit h is applied to the bootstrap sample to obtain a RL value. This bootstrap re-sampling procedure is repeated $B = 1,000$ times, and the average of the B RL values is used to approximate the ARL_0 value for the given h. Finally, h can be empirically selected so that assumed ARL_0 is reached. Finally, h be searched by a numerical algorithm so that the assumed ARL_0 value is reached.

Distance Based Control Chart Using SVM
The charting statistic of the RTC chart discussed above actually take discrete values, because the estimated "out-of-bag" correct classification probabilities $\{P_{OOB}(\mathbf{X}_i)\}$ are obtained from an ensemble of decision trees [4] and [13]. As an alternative, [13] suggested a distance-based control chart under the framework of SVM, which is denoted as DSVM. The DSVM method uses the distance between the support vectors and the process observations in the dataset $\mathcal{X}_{AC_n}$ as a charting statistic. Unlike charting statistic P_n of the RTC chart, this distance-based charting statistic is a continuous variable. Because the distance from a sample of process observations to the boundary surface defined by the support vectors can be either positive or negative, He, Jiang, and Deng suggested transforming the distance using the standard logistic function

$$g(a) = \frac{1}{1 + \exp(-a)}.$$

Then, the following average value of the transformed distances from individual observations in $\mathcal{X}_{AC_n}$ to the boundary surface can be defined to be the charting statistic:

$$M_n = \frac{\sum g(d(\mathbf{X}_i))I(\mathbf{X}_i \in \mathcal{X}_{AC_n})}{|\mathcal{X}_{AC_n}|},$$

where $d(\mathbf{X}_i)$ is the distance from the observation $\mathbf{X}_i$ to decision boundary determined by the SVM algorithms at time n.

In the above DSVM chart, the kernel function and the penalty parameter need to be selected properly. [13] suggested using the following Gaussian radial basis function (RBF): for any $\mathbf{X}, \mathbf{X}' \in R^p$,

$$K(\mathbf{X}, \mathbf{X}') = \exp\left(\frac{\|\mathbf{X} - \mathbf{X}'\|^2}{\sigma^2}\right)$$

as the kernel function, where p is dimension of the process observations, and the parameter σ^2 was chosen to be larger than 2.8. They also suggested choosing the penalty parameter to be 1. The control limit of the chart DSVM can be determined by a bootstrap procedure, similar to the one described above for the RTC chart.

Control Chart Based on the KNN Classification
Another approach to develop machine learning control charts is to use *one-class classification (OCC)* algorithms. [25] developed a nonparametric control chart based

on the so-called support vector data description (SVDD) approach [26], described below. By SVDD, the boundary surface of an IC data can be defined so that the volume within the boundary surface is as small as possible while the Type-I error probability is controled within a given level of α. Then, the boundary surface is used as the decision rule for online process monitoring as follows: a new observation is claimed to be OC if it falls outside of the boundary surface, and IC otherwise. See [6] for some modifications and generalizations. However, determination of this boundary surface is computationally intensive. To reduce the computation burden, Sukchotrat, Kim and Tsung (2009) suggested a control chart based on the KNN classification, denoted as KNN. In KNN, the average distance between a given observation $\mathbf{X}_i$ and its k nearest neighboring observations in the IC dataset is first calculated as follows:

$$K_i^2 = \frac{\sum_{j=1}^{k} \|\mathbf{X}_i - NN_j(\mathbf{X}_i)\|}{k},$$

where $NN_j(\mathbf{X}_i)$ is the jth nearest neighboring observation of $\mathbf{X}_i$ in the IC dataset, and $\|\cdot\|$ is the Euclidean distance. Then, the $(1-\alpha)$th quantile of all such distances of individual observations in the IC data can be computed. This quantile can be used as the decision rule for online process monitoring as follows. At the current time n, if the average distance from $\mathbf{X}_n$ to its k nearest neighboring observations (i.e., K_n^2) is less than the quantile, then $\mathbf{X}_n$ is claimed as IC. Otherwise, it is claimed as OC.

In the above KNN chart, the control limit (i.e., the $(1-\alpha)$th quantile of $\{K_i^2\}$ of individual observations in the IC data) can be refined by the following bootstrap procedure suggested by Sukchotrat et al. (2009). First, a total of $B = 1,000$ bootstrap samples are obtained from the IC dataset by the simple random sampling procedure with replacement. Then, the $(1-\alpha)$th quantile of $\{K_i^2\}$ of individual observations in each bootstrap sample can be computed. Then, the final control limit is chosen to be the mean of the B such quantiles. The KNN chart assumes that process observations at different time points are independent. Thus, its ARL_0 value equals $1/\alpha$.

2.2 Sequential Data De-Correlation

In this subsection, the sequential data de-correlation procedure for multivariate serially correlated data is described in detail. It is assumed that the IC process mean is $\boldsymbol{\mu}$ and the serial data correlation is stationary with the covariances $\boldsymbol{\gamma}(s) = \mathrm{Cov}(\mathbf{X}_i, \mathbf{X}_{i+s})$, for any i and s, that depend only on s.

For the first observation $\mathbf{X}_1$, its covariance matrix is $\boldsymbol{\gamma}(0)$. Then, its standardized vector can be defined to be

$$\mathbf{X}_1^* = \boldsymbol{\gamma}(0)^{-1/2}(\mathbf{X}_1 - \boldsymbol{\mu}).$$

After the second observation $\mathbf{X}_2$ is collected, let us consider the long vector $(\mathbf{X}_1', \mathbf{X}_2')'$. Its covariance matrix can be written as $\Sigma_{2,2} = \begin{pmatrix} \boldsymbol{\gamma}(0) & \boldsymbol{\sigma}_1 \\ \boldsymbol{\sigma}_1' & \boldsymbol{\gamma}(0) \end{pmatrix}$, where $\boldsymbol{\sigma}_1 = \boldsymbol{\gamma}(1)$. The Cholesky decomposition of $\Sigma_{2,2}$ is given by $\Phi_2 \Sigma_{2,2} \Phi_2' = \mathbf{D}_2$, where $\Phi_2 = \begin{pmatrix} \mathbf{I}_p & \mathbf{0} \\ -\boldsymbol{\sigma}_1' \boldsymbol{\gamma}(0)^{-1} & \mathbf{I}_p \end{pmatrix}$, and $\mathbf{D}_2 = \begin{pmatrix} \mathbf{d}_1 & \mathbf{0} \\ \mathbf{0} & \mathbf{d}_2 \end{pmatrix} = diag(\mathbf{d}_1, \mathbf{d}_2)$, $\mathbf{d}_1 = \boldsymbol{\gamma}(0)$, and $\mathbf{d}_2 = \boldsymbol{\gamma}(0) - \boldsymbol{\sigma}_1' \boldsymbol{\gamma}(0)^{-1} \boldsymbol{\sigma}_1$. Therefore, we have $\mathrm{Cov}(\Phi_2 \mathbf{e}_2) = \mathbf{D}_2$, where $\mathbf{e}_2 = [(\mathbf{X}_1 - \boldsymbol{\mu})', (\mathbf{X}_2 - \boldsymbol{\mu})']'$. Since $\Phi_2 \mathbf{e}_2 = \begin{pmatrix} \mathbf{I}_p & \mathbf{0} \\ -\boldsymbol{\sigma}_1' \boldsymbol{\gamma}(0)^{-1} & \mathbf{I}_p \end{pmatrix} \begin{pmatrix} (\mathbf{X}_1 - \boldsymbol{\mu})' \\ (\mathbf{X}_2 - \boldsymbol{\mu})' \end{pmatrix} = (\epsilon_1', \epsilon_2')'$, where

$$
\begin{aligned}
\epsilon_1 &= \mathbf{X}_1 - \boldsymbol{\mu}, \\
\epsilon_2 &= -\boldsymbol{\sigma}_1' \Sigma_{1,1}^{-1} (\mathbf{X}_1 - \boldsymbol{\mu}) + (\mathbf{X}_2 - \boldsymbol{\mu}),
\end{aligned}
$$

ϵ_1 and ϵ_2 are uncorrelated. Therefore, the de-correlated and standardized vector of $\mathbf{X}_2$ can be defined to be

$$
\mathbf{X}_2^* = \mathbf{d}_2^{-1/2} \epsilon_2 = \mathbf{d}_2^{-1/2} \left[-\boldsymbol{\sigma}_1' \Sigma_{1,1}^{-1} (\mathbf{X}_1 - \boldsymbol{\mu}) + (\mathbf{X}_2 - \boldsymbol{\mu}) \right].
$$

It is obvious that $\mathbf{X}_1^*$ and $\mathbf{X}_2^*$ are uncorrelated, and both have the identity covariance matrix $\mathbf{I}_p$.

Similarly, for the third observation $\mathbf{X}_3$, which could be correlated with $\mathbf{X}_1$ and $\mathbf{X}_2$, consider the long vector $(\mathbf{X}_1', \mathbf{X}_2', \mathbf{X}_3')'$. Its covariance matrix can be written as $\Sigma_{3,3} = \begin{pmatrix} \Sigma_{2,2} & \boldsymbol{\sigma}_2 \\ \boldsymbol{\sigma}_2' & \boldsymbol{\gamma}(0) \end{pmatrix}$, where $\boldsymbol{\sigma}_2 = ([\boldsymbol{\gamma}(2)]', [\boldsymbol{\gamma}(1)]')'$. If we define $\Phi_3 = \begin{pmatrix} \Phi_2 & \mathbf{0} \\ -\boldsymbol{\sigma}_2' \Sigma_{2,2}^{-1} & \mathbf{I}_p \end{pmatrix}$ and $\mathbf{D}_3 = \begin{pmatrix} \mathbf{d}_1 & \mathbf{0} & \mathbf{0} \\ \mathbf{0} & \mathbf{d}_2 & \mathbf{0} \\ \mathbf{0} & \mathbf{0} & \mathbf{d}_3 \end{pmatrix} = diag(\mathbf{d}_1, \mathbf{d}_2, \mathbf{d}_3)$, where $\mathbf{d}_3 = \Sigma_{3,3} - \boldsymbol{\sigma}_2' \Sigma_{2,2}^{-1} \boldsymbol{\sigma}_2$, then we have $\Phi_3 \Sigma_{3,3} \Phi_3' = \mathbf{D}_3$. This motivates us to consider $\Phi_3 \mathbf{e}_3$, where $\mathbf{e}_3 = [(\mathbf{X}_3 - \boldsymbol{\mu})', (\mathbf{X}_1 - \boldsymbol{\mu})', (\mathbf{X}_2 - \boldsymbol{\mu})']'$. It can be checked that $\Phi_3 \mathbf{e}_3 = (\epsilon_1', \epsilon_2', \epsilon_3')'$, where

$$
\epsilon_3 = -\boldsymbol{\sigma}_2' \Sigma_{2,2}^{-1} \mathbf{e}_2 + (\mathbf{X}_3 - \boldsymbol{\mu}).
$$

Since $\mathrm{Cov}(\Phi_3 \mathbf{e}_3) = \mathbf{D}_3$, $\mathbf{e}_3$ is uncorrelated with $\mathbf{e}_1$ and $\mathbf{e}_2$. Therefore, the de-correlated and standardized vector of $\mathbf{X}_3$ is defined to be

$$
\mathbf{X}_3^* = \mathbf{d}_3^{-1/2} \epsilon_3 = \mathbf{d}_3^{-1/2} (-\boldsymbol{\sigma}_2' \Sigma_{2,2}^{-1} \mathbf{e}_2 + (\mathbf{X}_3 - \boldsymbol{\mu})),
$$

which is uncorrelated with $\mathbf{X}_1^*$ and $\mathbf{X}_2^*$ and has the identity covariance matrix $\mathbf{I}_p$.

Following the above procedure, we can define the de-correlated and standardized vectors sequentially after a new observation is collected. More specifically, at the j-th observation time, the covariance matrix of the long vector $(\mathbf{X}_1', \mathbf{X}_2', \ldots, \mathbf{X}_j')'$ can be written as $\Sigma_{j,j} = \begin{pmatrix} \Sigma_{j-1,j-1} & \boldsymbol{\sigma}_{j-1} \\ \boldsymbol{\sigma}_{j-1}' & \boldsymbol{\gamma}(0) \end{pmatrix}$, where $\boldsymbol{\sigma}_{j-1} = ([\boldsymbol{\gamma}(j-1)]', \ldots, [\boldsymbol{\gamma}(2)]',$

$[\boldsymbol{\gamma}(1)]')'$. It can be checked that $\Phi_j \Sigma_{j,j} \Phi'_j = \mathbf{D}_j$, where $\Phi_j = \begin{pmatrix} \Phi_{j-1} & \mathbf{0} \\ -\boldsymbol{\sigma}'_{j-1}\Sigma^{-1}_{j-1,j-1} & \mathbf{I}_p \end{pmatrix}$, $\mathbf{D}_j = diag(\mathbf{d}_1, \mathbf{d}_2, \ldots \mathbf{d}_j)$, and $\mathbf{d}_j = \Sigma_{j,j} - \boldsymbol{\sigma}'_{j-1}\Sigma^{-1}_{j-1,j-1}$ $\boldsymbol{\sigma}_{j-1}$. Therefore, if we define

$$\epsilon_j = -\boldsymbol{\sigma}'_{j-1}\Sigma^{-1}_{j-1,j-1}\mathbf{e}_{j-1} + (\mathbf{X}_j - \boldsymbol{\mu}),$$

then $\Phi_j \epsilon_j = (\mathbf{e}'_1, \mathbf{e}'_2, \ldots, \mathbf{e}'_j)'$ and $\text{Cov}(\Phi_j \epsilon_j) = D_j$, which implies that $\mathbf{e}_j$ is uncorrelated with $\{\mathbf{e}_1, \ldots, \mathbf{e}_{j-1}\}$. Therefore, the de-correlated and standardized vector of $\mathbf{X}_j$ is defined to be

$$\mathbf{X}^*_j = \mathbf{d}_j^{-1/2}\epsilon_j = \mathbf{d}_j^{-1/2}(-\boldsymbol{\sigma}'_{j-1}\Sigma^{-1}_{j-1,j-1}\mathbf{e}_{j-1} + (\mathbf{X}_j - \boldsymbol{\mu})),$$

which is uncorrelated with $\mathbf{X}^*_1, \ldots, \mathbf{X}^*_{j-1}$ and has the identity covariance matrix $\mathbf{I}_p$.

By the above sequential data de-correlation procedure, we can transform the originally correlated process observations to a sequence of uncorrelated and standardized observations, each of which has the mean $\mathbf{0}$ and the identity covariance matrix $\mathbf{I}_p$. In reality, the IC parameters $\boldsymbol{\mu}$ and $\{\boldsymbol{\gamma}(s)\}$ are usually unknown and should be estimated in advance. To this end, $\boldsymbol{\mu}$ and $\{\boldsymbol{\gamma}(s)\}$ can be estimated from the IC dataset $\mathcal{X}_{IC} = \{\mathbf{X}_{-m_0+1}, \mathbf{X}_{-m_0+2}, \ldots, \mathbf{X}_0\}$ as follows:

$$\widehat{\boldsymbol{\mu}} = \frac{1}{m_0} \sum_{i=-m_0+1}^{0} \mathbf{X}_i \tag{1}$$

$$\widehat{\boldsymbol{\gamma}}(s) = \frac{1}{m_0 - s} \sum_{i=-m_0+1}^{-s} (\mathbf{X}_{i+s} - \widehat{\boldsymbol{\mu}})(\mathbf{X}_i - \widehat{\boldsymbol{\mu}})'.$$

2.3 *Machine Learning Control Charts for Monitoring Serially Correlated Data*

To monitor a serially correlated process with observations $\mathbf{X}_1, \mathbf{X}_2, \ldots, \mathbf{X}_n, \ldots$, we can sequentially de-correlate these observations first by using the procedure described in the previous subsection and then apply the machine learning control charts described in Subsect. 2.1. However, at the current time point n, to de-correlate $\mathbf{X}_n$ with all its previous observations $\mathbf{X}_1, \mathbf{X}_2, \ldots, \mathbf{X}_{n-1}$, will take much computing time, especially when n becomes large. To reduce the computing burden, [22] suggested that the observation $\mathbf{X}_n$ only need to be de-correlated with its previous b_{max} observations, based on the assumption that two process observations becomes uncorrelated if their observation times are more than b_{max} apart. This assumption basically says that the serial data correlation is short-ranged, which should be reasonable in many applications. Based on this assumption, a modified machine learning control chart for monitoring serially correlated data is summarized below.

- When $n = 1$, the de-correlated and standardized observation is defined to be $\widehat{\mathbf{X}}_1^* = \widehat{\gamma}(0)^{-1/2}(\mathbf{X}_1 - \widehat{\mu})$. Set an auxiliary parameter b to be 1, and then apply a machine learning control chart to $\widehat{\mathbf{X}}_1^*$.
- When $n > 1$, the estimated covariance matrix of $(\mathbf{X}_{n-b}', \ldots, \mathbf{X}_n')'$ is defined to be

$$\widehat{\Sigma}_{n,n} = \begin{pmatrix} \widehat{\gamma}(0) & \cdots & \widehat{\gamma}(b) \\ \vdots & \ddots & \vdots \\ \widehat{\gamma}(b) & \ldots & \widehat{\gamma}(0) \end{pmatrix} =: \begin{pmatrix} \widehat{\Sigma}_{n-1,n-1} & \widehat{\sigma}_{n-1} \\ \widehat{\sigma}_{n-1}' & \widehat{\gamma}(0) \end{pmatrix}.$$

Then, the de-correlated and standardized observation at time n is defined to be

$$\widehat{\mathbf{X}}_n^* = \widehat{\mathbf{d}}_n^{-1/2}\left[-\widehat{\sigma}_{n-1}'\widehat{\Sigma}_{n-1,n-1}^{-1}\widehat{\mathbf{e}}_{n-1} + (\mathbf{X}_n - \widehat{\mu})\right],$$

where $\widehat{\mathbf{d}}_j = \widehat{\Sigma}_{j,j} - \widehat{\sigma}_{j-1}'\widehat{\Sigma}_{n-1,n-1}^{-1}\widehat{\sigma}_{j-1}$, and $\widehat{\mathbf{e}}_{n-1} = [(\mathbf{X}_{n-b} - \widehat{\mu})', (\mathbf{X}_{n-b+1} - \widehat{\mu})', \ldots, (\mathbf{X}_{n-1} - \widehat{\mu})']'$. Apply a machine learning control chart to $\widehat{\mathbf{X}}_n^*$ to see whether a signal is triggered. If not, set $b = min(b+1, b_{max})$ and $n = n+1$, and monitor the process at the next time point.

3 Simulation Studies

In this section, we investigate the numerical performance of the four existing machine learning control charts AC, RTC, DSVM and KNN described in Subsect. 2.1, in comparison with their modified versions AC-D, RTC-D, DSVM-D and KNN-D discussed in Subsect. 2.3, where "-D" indicates that process observations are de-correlated before each method is used for process monitoring. In all simulation examples, the nominal ARL_0 values of all charts are fixed at 200. If there is no further specification, the parameter λ in the chart AC is chosen to be 0.2, as suggested in He et al. (2010), the moving window size w in the charts RTC and DSVM is chosen to be 10, as suggested in [10] and [13], and the number of nearest observations k in the chart KNN is chosen to be 30, as suggested in Sukchotrat et al. (2009). The number of quality variables is fixed at $p = 10$, the parameter b_{max} is chosen to be 20, and the IC sample size is fixed at $m_0 = 2,000$. The following five cases are considered:

- Case I: Process observations $\{\mathbf{X}_n, n \geq 1\}$ are i.i.d. with the IC distribution $N_{10}(\mathbf{0}, I_{10\times 10})$.
- Case II: Process observations $\{\mathbf{X}_n, n \geq 1\}$ are i.i.d. at different observation times, the 10 quality variables are independent of each other, and each of them has the IC distribution χ_3^2, where χ_3^2 denotes the chi-square distribution with the degrees of freedom being 3.

- Case III: Process observations $\mathbf{X}_n = (X_{n1}, X_{n2}, \ldots, X_{n10})'$ are generated as follows: for each i, X_{ni} follows the AR(1) model $X_{ni} = 0.1X_{n-1,i} + \epsilon_{ni}$, where $X_{01} = 0$ and $\{\epsilon_{n1}\}$ are i.i.d. random errors with the $N(0, 1)$ distribution. All 10 quality variables are assumed independent of each other.
- Case IV: Process observations $\mathbf{X}_n = (X_{n1}, X_{n2}, \ldots, X_{n10})'$ are generated as follows: for each i, X_{ni} follows the ARMA(3,1) model $X_{ni} = 0.8X_{n-1,i} - 0.5X_{n-2,i} + 0.4X_{n-3,i} + \epsilon_{ni} - 0.5\epsilon_{n-1,i}$, where $X_{1i} = X_{2i} = X_{3i} = 0$ and $\{\epsilon_{ni}\}$ are i.i.d. random errors with the distribution χ^2_3. All 10 quality variables are assumed independent of each other.
- Case V: Process observations follow the model $\mathbf{X}_n = A\mathbf{X}_{n-1} + \boldsymbol{\epsilon}_n$, where $\{\boldsymbol{\epsilon}_n\}$ are i.i.d. random errors with the $N_{10}(0, B)$ distribution, A is a diagonal matrix with the diagonal elements being 0.5, 0.4, 0.3, 0.2, 0.1, 0.1, 0.2, 0.3, 0.4, 0.5, and B is a 10×10 covariance matrix with all diagonal elements being 1 and all off-diagonal elements being 0.2.

In all five cases described above, each variable is standardized to have mean 0 and variance 1 before process monitoring. Obviously, Case I is the conventional case considered in the SPC literature with i.i.d. process observations and the standard normal IC process distribution. Case II also considers i.i.d. process observations, but the IC process distribution is skewed. Cases III and IV consider serially correlated process observations across different observation times; but the 10 quality variables are independent of each other. In Case V, process observations are serially correlated and different quality variables are correlated among themselves as well.

Evaluation of the IC Performance. We first evaluate the IC performance of the related control charts. The control limits of the four control charts AC, RTC, DSVM and KNN are determined as discussed in Subsect. 2.1. For each method, its actual ARL_0 value is computed as follows. First, an IC dataset of size $m_0 = 2,000$ is generated, and some IC parameters (e.g. $\boldsymbol{\mu}$ and $\boldsymbol{\gamma}(s)$) are estimated from the IC dataset. Then, each control chart is applied to a sequence of 2,000 IC process observations for online process monitoring, and the RL value is recorded. This simulation of online process monitoring is then repeated for 1,000 times, and the actual conditional ARL_0 value conditional on the given IC data is computed as the average of the 1,000 RL values. Finally, the previous two steps are repeated for 100 times. The average of the 100 actual conditional ARL_0 values is used as the approximated actual ARL_0 value of the related control chart, and the standard error of this approximated actual ARL_0 value can also be computed. For the four modified charts AC-D, RTC-D, DSVM-D and KNN-D, their actual ARL_0 values are computed in a same way, except that process observations are de-correlated before online monitoring.

From Table 1, we can have the following results. First, the IC performance of the charts AC, RTC, DSVM and KNN all have a reasonably stable performance in Cases I and II when process observations are assumed to be i.i.d. at different observation times and different quality variables are assumed independent as well. Second, in Cases III-V when there is a serial data correlation across different observation times and data correlation among different quality variables, the IC performance of the charts AC, RTC, DSVM and KNN becomes unreliable since their actual ARL_0

Table 1 Actual ARL_0 values and their standard errors (in parentheses) of four machine learning control charts and their modified versions when their nominal ARL_0 values are fixed at 200.

Methods	Case I	Case II	Case III	Case IV	Case V
RF	189(3.98)	194(4.20)	105(1.42)	119(2.05)	106(1.33)
RF-D	193(3.22)	182(3.49)	188(3.61)	193(3.70)	194(3.37)
RTC	203(4.66)	207(5.23)	252(5.97)	133(3.02)	269(6.01)
RTC-D	194(3.68)	196(3.64)	201(4.00)	188(3.49)	190(3.96)
DSVM	213(5.20)	195(4.77)	263(6.99)	118(2.87)	277(6.34)
DSVM-D	193(4.33)	198(3.50)	193(4.16)	190(3.72)	188(3.73)
KNN	196(4.77)	188(3.88)	156(3.70)	266(6.02)	134(4.03)
KNN-D	191(4.20)	194(3.69)	194(4.01)	187(3.20)	190(3.18)

values are substantially different from the nominal ARL_0 value of 200. Third, as a comparison, the IC performance of the four modified charts AC-D, RTC-D, DSVM-D and KNN-D is stable in all cases considered. Therefore, this example confirms that the IC performance of the machine learning control charts can be improved in a substantial way by using the suggested modification discussed in Subsect. 2.3.

Evaluation of the OC Performance. Next, we evaluate the OC performance of the related charts in the five cases discussed above. In each case, a shift is assumed to occur at the beginning of online process monitoring with the size 0.25, 0.5, 0.75 and 1.0 in each quality variable. Other setups are the same as those in Table 1. To make the comparison among different charts fair, the control limits of the charts have been adjusted properly so that their actual ARL_0 values all equal to the nominal level of 200. The results of the computed ARL_1 values of these charts in Cases I-V are presented in Fig. 1.

From the Fig. 1, it can be seen that the modified versions of the four control charts all have a better OC performance in Cases III-V when the serial data correlation exists. In Cases I and II when process observations are independent at different observation times, the OC performance of the modified versions of the four charts have a slightly worse performance than the original versions of the related charts. The main reason for the latter conclusion is due to the "masking effect" of data de-correlation, as discussed in [29]. Remember that the de-correlated process observations are linear combinations of the original process observations. Therefore, a shift in the original data would be attenuated during data de-correlation, and consequently the related control charts would be less effective in cases when serial data correlation does not exist.

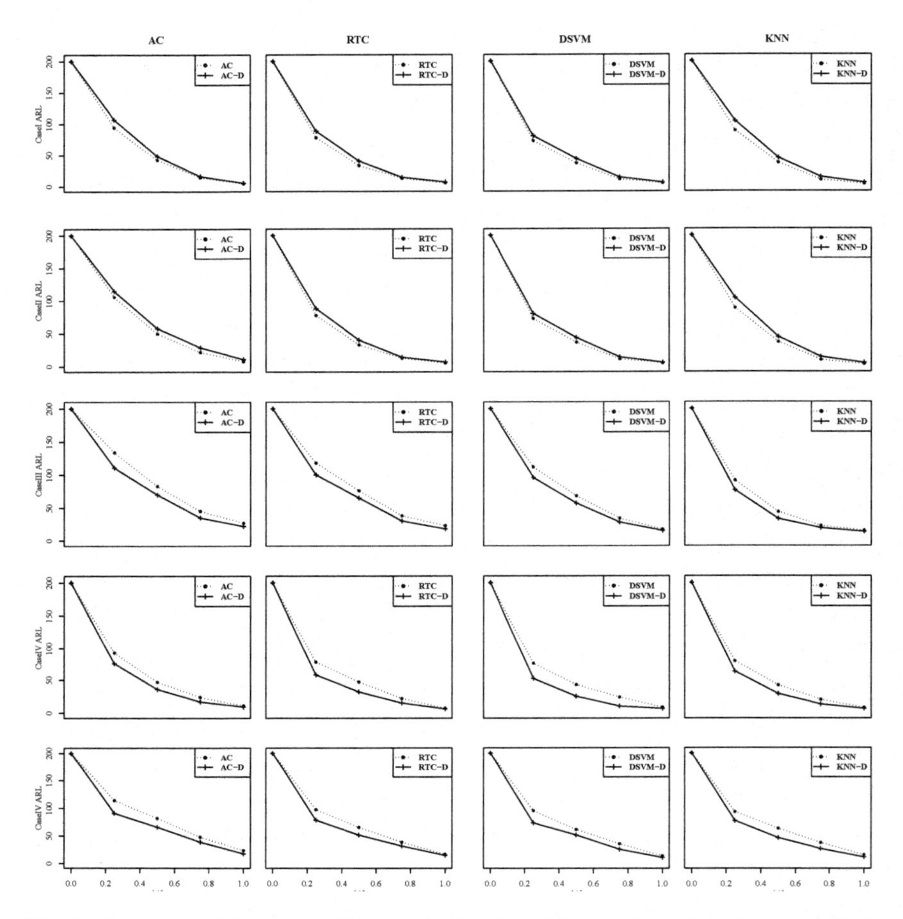

Fig. 1 Computed ARL_1 values of the original and modified versions of the four control charts AC, RTC, DSVM and KNN when their nominal ARL_0 values are fixed at 200, the parameters of the charts are chosen as in the example of Table 1, all quality variables have the same shift, and the shift size changes among 0.25, 0.5, 0.75 and 1.0.

4 A Real-Data Application

In this section, a dataset from a semiconductor manufacturing process is used to demonstrate the application of the modified machine learning control charts discussed in the previous sections. The dataset is available in the UC Irvine Machine Learning Repository (http://archive.ics.uci.edu/ml/datasets/SECOM). It has a total of 590 quality variables and 1,567 observations of these variables. A total of 600 observations of five specific quality variables are selected here. The original data are shown in Fig. 2. From the figure, it seems that the first 500 observations are quite stable, and thus they are used as the IC data. The remaining 100 observations are

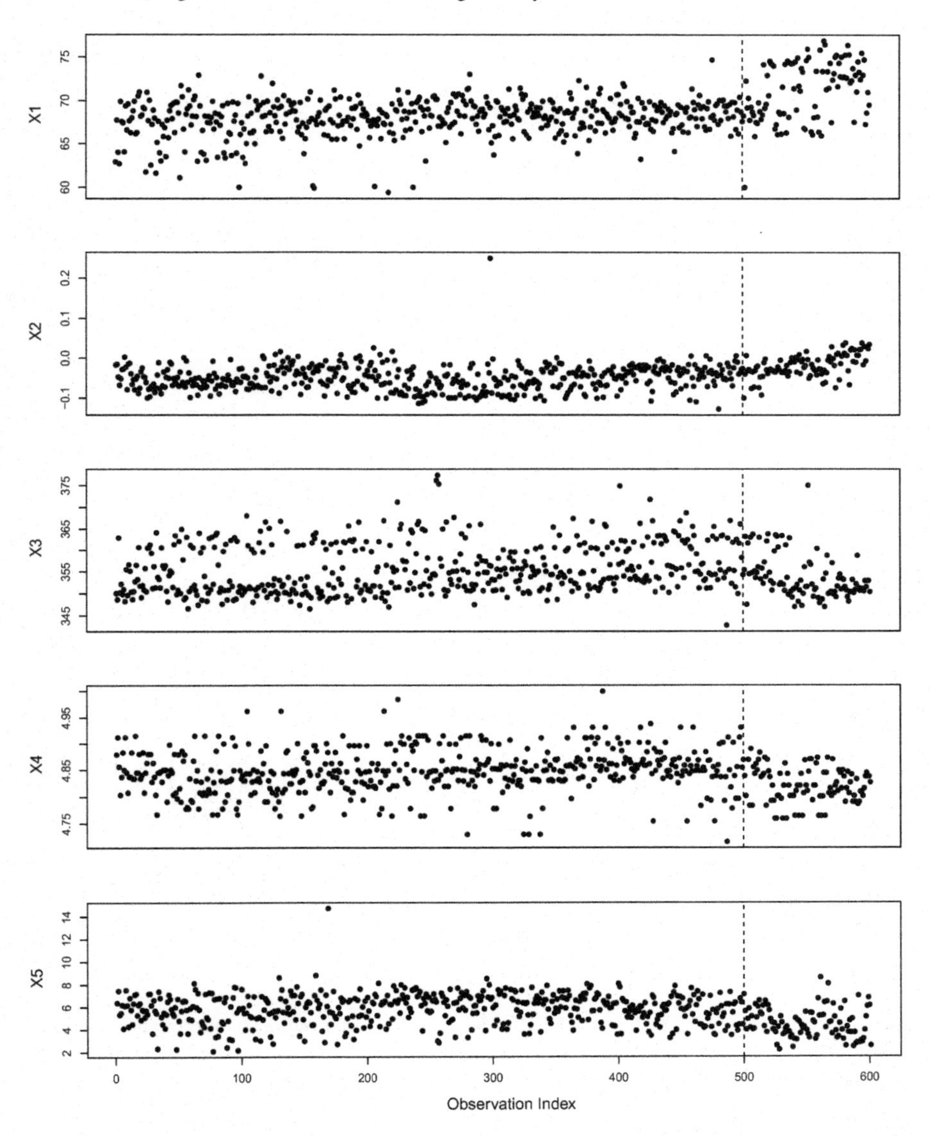

Fig. 2 Original observations of the five quality variables of a semiconductor manufacturing data. The vertical dashed line in each plot separates the IC data from the data for online process monitoring.

used for online process monitoring. In Fig. 2, the training and testing datasets are separated by the dashed vertical lines.

For the IC data, we first check for existence of serial data correlation. To this end, the p-values of the Durbin-Watson test for the five quality variables are 1.789×10^{-3}, 4.727×10^{-1}, 4.760×10^{-4}, 1.412×10^{-4}, and 9.744×10^{-2}. Thus, there

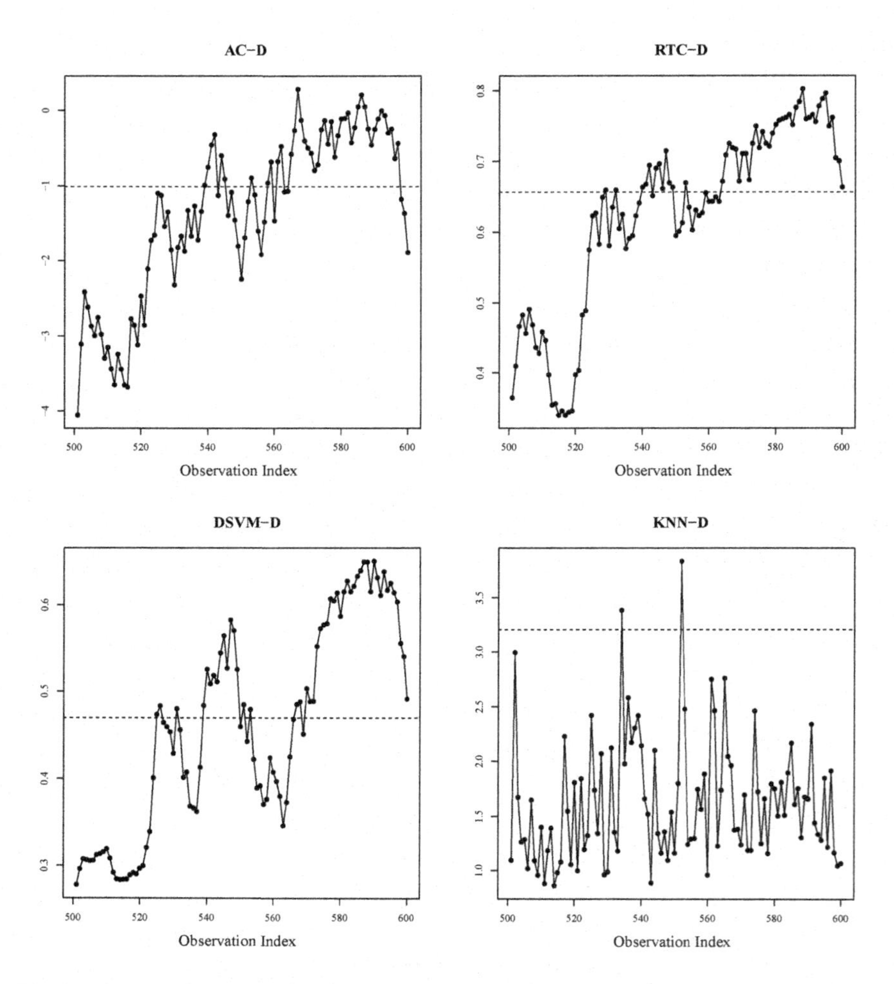

Fig. 3 Control charts AC-D, RTC-D, DSVM-D, and KNN-D for online monitoring of a semiconductor manufacturing data. The horizontal dashed line in each plot denotes the control limit of the related control chart.

is a significant autocorrelation for the first, third and fourth quality variables. The Augmented Dickey-Fuller (ADF) test for stationality of the autocorrelation gives p-values that are < 0.01 for all quality variables. This result suggests that the stationary assumption is valid in this data. Therefore, the IC data have a significant stationary serial data correlation in this example, and the modification for the machine learning charts discussed in Sects. 2 and 3 should be helpful.

Next, we apply the four modified control charts AC-D, RTC-D, DSVM-D and KNN-D to this data for online process monitoring starting from the 501st observation time. In all control charts, the nominal ARL_0 values is fixed at 200, and their control limits are computed in the same way as that in the simulation study of Sect. 3. All four control charts are shown in Fig. 3. From the plots in the figure, the charts AC-D, RTC-D, DSVM-D and KNN-D give their first signals at the 539th, 529th, 525th, and 534th observation times, respectively. In order to determine whether these signals are false alarms or not, the change-point detection approach based on the generalized maximum likelihood estimation (cf., [21], Sect. 7.5) is applied to the test data (i.e., the data between the 501st and 600th observation times). The detected change-point position is at 517. The Hotelling's T^2 test for checking whether the mean difference between the two groups of data with the observation times in [501,516] and [517,600] is significantly different from $\mathbf{0}$ gives the p-value of 4.426×10^{-3}. Thus, there indeed is a significant mean shift at the time point 517. In this example, it seems that all four charts can detect the shift and the chart DSVM-D can give the earliest signal among them.

5 Concluding Remarks

Recently, several multivariate nonparametric control charts based on different machine learning algorithms have been proposed for online process monitoring. Most existing machine learning control charts are based on the assumption that the multivariate observations are independent of each other. These control charts have a reliable performance when the data independence assumption is valid. However, when the process data are serially correlated, they may not be able to provide a reliable process monitoring. In this chapter, we have suggested a modification for these machine learning control charts, by which process observations are first de-correlated before they are used for monitoring serially correlated data. Numerical studies have shown that the modified control charts have a more reliable performance than the original charts in cases when the serial data correlation exists.

There are still some issues to address in the future research. For instance, the "masking effect" of data de-correlation could attenuate the shift information in the de-correlated data. One possible solution is to use the modified data de-correlation procedure discussed in You and Qiu (2017). By this approach, the process observation at the current time point is de-correlated only with a small number of previous process observations within the so-called "spring length" (cf., [9]) of the current observation time. Another issue is related to the assumption of short-range stationary serial data correlation that has been used in the proposed modification procedure. In some applications, the serial data correlation could be long-range and non-stationary (cf., [5]). Thus, the proposed modification could be ineffective for such applications.

References

1. Aggarwal CC (2018). Neural Networks and Deep Learning. Springer, New Yorker

2. Alwan LC, Roberts HV (1995) The problem of misplaced control limits. J Roy Stat Soc (Ser C) 44:269–278
3. Apley DW, Tsung F (2002) The autoregressive T^2 chart for monitoring univariate autocorrelated processes. J Qual Technol 34:80–96
4. Breiman L (2001) Random forests. Mach Learn 45:5–32
5. Brean J (1992) Statistical methods for data with long-range dependence. Stat Sci 4:404–416
6. Camci F, Chinnam RB, Ellis RD (2008) Robust kernel distance multivariate control chart using support vector principles. Int J Prod Res 46:5075–5095
7. Capizzi G, Masarotto G (2008) Practical design of generalized likelihood ratio control charts for autocorrelated data. Technometrics 50:357–370
8. Carvalhoa TP, Soares F, Vita R, Francisco R, Basto JP, Alcalá SGS (2019) A systematic literature review of machine learning methods applied to predictive maintenance. Comput Ind Eng 137:106024
9. Chatterjee S, Qiu P (2009) Distribution-free cumulative sum control charts using bootstrap-based control limits. Ann Appl Stat 3:349–369
10. Deng H, Runger G, Tuv E (2012) System monitoring with real-time contrasts. J Qual Technol 44:9–27
11. Göb R (2006) Data mining and statistical control - a review and some links. In: Lenz HJ, Wilrich PT (ed) Frontiers in Statistical Quality Control, vol 8, pp 285–308. Physica-Verlag, Heidelberg
12. Hastie T, Tibshirani R, Friedman J (2001) The Elements of Statistical Learning - Data Mining, Inference, and Prediction. Springer-Verlag, Berlin
13. He S, Jiang W, Deng H (2018) A distance-based control chart for monitoring multivariate processes using support vector machines. Ann Oper Res 263:191–207
14. Hwang W, Runger G, Tuv E (2007) Multivariate statistical process control with artificial contrasts. IIE Trans 2:659–669
15. Knoth S, Schmid W (2004) Control charts for time series: a review. In: Lenz HJ, Wilrich PT (ed) Frontiers in Statistical Quality Control, vol 7, pp 210–236. Physica-Verlag, Heidelberg
16. Lee HC, Apley DW (2011) Improved design of robust exponentially weighted moving average control charts for autocorrelated processes. Qual Reliab Eng Int 27:337–352
17. Li W, Qiu P (2020) A general charting scheme for monitoring serially correlated data with short-memory dependence and nonparametric distributions. IISE Trans 52:61–74
18. Zhang L, Mei T (2020) Nonparametric monitoring of multivariate data via KNN learning. Int J Prod Res. https://doi.org/10.1080/00207543.2020.1812750
19. Prajapati DR, Singh S (2012) Control charts for monitoring the autocorrelated process parameters: a literature review. Int J Prod Qual Manag 10:207–249
20. Psarakis S, Papaleonida GEA (2007) SPC procedures for monitoring autocorrelated processes. Qual Technol Quant Manag 4:501–540
21. Qiu P (2014) Introduction to Statistical Process Control. Chapman Hall/CRC, Boca Raton
22. Qiu P, Li W, Li J (2020) A new process control chart for monitoring short-range serially correlated data. Technometrics 62:71–83
23. Runger GC, Willemain TR (1995) Model-based and model-free control of autocorrelated processes. J Qual Technol 27:283–292
24. Sukchotrat T, Kim SB, Tsui K-L, Chen VCP (2011) Integration of classification algorithms and control chart techniques for monitoring multivariate processes. J Stat Comput Simul 81:1897–1911
25. Sun R, Tsung F (2003) A kernel-distance-based multivariate control chart using support vector methods. Int J Prod Res 41:2975–2989
26. Tax DM, Duin RPW (2004) Support vector data description. Mach Learn 54:45–66
27. Tuv E, Runger G (2003) Learning patterns through artificial contrasts with application to process control. Trans Inf Commun Technol 29:63–72
28. Weiß CH (2015) SPC methods for time-dependent processes of counts - a literature review. Cogent Math 2:1111116

29. You L, Qiu P (2019) Fast computing for dynamic screening systems when analyzing correlated data. J Stat Comput Simul 89:379–394
30. Xue L, Qiu P (2020) A nonparametric CUSUM chart for monitoring multivariate serially correlated processes. J Qual Technol. https://doi.org/10.1080/00224065.2020.1778430
31. Zhang C, Tsung F, Zou C (2015) A general framework for monitoring complex processes with both in-control and out-of-control information. Comput Ind Eng 85:157–168

A Review of Tree-Based Approaches for Anomaly Detection

Tommaso Barbariol, Filippo Dalla Chiara, Davide Marcato, and Gian Antonio Susto

Abstract Data-driven Anomaly Detection approaches have received increasing attention in many application areas in the past few years as a tool to monitor complex systems in addition to classical univariate control charts. Tree-based approaches have proven to be particularly effective when dealing with high-dimensional Anomaly Detection problems and with underlying non-gaussian data distributions. The most popular approach in this family is the Isolation Forest, which is currently one of the most popular choices for scientists and practitioners when dealing with Anomaly Detection tasks. The Isolation Forest represents a seminal algorithm upon which many extended approaches have been presented in the past years aiming at improving the original method or at dealing with peculiar application scenarios. In this work, we revise some of the most popular and powerful Tree-based approaches to Anomaly Detection (extensions of the Isolation Forest and other approaches), considering both batch and streaming data scenarios. This work will review several relevant aspects of the methods, like computational costs and interpretability traits. To help practitioners we also report available relevant libraries and open implementations, together with a review of real-world industrial applications of the considered approaches.

T. Barbariol · F. D. Chiara · D. Marcato · G. A. Susto (✉)
Department of Information Engineering, University of Padova, Padova, Italy
e-mail: gianantonio.susto@unipd.it

T. Barbariol
e-mail: tommaso.barbariol@dei.unipd.it

F. D. Chiara
e-mail: filippo.dallachiara@studenti.unipd.it

D. Marcato
e-mail: davide.marcato@lnl.infn.it

D. Marcato
Legnaro National Laboratories, INFN, Legnaro, Italy

K. P. Tran (ed.), *Control Charts and Machine Learning for Anomaly Detection in Manufacturing*, Springer Series in Reliability Engineering,
https://doi.org/10.1007/978-3-030-83819-5_7

1 Introduction

The problem of finding novel or anomalous behaviours, often referred as Anomaly or outlier Detection (AD), is common to many contexts: anomalies may be very critical in many circumstances that affect our everyday life, in contexts like cybersecurity, fraud detection and fake news [40, 56]. In science Anomaly Detection tasks can be found in many areas, from astronomy [62] to health care [65]; moreover, Anomaly Detection approaches have been even applied to knowledge discovery [16] and environmental sensor networks [35].

One of the areas that mostly benefits from the employment of AD modules is the industrial sector, where quality is a key driver of performance and success of productions and products. With the advent of the Industry 4.0 paradigm, factories and industrial equipment are generating more and more data that are hard to be fully monitored with traditional approaches; on the other hand, such availability of data can be exploited for enhanced quality assessment and monitoring [55, 78]. Moreover, products and devices, thanks to advancements in electronics and the advent of the Internet of Things, are increasingly equipped with sensors and systems that give them new capabilities, like for example the ability to check their health status thanks to embedded/cloud Anomaly Detection modules [4, 44, 89].

Despite the heterogeneous systems that may benefit from AD modules/capabilities, there are several desiderata that are typically requested for an AD module, since obviously looking for high detection accuracy is not the only important requisite. For example, in many contexts the delay between the occurred anomaly and its detection might be critical and the low latency of the model becomes a stringent requirement. One way to mitigate the detection delay problem is typically to embed the AD model in the equipment/device: this implementation scenario directly affects both the choice of the detection model and, in some cases, the hardware. As a consequence practitioners will typically have to find a compromise between detection accuracy and computational complexity of the model: while this is true for any Machine Learning module, it is typically more critical when dealing with AD tasks. Moreover, in presence of complex processes or products equipped with different sensors, data will exhibit high dimensionality and therefore practitioners will tend to prefer models that are able to efficiently handle multiple inputs. Summarizing, a good real-world AD solution: i) has to provide high detection performances; ii) has to guarantee low latency; iii) requires low computational resources; iv) should be able to efficiently handle high dimensional data.

The former list of desiderata for AD solutions is not exhaustive, and other characteristics can increase the model appeal in front of practitioners. In recent years, model interpretability is an increasingly appreciated property. Detecting an anomaly is becoming no longer enough and providing a reason why a point has been labelled as anomalous is getting more and more importance. This is particularly true in manufacturing processes where the capability of quickly finding the root cause of an anomalous behaviour can lead to important savings both in terms of time and costs. In addition, interpretability enhances the trust of users in the model, leading to widespread adoption of the AD solution.

Another attractive property is the ability of the model to handle data coming from non-stationary environments. Especially in early stages of AD adoption, available data are few and restricted to a small subset of possible system configurations; a model trained on such data risks to label as anomalous all the states not covered in the training domain, even if they are perfectly normal. In order to overcome this issue, the model should detect the changes in the underling distribution and continuously learn new *normal* data. In this process, however, the AD model should not lose its ability to detect anomalies.

Given the importance and diffusion of AD approaches, we deemed relevant to review and compare an important class of algorithms particularly suited for the aforementioned requirements. The subject of this investigation is the tree-based approaches to AD, i.e. approaches that have a tree structure in their decision making evaluation; the most popular representative of this class is the famous Isolation Forest algorithm, originally proposed by Liu [52, 53], an algorithm that is receiving increasing attention and sees application in many scenarios. In this work, for the first time to our knowledge, we try to systematically review all the AD methods in this emerging class, to discuss their costs and performances in benchmarks, to report industrial applications and to guide readers through available implementations and popularity of the various approaches in the scientific literature.

This review is conceived both for researchers and practitioners. The first ones will find a comparison between the many proposed variants, while the second ones will find useful information for practical implementation. Despite this work mainly copes with AD algorithms designed for tabular datasets, it is important to note that AD can be performed on a variety of data structures like images or audio signals and algorithms dedicated for different types of data format are also present in the literature. Nevertheless, it should be remarked that any data structures can be transformed into tabular data extracting appropriated features, making AD approaches for tabular data applicable potentially to any scenario.

2 Taxonomy and Approaches to Anomaly Detection

While many Anomaly Detection approaches have been developed, a simple taxonomy divides such methods into two categories:

- *Model-based* - It is the most traditional Anomaly Detection category and employs a predefined model that describes the normal or *all* the possible anomalous operating conditions. These approaches usually rely on physics or domain-knowledge heuristics; unfortunately they are often unfeasible and costly to be developed since they require extended knowledge of the system under exam.
- *Data-driven* - The approaches examined in this chapter rely instead on data availability. More precisely, such approaches make use of two ingredients: data sampled

from the analyzed system and algorithms able to automatically learn the abnormality level of those data. Such category of approaches is often referred as data-driven anomaly detection and its advantages are the great flexibility and absence of strong assumptions that limit the model applicability.

Additionally, when looking at anomalous behaviours, two problem settings can be defined: the supervised and the unsupervised one. The former consists in classifying data, based on previously tagged anomalies. It is called supervised since training data are collected from sensor measurements and have an associated label that identifies them as anomalous or not; in industrial context, the supervised scenario is typically named Fault Detection. Unfortunately, supervised settings are seldom available in reality [13]: labelling procedures are very time consuming and typically require domain experts to be involved.

On the contrary, the unsupervised scenario is the most common in real world applications. In this case, data are not equipped with labels and therefore the learning algorithm lacks a ground truth of what is anomalous and what is not. Given that, the goal of the algorithm is to highlight the most abnormal data, assigning to each one an Anomaly Score (AS). In this chapter the focus will be on the unsupervised setting since it is the most applicable in real-world scenarios.

To be more precise, the unsupervised setting can be further divided into two subcategories, based on the nature of the available data. The fully unsupervised one relies on training set composed of both anomalies and normal data. However in some applications obtaining training data with anomalies is quite complex, therefore in such cases data are composed only of normal instances: in this scenario semi-supervised approaches, sometimes named one-class setting, are the most natural ones to be adopted.

2.1 Formal Definition of Anomaly

The definition of anomaly is far from trivial, and depending on it, methods that try to detect anomalous data behave differently. The most widely accepted definition is quite general, and it was given by Hawkins in [34]:

> *"Observation which deviates so much from other observations as to arouse suspicion it was generated by a different mechanism".*

This statement can refer to multiple different anomalies, and does not give a clear indication on which way to follow to detect them. According to this definition it may be inferred that: i) a model needs to measure (in a not-specified way) the deviation between points; ii) each observation has an associated probability to be an outlier; iii) a different mechanism is present in the case of anomalous samples, suggesting that, on a data perspective, outliers follow a distribution that is different from the one of the inliers. The reported definition does not speak about the numerosity of outliers, but there is an hint that outliers are fewer than inliers in number. These indications

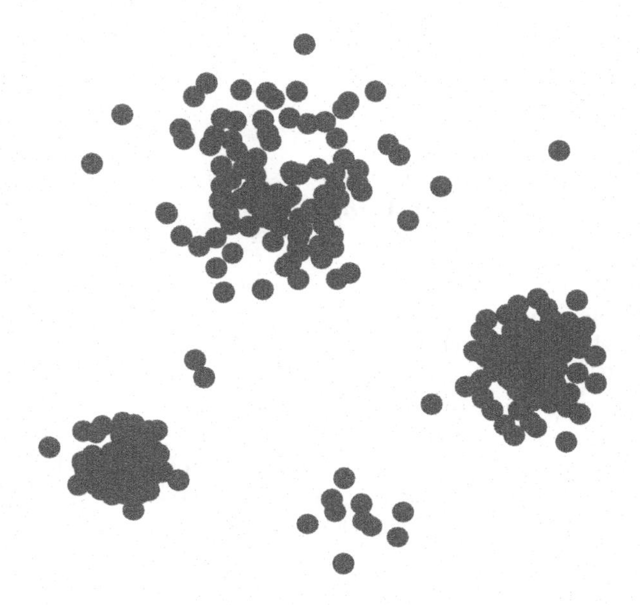

Fig. 1 Example of anomalies in a simple dataset composed of dense and scattered normal clusters. Anomalies can be locally defined w.r.t a specific cluster, or global. They can be scattered in the domain or they can group together in anomalous clusters. Figure adapted from [6]

give wide space to interpretations and as it will be clear later on, they encourage very different approaches.

Anomalies are traditionally divided into 4 categories even if some authors suggest different classes. Generic datasets looks like Fig. 1: they are made up of normal dense and sparse clusters, surrounded by global sparse and dense anomalies. These can be defined global anomalies if they look anomalous w.r.t all the normal points, or local if their abnormality is w.r.t a single normal cluster.

2.2 *Static and Dynamic Problems*

Depending on the application and the available data, different problem statements can be defined. The anomaly detection problem is defined *static* if the analysis is performed on time-independent data (*static* datasets, where the order of the observations do not matter), while it is *dynamic* if it is performed on time-dependent data (time-series data or dynamic datasets). Another very basic discrimination is between univariate or multivariate anomaly detection.

A more subtle distinction concerns the way in which the algorithm training is performed. The most traditional one is the batch training where the model is trained only once, using all the available training data. This approach might be unpractical when the dataset is too large to fit into the memory, or when sampled data do not cover sufficiently well the normal operating domain: in this case the model needs to be continuously updated as new data are collected; this is even more important in situations where the normal data distribution undergoes a drift, and samples that were used in the first part of the training now become anomalous. Therefore in such case, the model has to adapt by learning the new configuration and by forgetting the outdated distribution.

2.3 Classes of Algorithms

Depending on the interpretation that each author gives to the definition in Sect. 2.1, there exists different ways to measure anomaly. The only thing that brings together all the approaches is the use of an Anomaly Score (AS) assigned to each point. This score should serve as a proxy of the probability to be an outlier. Obviously, each method assigns a different anomaly score to the same point since it is based on different detection strategies.

There exists a variety of anomaly detection algorithms, but they can be categorized into 5 classes. The distinction between classes is not strongly fixed, and some methods could be categorized at the same time in different classes. The most intuitive class of algorithms is the *distance-based* one. It is based on the assumption that outliers are spatially far from the rest of the points [3]. Also the *density-based* class is quite intuitive since assumes outliers living in rarefied areas [28]. The *statistics-based* ones are conceptually simple, but often make use of heavy assumptions on the distribution that generated training data [38]. *Clustering-based* employ clustering techniques in order to find clustered data, moreover, they are strongly susceptible to hyper-parameters and often rely on density or distance measures [39].

Unfortunately, these approaches are expensive to compute or rely on too strong assumptions. Density and distance-based methods are hard to compute especially in high dimensional settings and when many data are available; such approaches are hardly applicable in scenarios where detection has to be performed online and on fast evolving data streams. Moreover, statistical methods are often restricted to ideal processes, rarely observed in practice.

Quite recently a new class of methods emerged: the *isolation-based*. This class is very different from the previous ones: it assumes that outliers are few, different and, above all, easier to be separated from the rest of data. This draws the attention from normal data to anomalies and allows to obtain much more efficient anomaly detectors. The primary goal of these methods is to quickly model the anomalies by isolating them, rather than spending resources on the modeling of the normal distribution. The seminal, and most popular, approach in this class is the Isolation Forest algorithm [53] that will be extensively discussed in Sect. 3. The original idea is based on tree-methods, but it has been recently extended to Nearest Neighbors algorithm [7].

2.4 Tree-Based Methods

Tree-based methods, as suggested by the class name, rely on tree structures where the domain of the available data is recursively split in a hierarchical way, in non-overlapping intervals named leaves. In the Anomaly Detection context, these models are seen as a tool rather than a separate detection approach. As a matter of fact, they are employed in both density-based and isolation-based approaches. The former approach is perhaps the most intuitive since at high densities one expects normal data clusters, vice-versa in low density regions. However this approach is in contrast with the simple principle of never solving a more difficult process than is needed [68]. Density estimation is a computationally expensive task since it focuses on normal data points, that are the majority. However, the ultimate goal of anomaly detection problem is to find anomalies, not to model normal data. Outliers are inferenced only at a second step. On the contrary, the isolation approach is less simple to formulate but also less computationally expensive. It directly addresses the detection of anomalies since points that are quickly isolated are more likely to be outliers.

The literature concerning anomaly detection using tree-based methods is quite vast, but in the present review we decided to focus on methods that naturally apply to the unsupervised setting, due to its relevance in industry. Not only we decided to exclude the supervised approaches, but also we excluded the ones that artificially create a second class of outliers like in [17]. These approaches often try to fit supervised models into unsupervised settings at cost of inefficient computations, especially at high dimensionality.

2.5 Structure of the Chapter and Contribution

The aim of this review will be to describe the most salient features of unsupervised tree-based algorithms for AD. Great attention will be devoted to the algorithms that primarily try to isolate anomalies, and then, as a by-product, estimate density. As stated above, to the best of our knowledge this is the first work that reviews the methods that originated from Isolation Forest, or that are closely related to it.

Throughout this chapter, we will refer to the Isolation Forest algorithm [53] as the *original* algorithm, or by using the acronym IF. Moreover, the term *outlier* and *anomaly* will be considered synonyms. Normal data will be often named inliers and must not be confused with Gaussian data.

This chapter is divided into 5 sections. After the first two introductory sections, the third reviews the tree-based approaches: in the first part of such section the Isolation Forest original algorithm is extensively discussed together with all the variants applied to time independent datasets; this part prepares the ground for the more complex time dependent datasets and their algorithms presented in Sect. 3.2. After this, some paragraphs are devoted to distributed and interpretable models. The fourth section compares the performances of the methods, looking at the results declared in the papers. A practical comparison between the methods fall outside the scope of this chapter, but case studies and multiple source code repositories are listed for the interested reader. In the last Section, conclusions and ideas for future research directions are discussed.

3 Isolation Tree Based Approaches

Isolation forest (IF) is the seminal algorithm in the field of isolation tree-based approaches and it was firstly described in [53]: in recent years IF has received an increasing attention from researchers and practitioners as it can be noted in Fig. 2, where the evolution of citations of the algorithm in scientific papers has increased exponentially over the years.

As the name suggests, IF is an ensemble algorithm that resembles in some aspects the popular Random Forest algorithm revised in the unsupervised anomaly detection settings. Indeed, IF is a collection of binary trees: while in the popular work of Breiman [9] we are dealing with decision trees, here the ensemble model is composed by *isolation trees*, that aim at isolating a region of the space where only a data point lies. IF is based on the idea that, since anomalies are by definition few in numbers, an isolation procedure will be faster in separating an outlier from the rest of the data than when dealing with inliers.

More in details, the algorithm consists in two steps: training and testing. In the training phase, each isolation tree recursively splits data into random partitions of the domain. As said, the core idea is that anomalies on average require less partitions to be isolated. Therefore inliers live in a leaf in the deepest part of the tree, while outliers in a leaf close to the root. More formally, the anomaly score is proportional to the average depth of the leaf where each datum lies. For the sake of clarity, we report the training and testing pseudo-codes (Algorithm 1, 2 and 3 - adapted from [53]).

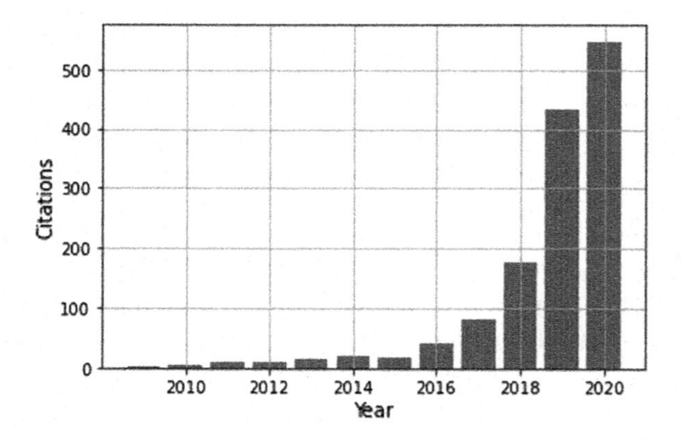

Fig. 2 Combined citations of IF original paper [53] and its extended version [52] by the same authors. Source: Scopus. Retrieved on the 30th of March 2021

Algorithm 1: IsolationForest(X, n, ψ)

Input: X – data in $\mathbb{R}^d$, n – number of trees, ψ – sample size
Output: list of Isolation Trees
$forest \leftarrow$ empty list of size n;
$h_{max} \leftarrow \lceil \log_2 |X| \rceil$;
for $i = 1$ *to* n **do**
  $\hat{X} \leftarrow sample(X,\ \psi)$;
  $forest[i] \leftarrow IsolationTree(\hat{X},\ 0,\ h_{\max})$;
end
return *forest*

Algorithm 2: IsolationTree(X, h, $h_{\max}$)

Input: X – data in $\mathbb{R}^d$, h – current depth of the tree, $h_{\max}$ – depth limit
Output: Isolation Tree (root node)
if $h \geq h_{max}$ *or* $|X| \geq 1$ **then**
  return *Leaf{*
    $size \leftarrow |X|$
  };
else
  $q \leftarrow$ randomly select a dimension from $\{1,\ 2,\ \ldots,\ d\}$;
  $p \leftarrow$ randomly select a threshold from [min $X^{(q)}$, max $X^{(q)}$];
  $X_L \leftarrow filter(X,\ X^{(q)} \geq p)$;
  $X_R \leftarrow filter(X,\ X^{(q)} < p)$;
  return *Node{*
    $left \leftarrow IsolationTree(X_L,\ h+1,\ h_{max})$,
    $right \leftarrow IsolationTree(X_R,\ h+1,\ h_{max})$,
    split_dim $\leftarrow q$,
    split_thresh $\leftarrow p$
  };
end

The training phase starts subsampling the dataset composed of n data points, in t randomly drawn subsets of ψ samples. Then, for each subset a random tree is built. At each node of the random tree a feature is uniformly drawn. The split point is uniformly sampled between the minimum and maximum value of the data along the selected feature, while the split criterion is simply the inequality w.r.t the feature split point. The splitting procedure is recursively performed until a specific number of points are isolated or when a specific tree depth is reached. In principle, the full isolation tree should grow until all points are separated, unfortunately risking to grow trees with depth close to $n-1$. However data that lie deep in the tree are the normal ones, not the target of the detector. For efficiency reasons, this is not practical and since anomalies are easy to be isolated, the tree only needs to reach its average depth $\lceil * \rceil \log_2 n$.

Algorithm 3: PathLength(x, T, h)

Input: x – instance in $\mathbb{R}^d$, T – node of IsolationTree, l – current length (to be initialized to 0 when first called on the root node)
Output: path length of x
if *T is a leaf node* **then**
  return $h + c(T.size)$;
end
$q \leftarrow T.split\_dim$;
if $x^{(q)} < T.split\_tresh$ **then**
  return $PathLength(x,\ T.left,\ l+1)$;
else
  return $PathLength(x,\ T.right,\ l+1)$;
end

The testing phase is different and consists in checking the depth $h(\cdot)$ reached by the data point x in all the isolation trees, and taking the average.

The anomaly score $s(x, n)$ is defined as:

$$s(x,n) = 2^{-\frac{E(h(x))}{c(n)}}$$

where $c(n)$ is a normalizing factor and $E(h(x))$ is the average of the tree depths. Note that when x is an anomaly, $E(h(x)) \rightarrow 0$ and therefore $s(x) \rightarrow 1$, while when $E(h(x)) \rightarrow n-1$, $s(x) \rightarrow 0$. When $E(h(x)) \rightarrow c(n)$, $s(x) \rightarrow 0.5$.

The computational complexity of IF is $O(t\psi \log \psi)$ in training while $O(nt\psi \log \psi)$ in testing, where we recall that ψ is the subsampling size of the dataset. It is interesting to note that in order to have better detection results, ψ needs to be small and constant across different datasets.

Isolation forest has many advantages compared to the methods belonging to other classes. Firstly it is very intuitive and requires a small amount of computations. For this reason it is particularly suited for big datasets and for applications where low latency is a strict requirement. The use of random feature selection and bagging allows to efficiently handle high dimensional datasets. In addition, the use of tree collections makes the method highly parallelisable. Unfortunately the algorithm has some issues. The most severe is the *masking effect* created by the axis parallel partitions and anomalous clusters, that perturbs the anomaly score of some points. A closely related issue is the algorithm difficulty to detect the local anomalies.

Trying to make a summary: the standard isolation forest defines anomalies as few and different, and approaches their detection not modelling normal data but trying to separate them as fast as possible with the aid of bagged trees. The split criterion is based on randomly selected feature and split point, that create axis-parallel partitions. These characteristics will be challenged by the following methods but the structure of the algorithms will remain quite the same.

In the following subsections the focus will be on static approaches (Sect. 3.1), dynamic (Sect. 3.2), distributed (Sect. 3.3) and finally interpretable and feature selection methods (Sect. 3.4).

3.1 Static Learning

Static learning methods can be generally divided into two sub-categories, using two approaches. The first one groups i) methods that directly originate from the seminal work [53] slightly modifying the Isolation Forest, and ii) methods that start from a different but similar algorithms like the Half-Space (HS) trees or Random Forests (RF). The former group focuses on the importance of fast isolation, while the latter on the density approximation. The second grouping approach subdivides the static methods based on how is computed the anomaly score. The majority relies on the mean leaf depth but a growing number of algorithms employs some variation of the leaf mass.

Filtering and Refinement: A Two-Stage Approach for Efficient and Effective Anomaly Detection

The general intuition that an AD algorithm should focus more on the detection of anomalies than the modelisation of normal data, has been developed also in [96]. In this case, the algorithms is based on two stages: filtering and refinement. At the first stage, the majority of normal data are filtered using a computationally cheap algorithm, while at the second stage, the remaining data are processed by a more refined but expensive tool. Filtering is performed by a tree based method quite similar to IF except for the splitting criterion: here the choice is deterministic and based on feature entropy and univariate densities computed by histograms. After that, distance-based methods are applied to the most abnormal data points and two anomaly scores are proposed to detect sparse global anomalies and clustered local anomalies.

The time complexity of the filtering stage is close to the IF, while the refinement stage is $O(s^2)$ where s is the number of filtered samples. It is easy to see that the filtering stage is very important to get competitive computational performances.

On Detecting Clustered Anomalies Using SCiForest (SCiForest)

SCiForest [54] takes the assumptions of IF and tries to improve it, with special attention to clustered anomalies. Indeed Isolation Forest performs quite poorly on them. The two most important novelties that this method introduces are: i) the use of oblique hyper-planes, instead of axis-aligned, and ii) the use of a split criterion that replaces the random split. At each partition multiple hyper-planes are generated, but only the one that maximises a certain criterion is selected.

The intuition behind this criterion is that clustered anomalies have their own distribution and the optimal split separates normal and anomalous distributions minimizing the dispersion. Therefore the split criterion is formalized as:

$$\mathrm{Sd}_{\mathrm{gain}} = \frac{\mathrm{std}(X) - \mathrm{average}\big(\mathrm{std}(X^{\mathrm{left}}), \mathrm{std}(X^{\mathrm{right}})\big)}{\mathrm{std}(X)}$$

where std(·) computes the standard deviation.

Due to the new computations, the complexity of the SCiForest increases reaching $O(t\tau\psi(q\psi + \log\psi + \psi))$ in the training stage, and $O(qnt\psi)$ in the evaluation stage, where t is the number of trees in the forest, τ the number of hyper-plane trials and q the number of features composing each hyper-plane dimensions.

Mass Estimation (MassAD)

The authors of Mass estimation (MassAD) started in their papers [83, 84] recalling the classic definition of the mass i.e. the number of points in a region. However their definition of the mass of a point is slightly more complex since they consider all the overlapping regions that cover that point. Doing that, they obtain a family of functions that accentuates the fringe points in a data cloud. Despite its usefulness, this sort of anomaly score is too computationally expensive. To overcome this issue they propose an algorithm that approximates it, employing Half Space trees. These can be thought as a simplification of isolation trees, indeed only the splitting feature is taken at random, while the splitting value is half of the range along that feature. They propose two variants of the same algorithm: one grows leaves of the same depth, while the other lets them to have different depths. The latter not only estimates the score using the leaf mass, but also improves it with a factor dependent on the tree depth.

The authors report a time complexity $O(t(\psi + n)h)$ that includes both training and testing. The space complexity is $O(t\psi h)$.

Ordinal Isolation: An Efficient and Effective Intelligent Outlier Detection Algorithm (kpList)

The method proposed in [15] eliminates the randomness of isolation forest, partitioning the space by means of successive uniform grids and at each depth the grid doubles its resolution.

The time complexity is $O(n \log n)$.

Improving iforest with Relative Mass (ReMass IF)

ReMass IF [6] starts from quite similar premises to the SCiForest's, but focuses on the poor performances of IF on local anomalies. Unlike SCiForest, it does not suggest to modify the training algorithms but the anomaly score: it proposes to substitute the tree depth with a new function, the *relative mass*.

The mass of a leaf $m(\cdot)$ is defined as the number of data points inside the leaf while the relative mass of the leaf is the ratio between the mass of the parent and the mass of the leaf. More precisely, the anomaly score for each tree is defined in this way:

$$s_i(x) = \frac{1}{\psi}\frac{m(X_{\mathrm{parent}})}{m(X_{\mathrm{leaf}})}$$

Note the authors suggest to modify only the anomaly score formula, keeping the rest of the algorithm untouched. This helps improving the anomaly score, while preserving the low computational complexity.

The time and space complexities are the same as IF.

Extended Isolation Forest (EIF)
In the paper [32] the authors observe the masking effect created in the IF algorithm by the axis-aligned partitions. The intersection of multiple masks can even create some fake normal areas of the domain, leading to completely wrong anomaly detection. In order to overcome the described issue, the authors suggest a very simple but effective strategy already employed in SCiForest: the use of oblique partitions. However in this case the authors do not use a repeated split criterion, loosing its benefits but also the additional computational overload.

The time complexity is similar to the SCiForest, except for the saving of the τ repetitions.

Identifying Mass-Based Local Anomalies Using Binary Space Partitioning
The approach presented in [26] is very similar to [84] and mainly differs in the way the anomaly score is computed: it does not rely only on the mass and depth of the leaf, but is weighted by the deviation between the selected split point and the corresponding feature value of the tested point.

The authors report time and space complexities similar to MassAD, i.e. a time complexity $O(th(\log n + \psi))$.

Functional Isolation Forest (Functional IF)
IF naturally born for static data but can also be employed for functional data. In this case anomalies can be subdivided into shift, amplitude and shape anomalies. These can be transient or persistent depending on their duration. In [76] the authors formalise this approach adapting the IF algorithm to the new setting. The set of features are not given with the dataset, but are extracted projecting the functional sample over a dictionary chosen by the user. This choice is arbitrary and highly affects the resulting performances. The projection is not performed using a classical inner product because it does not account for shape anomalies; on the contrary the authors suggest to employ the Sobolev scalar product.

One Class Splitting Criteria for Random Forests (OneClassRF)
The Random Forest algorithm naturally applies to the supervised setting, however some attempts to adapt it to the unsupervised one has been made. As previously discussed in the former section, the most intuitive solution is to artificially sample the domain in order to create outlier data [17], but this has been shown to be inefficient. Another approach described in [27] extends the split criterion based on the Gini index to the unsupervised scenario. Intuitively, the criterion tries to generate two children: the first that isolates the minimum number of samples in the maximum volume, while the second the contrary. In practice, the authors suggest two strategies to adapt the Gini index in absence of a second class: one considers the outliers uniformly distributed, while the other at each split estimates the outliers as a fixed fraction of inliers.

A Novel Isolation-Based Outlier Detection Method (EGiTree)
Sciforest is not the only one that tries to improve the algorithm using split criteria: EgiTree [75] (a very similar approach was presented in [50]) employs heuristics based on entropy to effectively select both attribute and split value. Indeed, the goal is to take the randomness out of the algorithm. The authors start observing that in practice, looking at the features individually, two kind of anomalies exist: anomalies that are outside the normal range, and other that are inside the normal feature range but have abnormal feature combinations. In the first case anomalous data are easier to be detected since a gap is easily identifiable, and the disorder is lower. In the second, it is difficult to find a gap simply looking at the projections of data over the axes. From these observations the authors defined two heuristics. If the feature that exhibits lower entropy has an entropy value i) less than a threshold, it is partitioned along the biggest gap between the feature data ii) greater that this threshold, it is partitioned along the mean feature value, creating a balanced partition. At each splitting iteration a partition cost is computed. When the first heuristic is used, the partition cost is roughly inversely proportional to the gap, and takes the form:

$$\mathrm{cost}(X) = 1 - \frac{\mathrm{maxgap}(X_{\mathrm{feature}})}{\max(X_{\mathrm{feature}}) - \min(X_{\mathrm{feature}})}$$

On the contrary, when the second heuristic is employed, the partition cost is maximal and equal to 1. The total partition cost of a data point is the sum on the partition costs of each node traversed by the datapoint, and the anomaly score related to a single tree is the inverse of the total partition costs.

LSHiForest: A Generic Framework for Fast Tree Isolation Based Ensemble Anomaly Analysis (LSHiForest)
The algorithm presented in [98] combines the isolation tree approach with the Locality Sensitive Hashing (LSH) forest, where given a certain distance function d, neighbours samples produce the same hash with high probability while samples far from each other produce the same hash with low probability. The probabilities can be tuned by concatenating different hash functions, so that an isolation tree can be constructed by concatenating a new function at each internal node. The path from the root to a leaf node is the combined key of the corresponding data instance. Since d is generic, this extension allows to incorporate any similarity measure in any data space. Moreover, the authors show that their framework easily accommodates IF and SCiForest when particular hash functions are selected. They adapted the method in this way: i) the sampling size is not fixed but variable, ii) the trees are built using the LSH functions, iii) the height limit and the normalisation factor are changed consequently and iv) the individual scores are combined after the exponential rescaling. The average-case time complexity in the training stage is $\Theta(\psi \log \psi)$, while in the evaluation phase it is $\Theta(\log \psi)$.

Hybrid Isolation Forest (HIF and HEIF)
The authors of [64] observe that IF behaves differently if the dataset has a convex or concave shape. For example they analyze the detection performances on a dataset

composed of a toroidal normal cluster and some scattered anomalies that lie inside and just outside the torus. It turns out that IF struggles to detect inner anomalies, giving them a score too close to the normality. To overcome this issue the authors propose two approaches, one of which is unsupervised: at each leaf node the centroid of leaf training data is computed and recorded, then in the testing phase the distance between the point and the corresponding centroid is measured. This new score is linearly combined with the traditional leaf depth, obtaining a more robust score. Unfortunately this approach employs euclidean distance that is not scale invariant and therefore requires some unpractical normalizations. In [37] the approach described in the previous paper is enhanced by the using of the Extended Isolation Forest [33], obtaining a better detector.

The authors claim the time complexity of this algorithm is slightly higher than IF due to the additional computations, but anyway comparable. To verify their hypotheses they perform some simple simulations.

A Novel Anomaly Detection Algorithm Based on Trident Tree (T-Forest)
The method [97] is based on a very simple tree structure: the trident tree. As the name suggests, this structure is not a binary tree but it generates three children at a time. Like IF, a random feature is selected and a split criterion is applied. The split criterion is simple: data that are three std to the left of the mean are sent to the left child, the contrary for the right child, and the part in the middle of the distribution is assigned to the central child. The anomaly score is then computed in a similar way to ReMass IF, i.e. using the mass instead of the leaf depth.

The authors report a time complexity of $O(t\psi \log n)$ and space complexity of $O(t\psi n)$ for the training phase, while the time complexity of the evaluation phase is $O(mt \log (n\psi))$.

Hyperspectral Anomaly Detection with Kernel Isolation Forest
In the context of computer vision, a small modification to the original algorithm has been shown in [49]. Here the goal is to find anomalous pixels inside a hyper spectral image. A kernel is employed in order to extract non linear features to be used in the IF. Then the principal components are selected, a global method is trained on the whole image and the most anomalous pixels are detected. The connected components of anomalous pixels are subjected to a local procedure, where a new model is trained and tested on the pixels. This method is applied recursively until a sufficiently small anomalous area is detected.

The authors calculate a time complexity of $O(t\psi(\psi + n_{\text{pixels}}))$.

A Novel Anomaly Score for Isolation Forests
The classic mean leaf depth as proxy of the data anomaly is questioned in [66]. According to the authors, the information encoded in the structure of the original isolation tree, is not fully exploited by its anomaly score. After this premise, they suggest three different alternatives that do not change the learning algorithms but only how is computed the anomaly score. Instead of the standard path length that adds a unit at each traversed node, they propose a weighted path. They suggest multiple strategies to obtain these weights. The first relies on the concept of neighborhood:

more isolated points will have smaller neighbors, therefore at each node the weight will be the inverse of data passing through it. The second strategy starts from 3.1 where a split criterion based on Gini impurity was employed. In this context the authors suggest to weight the node using the inverse of the split criterion value, since it measures how well the split was performed. The last strategy simply takes the product between the former two.

Distribution Forest: An Anomaly Detection Method Based on Isolation Forest (dForest)
The variant proposed in [95] does not rely on axis parallel or oblique partitions but on elliptic ones. In this case, at each node multiple random features are selected and the covarance is computed. Then data are divided using the Mahalanobis distance: points that lye inside the hyper ellipsoid are sent to the left child, while the ones outside of the elliptic boundary are sent to the right leaf. The split value is chosen such that a fixed portion of data are outside the ellipse.

The time complexity differs from the IF one in the last step of covariance computing. As the subset of selected features increases, this diversity becomes more marked.

Research and Improvement of Isolation Forest in Detection of Local Anomaly Points (CBIF)
Multiple approaches have been proposed to overcome the limitations of IF in detecting local outliers. In [25] the authors suggest the combination of clustering based algorithms and the original IF. Despite its efficacy, the choice of using a more expensive model to enhance a cheaper one, seems counter intuitive.

K-Means-Based Isolation Forest (k-means IF and n-ary IF)
In the papers [42, 43] the authors investigate the impact of the branching process in the original isolation forest algorithm. More precisely they are interested in how the algorithm behaves changing the number of children each node has to grow. They try to improve the original algorithm by means of K-means clustering over the selected splitting feature and using a score that measures the degree of membership of the point to the each traversed node.

OPHiForest: Order Preserving Hashing Based Isolation Forest for Robust and Scalable Anomaly Detection (OPHiForest)
The work shown in [93] improves on the core ideas of the LSHiForest by proposing a learning to hash (LTH) method to select the hashing function which best preserve similarities in the dataset in the projected space. The order preserving hashing algorithm (OPH) is chosen for such task as it shows excellent performances in nearest neighbour search. This algorithm is able to find the hash function which minimizes the order alignment errors between original and projected data samples. Moreover, an improved two-phases learning process for OPH is presented to enable faster computation. An isolation forest is built based on the hashing scheme, where the specific hash function to use at each node is not random, but it is fine-learnt by OPH. Finally, the evaluation phase is similar to the LSHiForest.

While this method hash higher training time complexity ($\Theta(Hb_r\psi^2 ta \log\psi)$) with respect to LSHiForest due to the learning process, it shows similar performance in the evaluation phase ($\Theta(tna \log\psi)$).

PIDForest: Anomaly Detection via Partial Identification (PIDForest)
Classical IF relies on the concept of isolation susceptibility, which intuitively can be outlined as the average number of random slices that are needed to fully isolate the target data. This definition of anomaly has some great advantages, but also some pitfalls. In particular, in high dimensional data, many attributes are likely to be irrelevant and isolation may be sometimes very demanding.

PIDForest [29] is based on an alternative definition of anomaly. The authors assert that an anomalous instance requires less descriptive information to be uniquely determined from the other data. Then, they define their partial identification score (PIDScore) in a continuous setting as a function of the maximum sparsity over all the possible cubical subregions containing the evaluated data point x. Say X full data, and C a subcube of the product space and ρ a sparsity measure, PIDScore can be formalized as

$$\text{PIDScore} = \max_{C \ni x} \rho(X, C) = \max_{C \ni x} \frac{\text{vol}(C)}{|C \cap X|}$$

PIDForest builds a heuristic that approximates the PIDscore. The strategy is to recursively choose an attribute to be splitted in k intervals, similarly to k-ary variants of IF (authors suggest default hyperparameter $k < 5$). Intuitively, we would like to partition the space into some sparse and some dense regions. For this purpose, a possible objective is to maximize the variance over the partitions in terms of sparsity, that can be treated as a well-studied computational problem related to histograms and admits efficient algorithms for its solution. For each attribute the optimal splits are computed and the best attribute is chosen as coordinate for partitioning. Then, the iteration is repeated on each partition, until a data point is fully isolated or a maximum depth parameter is exceeded. Now the resulting leaf is labelled with the sparsity of the related subregion. In the testing phase, a data point can be evaluated on each tree of the PIDForest and the maximum score (or a robust analog, like 75% percentile) gives an estimate of the PIDScore.

In the words of its authors, the fundamental difference between IF and PIDForest is that the latter zooms in on coordinates with higher signal, being less susceptible to irrelevant attributes at the cost of more expensive computation time. Each PIDTree takes $O(k^h d\psi \log\psi)$ as training time, while testing is pretty much equivalent to IF.

An Optimized Computational Framework for Isolation Forest
As many other methods, also [57] tries to improve the stability and accuracy of IF, designing new split criteria. It starts observing that the separability of two distributions, the anomalous and normal one, is proportional to two factors: the distance between peaks and the dispersion of the distributions. The authors developed a simple index, named *separability index* roughly similar to the one described in [54] but considering also the distance between distributions a feature at a time. Another difference relies in the choice of the best splitting value: instead of trying multiple

random values and taking the best out of them, an optimization procedure based on the gradient of separability function is chosen.

The time complexity is $O(kn(\log n)^2)$.

Anomaly Detection Forest (ADF)
In [77] the authors observe that IF is specifically suited for the unsupervised setting previously discussed in Sect. 2, where the training set is composed both of normal and anomalous samples. However, in case of normal-only training data, this model does not create leaves representing the anomalous feature space, and tends to give high scores to inliers. To overcome this issue the authors introduce a new structure based on two new concepts: i) the *anomaly leaves* that should model the feature values not contained in the range of training samples, and ii) the *isolation level* that is the node size below which the anomaly leaves are created. The authors also define a special kind of internal node, named *anomaly catcher*: when the node size is less than the *isolation level* threshold, it generates a generic child and an anomaly leaf. Two kinds of split criteria are used: one for the partition of generic internal nodes, and one for the partition of anomaly leaves. The first is quite similar to the uniformly random criteria of IF, with the difference it tries to guarantee a less unbalanced split. The second generates the empty (anomaly) leaf by splitting the feature between the extreme value of the dataset and the extreme value of the node space. These modifications require a small adjustment on the anomaly score since in this new settings the original normalising factor does not make sense anymore. As a consequence the *observed* average path length has been preferred.

The time complexity in the testing phase is similar to IF, but the one in the training phase is higher due to additional sortings done in the split value computation.

usfAD: A Robust Anomaly Detector Based on Unsupervised Stochastic Forest (usfAD)
The work presented in [5] addresses the issue of different units/scales in data, starting by showing some examples where different non-linear scales lead to completely different anomalies. To solve this issue the authors propose *usfAD*, a method that combines Unsupervised Stochastic Forest (USF) with IF, and naturally born for the semi-supervised task. This hybrid model recursively splits the subsample until all the samples are isolated. However it is different from the IF since it grows balanced trees with leaf of the same depth. This is accomplished using a splitting rule that uses the median value as split point. The core idea is that the median, since relies on ordering, is more robust to changes in scale or units. After the tree growth, normal and anomalous regions are associated to each node: the former consists in the hyper-rectangle containing the training points, while the latter is the complementary region. All these modifications lead to a quite different testing phase. The anomaly score of a test point is the depth of the first node where it falls outside the normal region.

The time complexity is slightly higher than IF: the training is $O(nth + t2^h d)$ while the testing $O(t(h + d))$. Moreover, it needs $O(t2^h d)$ memory space.

Fuzzy Set-Based Isolation Forest (Fuzzy IF)
Attempts to improve the IF algorithm have been made also by using Fuzzy Sets approaches [41]. The anomaly score is simply measured by the so called *degree of*

membership, i.e. at each node a function of the distance between the point and the centroid is incrementally added.

Integrated Learning Method for Anomaly Detection Combining KLSH and Isolation Principles
In the paper [70] a method based on Kernelized Locality-Sensitive Hashing (KLSH) combined with IF is proposed, with the aim of improving the detection of local anomalies. A gaussian kernel function is used to map features to a higher dimensional feature space to map local anomalies in the original space into global anomalies, which are easy to isolate and detect. IF is then used to isolate anomalies in the kernelized dataset. Furtheremore, two improvements on IF are proposed: a random non-repeating subsampling technique and a mean optimization strategy to optimally select the segmentation attributes and values.

RMHSForest: Relative Mass and Half-Space Tree Based Forest for Anomaly Detection (RMHSForest)
The algorithm presented in [59] tries to combine the advantages of the Half-Space tree described also in [83] and the anomaly score proposed in the ReMass-IF [6] algorithm. In this context the authors employ Half-Spaces for the tree construction and modify the ReMass score function adding the depth and taking a logarithmic function of the relative mass.

The time and space complexity are respectively $O(t(\psi + n)h)$ and $O(t\psi h)$.

Anomaly Detection by Using Random Projection Forest (RPF)
Many works in anomaly detection with tree-based methods still refer to Random Forests. One of them is [14] where a revised splitting rule based on the Kullback-Leibler divergence and oblique projections instead of axis aligned are used to model the dataset density distribution.

Randomized Outlier Detection with Trees (GIF)
The method proposed in [11] focuses on two aspects. Firstly it proposes a theoretical framework that interprets the isolation forest variants from a distributional point of view. More precisely, it interprets isolation as a density estimation heuristic in which the algorithm reckons the weights of a mixture distribution, where the dominant component characterizes normal data, while the minor ones can be considered as anomalous. The authors conclude that any tree-based algorithm with sufficiently many fine-grained splits can guarantee some approximation quality of the underlying probability distribution.

Afterwards, starting from these premises a new method is developed, named Generalized Isolation Forest. The proposed algorithm makes use of non-binary partitions and the data are divided based on the maximization of a custom inner kernel function, in order to produce regions that are small and dense enough, as the theoretical dissertation suggests. As in the original IF, the tree is not required to be fully grown, but the partitioning process stops when a sufficient level of the distribution approximation is reached. Then, a density function, like frequency of observations, is used in testing phase, instead of path length.

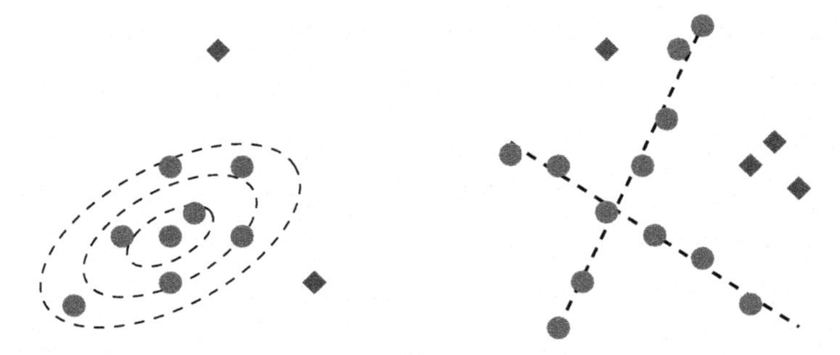

Fig. 3 Statistical vs pattern anomalies. The latter are very complex since they are defined as points not belonging to a specific but unknown pattern. Figure adapted from [47]

Despite its name, GIF turns to be pretty much different than original IF. Nevertheless, it promises interesting performances and a solid foundation, the downside resides on the need of fine-tuning of many hyperparameters and an arguably expensive computational time for big datasets.

PIF: Anomaly Detection via Preference Embedding (PIF)
As discussed above, anomaly definition varies across the papers: in this approach [47] the authors point out the difference between general statistical anomalies and pattern anomalies (Fig. 3). The former are typically defined as samples falling in regions where the density is low, while the latter are samples that deviate from some structured pattern. Finding and fitting these structures to find the anomalies is extremely expensive, therefore the authors suggest a new method, named Preference IF, to directly tackle the problem.

The proposed method consists mainly in two steps: i) the embedding of data in a new space, named preference space and ii) the adoption of tree based isolation approach specifically suited for this new space. In particular, the adoption of a nested Voronoi tessellation and the Tanimoto distance allows much better performances than simply using the original IF in the preference space.

The complexity of this algorithm is $O(\psi t b \log \psi)$ in the training phase, and similarly $O(n t b \log \psi)$ in the testing one, where b is the branching factor of the PI-tree.

An Explainable Outlier Detection Method Using Region-Partition Trees
Another tree-based approach employed in anomaly detection is described in [67], where Region-Partition trees are employed. It is quite different from the IF. A tree with maximum height h has the same amount of randomly selected features used at each depth level to split data. At each level a feature is selected and divided into k intervals. This tabular structure is used to build the actual tree, starting from the empty root intervals and adding recursively a new a data point, creating a new child when needed. This training procedure is performed only with normal data, therefore all nodes correspond only to regions where the distribution is expected to be normal. The detection of anomalies is quite simple since if a new sample arrives to a leaf

node it is marked normal, if instead it gets stuck in a node it is labelled as outlier. In order to get an intuition about the cause that generated that anomaly, the authors suggest to count the number of times a feature is responsible to the internal node stop. To get the anomalous range, the intersection between the anomalous feature intervals for each tree can be easily performed.

3.2 Dynamic Datasets

Dynamic datasets are made up of infinite data streams [81]. This poses new challenges that previously described methods cannot tackle. The most simple challenge is the continuous training: since the stream is infinite, the training data may be insufficient to fully describe it. In order to do that, the model should continuously learn from the incoming data stream, and at the same moment detect anomalous points. The most complex setting is represented by the distributional drift. In this case, the stream is not stationary and its distribution experiences time-dependent variations. Here, the model must adapt to the evolving data stream, but at the same time discern anomalies from new normal data points.

The weak point of these methods is the assumption about the rarity of anomalies. If they are too numerous, they can be confused with a change in the normal distribution of data points, leading to an erroneous adaptation of the model. Doing this the model will consider them as normal and it will not raise the necessary alarm.

An Anomaly Detection Approach Based on Isolation Forest Algorithm for Streaming Data Using Sliding Window (iForestASD)
The method proposed in [20] is an adaptation of the original algorithm to the streaming settings. It is very simple: it splits the stream in windows, and checks each window to detect anomalies. If the ratio between normal data and anomalies is too high (exceeds the expected anomaly ratio), it assumes a concept drift is happening. In this case the model is re-trained on the new window. Obviously, the threshold on the anomaly ratio and the width of the window are delicate hyper-parameters that highly depend on the specific application.

Fast Anomaly Detection for Streaming Data (Streaming HS)
In the paper [81] an adaptation of [84] to data streams is presented. Unlike other tree based approaches, it is not built starting from training data but its structure is induced only by the feature space dimension. It doesn't need split point evaluation, and therefore it is fast and able to continuously learn from new data. Contrary to the basic Isolation Forest, this method employs the concept of mass to determine the anomaly score. In practice, this method works segmenting the stream in windows, and working with two of them: *reference* and the *latest* windows (despite their name, they are immediately consecutive). The reference window is used to record the mass profile, while the latest one is used for testing. When this is done, the latest becomes the reference and a new mass profile is recorded.

The authors show a time complexity of $O(t(h + \psi))$ in the worst case.

RS-Forest: A Rapid Density Estimator for Streaming Anomaly Detection (RS-Forest)
The method published in [91] relies on a type of tree quite similar to the isolation one, that is let growing until a maximum depth is reached independently on the data. Before the tree construction, the feature space is enlarged in order to cope with possible feature drifts. The score is based on the density of the points in the leaf volume and the model update relies on dual node profiles similarly to the previous method.

The time complexity of the method is $O(\frac{n}{\psi}m2^{(h+1)})$ while space complexity is constant since is $O(2^h)$.

An Online Anomaly Detection Method for Stream Data Using Isolation Principle and Statistic Histogram (AHIForest)
Isolation Forest is a light-weighted method only designed for batch data, furthermore it suffers of slow convergence having no knowledge on the distribution of data.

In the paper [21] both these limitations are inspected, and a new method called AHIForest is proposed. This is identical to Isolation Forest, except for the selection step of the splitting point. Indeed, after a dimension is randomly picked, the choice of the threshold is not drawn from a uniform distribution, but it is based on the histogram that approximates the distribution of data projected on the given axis. The idea of histogram-based has the advantage to fasten the convergence, at the price of adding a new critical parameter, the bin size, to be carefully chosen.

On the other hand, AHIForest deals with streaming data using a sliding window strategy. Firstly, a forest is built by data sampled from the first window, and new observations are judged in real time. Then, if anomaly rate exceeds a pre-defined threshold (as in iForestASD) or the buffer is full, the forest is updated, growing new trees from the last window and pruning old estimators.

The time complexity of this algorithm is $O(MN)$, where N is the number of individual detectors and M is the maximum number of leaves of each tree. Space complexity is $O(MN)$, too.

Robust Random Cut Forest Based Anomaly Detection on Streams (RRCF)
RRCF [30] presents exactly the same structure and anomaly scoring of isolation forest, except for the mechanism how the splitting dimension is chosen. The intuition of the authors was to pick what they called "robust random cuts" proportionally to the span of data in each dimension, instead of uniformly at random, as in the original version. In this manner, the method loses the property of scaling invariance of the Isolation Forest, but achieves some sense of self-consistency after addition or deletion of data, i.e. any tree preserves the same distributional properties independently if constructed given the whole dataset in a batch fashion and if dynamically grown from a stream of data.

Naturally, RRCF is a straightforward solution to propose the same key-ideas of Isolation Forest in a suitable way for online anomaly detection, without the need of rebuilding the model from scratch.

When a new data instance arrives, it runs across the tree, starting from the root. At each node, a candidate cut is randomly proposed in the subregion represented by the node, following the same mechanism as described above. If the new point is fully isolated by the candidate cut, this is kept and a new leaf is there inserted, otherwise the cut is discharged and the data instance moves the next node through the branch. Instance by instance, new branches grow up, making evolve the shape of trees in RRCF with respect to concept drift phenomena of data stream.

Fast Anomaly Detection in Multiple Multi-dimensional Data Streams (Streaming LSHiForest)
The model presented in [79] extends the LSHiForest algorithm for streaming data exploiting a dynamic isolation forest. The procedure can be split into three main phases: i) a dataset of historical data points is used to build a LSHiForest data structure, as presented in the original paper, then ii) the data points collected from multiple data streams are preprocessed to find "suspicious" samples, which are outlier candidates. Finally, iii) the suspicious data are updated into the LSHiForest structure and the anomaly scores of the updated data points are recalculated. To effectively extract suspicious points from the streaming data, Principal Component Analysis (PCA) and the weighted Page-Hinckley Test (PHT) are applied to a sliding window, to cope with the challenges of high dimensionality and concept drift. An update mechanism is proposed to iteratively update the LSHiForest by replacing the previous data points observed on a stream with suspicious ones.

Isolation Mondrian Forest for Batch and Online Anomaly Detection (Isolation Mondrian Forest)
The Mondrian Forest [60] represents a family of random hierarchical binary partitions, based on the Mondrian Process, that tessellates the domain in a tree-like data structure. This random process recursively generates a series of axis-aligned slices that recall the abstract grid-based paintings by Piet Mondrian (1872–1944). Each slice is associated with a split time and the partitioning process can be eventually stopped after a given budget time. In the past few years, the interest upon Mondrian Forest raised up in machine learning, both for regression and classification purposes. Only recently, an application of Mondrian Forest has been proposed in anomaly detection, that exploits the similarities with the data structure from Isolation Forest, and uses the same depth-based anomaly score of the latter.

The advantage of Mondrian Forest lies on its nice self-consistency property, in particular a Mondrian Tree can be infinitely extended performing new partitions on its sub-domains, preserving the same distributional properties. Therefore a Mondrian Forest can be dynamically updated by the arrival of new data and this makes the algorithm particularly suitable for online/streaming applications.

The update mechanism is very similar to the one described for Robust Random Cut Forest. Again, each novel data point travels through the trees in the forest, and a candidate split is picked at each node. In this case, the candidate is maintained if its split time is lower than the one of the node; this can be interpreted as a split that is occurred before; otherwise it is discarded and the data instance moves to the next node until it reaches a leaf, which is associated with an infinite time.

The training and evaluation time complexities for online processing of m new points are $O(td(n+m)\log(n+m))$ and $O(t(n+m))$ respectively.

Interpretable Anomaly Detection with Mondrian Polya Forests on Data Streams (Mondrian Polya Forest)

Mondrian Polya Forest (MPF) [18] is another example of a method based on the Mondrian Process, that seems one of the most promising research paths in tree-based approaches for anomaly detection.

As the previously described Isolation Mondrian Forest, Mondrian Polya Forest grows a tree partition based on Mondrian Process. The difference lies on the evaluation procedure, that estimates the density function of data instead of inferring their isolation scores. As the name suggests, MPF makes use of Polya Trees for modeling the distribution of the mass in the nested binary partition constructed by each Mondrian Tree, each cut is then associated with a beta-distributed random variable, which reflects the probability of a data point to lie in one of the sub-partitions in the hierarchical structure of the tree. In this setting, an anomaly is identified by the fact it occurs in a region with lower density than normal data.

Alike Isolation Mondrian Forest, this method takes advantages of the properties of the Mondrian Forest, then it is able to efficiently update by nesting of new slices, instead of rebuilding the tree structure from scratch, when new data instances are available. A streaming version of the method has been proposed by the same authors of MPF with interesting results.

Anomalies Detection Using Isolation in Concept-Drifting Data Streams

In the paper [85] the authors review several isolation-based techniques in streaming anomaly detection, in particular iForestASD and Half-Space Trees, both previously introduced. The main differences are remarked, like the strong dependency by data of the first, versus the lack of knowledge in the building phase of the latter, and the different approaches for handling drift.

Moreover, a couple of new strategies are proposed, based on iForestASD. ADWIN (ADaptive WINdowing) is a well-known solution, that maintains a variable-length window of data, which increases with incoming observations. The algorithm compares any subset of the window until it detects a significant difference between data, then the new information is kept while the old one is erased. Slight variants are PADWIN (Prediction based ADWIN) and SADWIN (Scores based ADWIN), which take predictions and scores as input of ADWIN for detecting drift, respectively. KSWIN (Kolmogorov-Simirnov WINdowing) is a more innovative approach, based on Kolmogorov–Simirnov statistic test. KS is a non-parametric test, originally suitable for one-dimensional data only. The authors propose to overcome this restriction by declaring the occurrence of drift if it is detected in at least one dimension.

Empirical experiments show the inefficiency of vanilla iForestASD in real-world scenarios and the need of explicit concept drift detection methods, such as the proposed ones.

3.3 Distributed Approaches

Wireless Senor Networks (WSNs) pose new and more challenging constraints to Anomaly detection. Indeed sensor nodes are usually quite cheap but have multiple constrains on energy consumption, communication bandwidth, memory and computational resources. Moreover they are often deployed in harsh environments that can corrupt sensor measurements and communication [19]. Despite the distributed nature of the network, Anomaly Detection on such applications should minimize the communication burden as much as possible, since data transmission is the most energy intensive process.

Distributed Isolation for WSN
The authors of [19] suggest the adaptation of IF to this distributed problem, considering the spatial correlation between neighbor sensor nodes in a local and global manner. They chose this base algorithm due to its already mentioned properties, that fits perfectly in this settings. However in the WSN context, data can be anomalous w.r.t. the single sensor node or w.r.t. the whole network. The local detector consists in a collection of isolation trees trained on a group of neighbouring nodes while the global one is made up of local detectors. When an anomaly is locally detected, it is marked as an error if is not detected by neighbor sensors, otherwise it is considered an event.

The space and time complexity is $O(km)$, where k is the number of trees on a local node, and m the number of leaves.

Robust Distributed Anomaly Detection Using Optimal Weighted One-Class Random Forests (OW-OCRF)
A similar approach has been exploited in [87], where a one-class random forest has been chosen as base detector. Here each sensor node builds his own model, but it is also augmented with the models belonging to neighbouring devices. In addition, a strategy to weight the most effective neighbor models has been implemented, based on the minimization of the model uncertainty. Uniform voting is reasonable in circumstances where all the learners arise from the same distribution, but when models come from heterogeneous data distributions this strategy shows its weaknesses. Larger weights are assigned to trees that are in accordance with the majority, while trees that increase the overall uncertainty are penalized. The optimization of these weights is performed in a fully unsupervised fashion. In presence of distributional drifts, the overall model can be easily adapted to the new conditions, optimising new weights or substituting the trees with lower weight importance. The communication between the node is employed just at early stages for the sharing of the detecting models, not for the sampled data sharing.

The time and space complexity of this approach are $O(th)$ and $O(t2^{h+1})$ respectively.

3.4 Interpretability and Feature Selection

The detection of anomalies is an important activity in manufacturing processes but it is useless if a corresponding action does not take place. That action is expected to be proportional to the gravity of the anomaly (encoded by the anomaly score), and to the *cause* that generated it. For doing that a tool to interpret that anomaly is needed. It is easy to understand that if unsupervised anomaly detection is challenging, interpretable models face even more complex issues. In real word scenarios anomalies are unlabelled and lack of proper interpretations.

Moreover [12] observes that interpretable models enhance the trust of the user in the anomaly detection algorithms, leading to a more systematic use of these tools.

Interpretable Anomaly Detection with DIFFI: Depth-based Feature Importance for the Isolation Forest (DIFFI)

IF is a highly randomised algorithm and therefore the logic behind the model predictions is very hard to grasp. In the paper [12], a model specifically designed for the interpretability of IF outcomes is presented. In particular the authors developed two variants: i) a global interpretability method able to describe the general behaviour of the IF on the training set, and ii) a local version able to explain the individual IF predictions made on test points. The central idea of this method, named DIFFI, relies on the following two intuitions: the split of an important feature should a) induce faster isolation at low levels (close to the root) and b) produce higher unbalance w.r.t splits of less important features. This is encoded in a new index named *cumulative feature importance*. With this in mind, the authors formulate the global feature importance as the weighted ratio between the cumulative feature importance computed for outliers and inliers. The local interpretation of single detected anomalies is sightly different but relies on the same intuitions. DIFFI can also be exploited for unsupervised feature selection in the context of anomaly detection.

Anomaly Explanation with Random Forests

Authors in [46] developed an algorithm able to explain the outcome of a generic anomaly detector by using sets of human understandable rules. More specifically, the proposed model consists in a special random forest trained to separate the single anomaly from the rest of the dataset. This algorithm provides two kinds of explanations: the *minimal* and the *maximal*. The first is performed isolating the anomaly using the minimal number of necessary features. On the contrary the maximal explanation looks for all the features in which the anomaly is different, employing a recursive feature reduction. Once the forest are trained and the explanations are obtained, the decision rules are extracted in a human readable manner.

The time complexity of the algorithm is $O(n_t T_{\text{sel}} T_{\text{train}})$ where T_{sel} is linear with the number of normal samples in the data. For the minimal explanation n_t is the number of trees trained for each anomaly and $T_{\text{train}} = O(d|T|^2)$, where d is the number of features and T is the size of the training set. For the maximal explanation $n_t = O(d-1)$ while $T_{\text{train}} = O(d^2|T|^2)$.

Isolation-Based Feature Selection for Unsupervised Outlier Detection (IBFS) In settings where the dimensionality of data is very high, even the most efficient anomaly detection algorithm may suffer. Methods of feature selection are used to the purpose of reducing the computational and memory cost. Isolation-based feature selection (IBFS) described in [94] computes an unbalanced score each time a node is split, based on the resulting entropy weighted by fraction of data samples in each leaf. Adding all the scores of the traversed nodes, it is possible to obtain a global features score that highlights the best features for anomaly detection.

4 Experimental Comparison

4.1 Methods Comparison and Available Implementations

Unfortunately a small subset of authors provided an open implementation of the methods presented in the previous Section: for this reason it is hard to have a comprehensive overview of performances for the overall plethora of isolation-based and tree-based methods. Most of the authors report some performance scores (commonly ROC AUC) for their proposed methods, using as benchmarks only the original IF and few of the most popular close variants, such as Split-Criteria IF, Robust Random Cut Forest or Extended IF and other density or distance-based anomaly detection approaches.

To the best of our knowledge, an extended comparison between all the variants of tree-based AD has never been realised. To cope with this issue, we have worked to collect results available in literature on various AD benchmarks, in order to provide an easier comparison between the different approaches also in terms of accuracy.

We selected a subset of the datasets where we could have a consistent amount of outcomes, that turn out to be all from UCI Machine Learning Repository [23]. For this reason, we excluded all the methods that are intended to work on a specific scope, for instance image detection or functional-based anomaly detection. A schematic description of datasets is in Table 1.

Moreover, we limited to the static approach, since testing for streaming algorithms allows a variety of different setups and it is hard or impossible to achieve a fair comparison with existing results.

Table 2 contains the result from our survey. In all the cases it was possible, we used the ROC AUC scores from the original papers, thus we suppose each method is tuned at the best of author's expertise. For many of the algorithms that were publicly available, we filled the eventually missing scores by running the tests ourselves. In this case, we consider those hyperparameters indicated in the original IF paper [53] (number of trees = 100, sample size = 256) as the most appropriate universal setup. Finally, we avoided to fill missing outcomes for such methods, like GIF, where the authors provided a fine-tuning of their algorithms, since our last assumptions would not replicate the same performances.

Table 1 Description of test data sets

Dataset	Size	Dim.	% anomalies
HTTP	567497	3	0.4%
SMTP	95156	3	0.03%
Forest cver	286048	10	0.9%
Shuttle	49097	9	7%
Mammography	11183	6	2%
Satellite	6435	36	32%
Pima	768	8	35%
Breastw	683	9	35%
Arrhytmia	452	274	15%
Ionosphere	351	32	32%

Table 2 ROC AUC score of a selection of methods from literature.

	Code available	HTTP	SMTP	Forest C.	Shuttle	Mammogr.	Satellite	Pima	Breastw	Arrhytmia	Ionosph.
IF	✓	**1.00**	0.88	0.88	**1.00**	0.86	0.71	0.67	**0.99**	0.80	0.85
SCIForest	✓	**1.00**	-	0.74 [98]	**1.00** [98]	0.59 [11]	0.71 [98]	0.65 [98]	0.98 [98]	0.72 [98]	0.91 [93]
EIF	✓	*0.99*	0.85 [37]	0.92	0.99 [37]	0.86	0.78	0.70	***0.99***	0.80 [37]	0.91
RRCF	✓	0.99 [29]	0.89 [29]	*0.91*	0.91 [18]	0.83 [18]	0.68 [18]	0.59 [18]	0.64 [18]	0.74 [18]	0.90 [18]
IMF	✓	***1.00***	0.87	*0.90*	*0.99*	*0.74*	0.74	0.64	*0.97*	*0.80*	0.86
MPF		**1.00**	0.84	0.77	0.51	0.87	0.70	0.66	0.97	0.81	0.88
PIDForest	✓	0.99	**0.92**	0.84	**0.99**	0.84	0.70	0.70 [18]	0.99	-	0.84 [18]
OPHiForest		-	-	-	0.99	-	0.77	0.72	0.96	0.78	0.93
LSHiForest	✓	-	-	0.94	0.97	-	0.77	0.71	0.98	0.78	0.91
HIF	✓	-	0.90	-	**1.00**	**0.88**	0.74	0.70	*0.98*	0.80	*0.86*
HEIF		-	0.90	-	0.99	0.83	0.73	0.72	-	0.80	-
OneClassRF	✓	0.98	**0.92**	0.85	0.95	-	-	0.71	-	0.70	0.90
T-Forest		0.99	-	-	0.99	-	0.68	0.71	-	**0.84**	0.94
EGiTree		-	-	**0.97**	0.94	-	0.73	-	-	-	0.94
GIF	✓	-	-	0.94	-	0.87	**0.86**	**0.84**	-	-	-
dForest		**1.00**	-	-	**1.00**	-	-	0.75	**0.99**	-	**0.97**
ReMass IF		**1.00**	0.88	0.96	**1.00**	0.86	0.71	-	**0.99**	0.80	0.89
HSF	✓	**1.00**	0.90	0.89	**1.00**	0.86	-	0.69	**0.99**	**0.84**	0.80

Selected algorithms are: Isolation Forest (IF) [53], Split-Criteria Isolation Forest (SCIForest) [54], Extended Isolation Forest (EIF) [33], Robust Random Cut Forest (RRCF) [30], Isolation Mondrian Forest (IMF) [60], Mondrian Polya Forest (MPF) [18], Partial Identification Forest (PIDForest) [29], Order Preserving Hashing Based Isolation Forest (OPHIF) [93], Locality Sensitive Hashing Isolation Forest (LSHiForest) [98], Hybrid Isolation Forest (HIF) [64], Hybrid Extended Isolation Forest (HEIF) [37], One-class Random Forest (OneClassRF) [27], Trident Forest (T-Forest) [97], Entropy-based Greedy Isolation Tree (EGiTree) [50], Generalized Isolation Forest (GIF) [11], Distribution Forest (dForest) [95], Re-Mass Isolation Forest (ReMass IF) [6], Half-Spaces Forest (HSF) [84].

Scores are referenced when are provided by a different source than the original paper for the method. If scores were not available in the original paper, we have performed the missing experiments when an implementation of the algorithm was available or by using our own implementation: such cases were reported by using Italic entries; in these circumstances we always performed testing with 100 trees and sample size equals to 256. If scores were not available in the original paper and no implementation of the algorithm were available, the entries were left blank '-'. Finally, we highlighted by bold entries the best performances for each dataset.

Table 3 Source code repositories

Model	Repository	Language
DIFFI [12]	github.com/mattiacarletti/DIFFI	Python
EIF [33]	github.com/sahandha/eif	Python
Functional IF [76]	github.com/GuillaumeStaermanML/FIF	Python
GIF [11]	github.com/philippjh/genif	Python
HIF [71]	github.com/pfmarteau/HIF	Python
IF [53]	scikit-learn.org	Python
iForestASD [20, 85]	github.com/Elmecio/IForestASD_based_methods_in_scikit_Multiflow	Python
Isolation Mondrian Forest [60]	github.com/bghojogh/iMondrian	Python
LSHiForest [98]	github.com/xuyun-zhang/LSHiForest	Python
MassAD [84]	sourceforge.net/projects/mass-estimation	MATLAB
OneClassRF [27]	github.com/ngoix/OCRF	Python
PIDForest [29]	github.com/vatsalsharan/pidforest	Python
RRCF [30]	github.com/kLabUM/rrcf	Python
SCiForest [54]	github.com/david-cortes/isotree	Python

From the collected scores, we have not found a method consistently outperforming all the others, and it's not clear how to build a hierarchy between all the variants. We can conclude that IF is an efficient baseline that shows good performance in many cases, however each method may be the most appropriate in any specific real-life scenario and the final choice it can only be up to the practitioner.

One of the limitations of the proposed comparison is related to the choice of the Area Under Receiver Operating Characteristic Curve (ROC AUC) as main performance indicator used in most of the reviewed papers. In fact, [73] already highlights the inefficiency of ROC AUC if data are strongly unbalanced, and suggests the usage of other metrics, such as Area Under Precision-Recall Curve (PR AUC): ROC curve is drawn by plotting the true positive rate (or recall) against the false positive rate; however, when positive labelled data are rare, ROC AUC can be misleading since even a poor skilled models can achieve high scores. For such reasons, the validity of the reported results for strongly unbalanced datasets like HTTP, SMTP or Forest Cover should be considered with some skepticism.

On the contrary, PR curve represents the precision over the recall for a binary classifier, and it would be more informative when normal instances outnumber anomalies. A slightly different alternative is Precision-Recall-Gain (PRG) curve [24]. Specifically, PRG AUC maintains the pros of the PR AUC, but allows to evaluate the model against a baseline binary classifier, i.e. the *always-positive* classifier, as ROC AUC does with the random classifier model.

In order to promote reproducibility, and to help practitioners in developing realword applications, we provide a list of the available source codes about the previously discussed methods (Table 3). Unfortunately, as stated above, we were able to retrieve just a portion of the reviewed methods but we hope as anomaly detection becomes a more mature field, authors will be more used to share their code for enhancing adoption and comparisons of the proposed approaches.

4.2 Industrial Case Studies

Tree based AD approaches have been extensively employed in industry because of their nice properties. Some of the relevant industrial applications of tree based methods are summarized in Table 4. Despite the existence of multiple tree-based algorithms, the large majority of applications concerns the original Isolation Forest and a big part of them are applications in the power industry. Fraud detection and cybersecurity examples, while being really popular in the literature, were not considered in this list since they are not strictly industrial applications.

In some of the reported cases, authors used the reported anomaly detection method as part of a more complex pipeline that typically involve a feature extraction procedure when dealing with non-tabular data for example: for the sake of simplicity, we didn't report such 'evolutions' of the methods in our classification.

In this review, many approaches have been listed and it might be hard to get a feeling on their actual importance for the research community, also given the fact that many approaches have been only recently submitted. To mitigate such issue, some statistic related to the method citations have been collected in Table 5 and 6. Given that citations are only a proxy of the relevance of an AD method and that it is somehow

Table 4 Industrial applications of tree-based approaches for anomaly detection

Work	Year	Sector/equipment type	Method
Ahmed ed at. [1]	2019	Smart grid	IF
Alsini et al. [2]	2021	Construction industry	IF
Antonini et al. [4]	2018	IoT audio sensors	IF
Barbariol et al. [8]	2020	Multi-phase flow meters	IF
Brito et al. [10]	2021	Rotating machinery	DIFFI
Carletti et al. [13]	2019	Home appliances manufacturing	DIFFI
De Santis et al. [74]	2020	Power plants	EIF
Du et al. [22]	2020	Sensor networks	IF
Hara et al. [31]	2020	Hydroelectric generators	IF
Hofmockel et al. [36]	2018	Vehicle sensors	IF
Li et al. [48]	2021	Machine tools	IF
Lin et al. [51]	2020	Power plants	IF
Luo et al. [58]	2019	Electricity consumption	IF
Kim et al. [45]	2017	Energy & smart grids	IF
Maggipinto et al. [61]	2019	Semiconductor manufacturing	IF
Mao et al. [63]	2018	Power consumption	IF
Puggini et al. [69]	2018	Semiconductor manufacturing	IF
Riazi et al. [72]	2019	Robotic arm	IF
Susto et al. [80]	2017	Semiconductor manufacturing	IF
Tan et al. [82]	2020	Marine gas turbines	IF
Tran et al. [86]	2020	Fashion industry	IF
Wang et al. [88]	2019	Power transformers & Gas-insulated swithchgear	IF
Wetzig et al. [90]	2019	IoT-gateway	Streaming HS
Wu et al. [92]	2018	Energy & Smart grid	IF
Zhang et al. [99]	2019	Cigarette production	IF
Zhong et al. [100]	2019	Gas turbine	IF

Table 5 Static methods. The '*' highlights methods published in 2020 or 2021 for which the reported statistics at the time of the writing of this work (March 2021) is of course not reliable

Paper	Acronym	Citations in 2020	Total citations	Annual rate
Liu et al. (2010) [54]	SCIForest	10	49	4.1
Aryal et al. (2014) [6]	ReMass	7	20	2.5
Li et al. (2020) [49]		*7	*7	*3.5
Ting et al. (2010) [84]	MassAD	6	53	4.4
Zhang et al. (2017) [98]		6	26	5.2
Karczmarek et al. (2020) [43]	K-Means IF	*5	*7	*3.5
Liu et al. (2018) [57]		4	9	2.2
Marteau et al. (2017) [64]	HIF	4	6	1.2
Staerman et al. (2019) [76]	Functional IF	3	5	1.7
Goix et al. (2017) [27]	OneClassRF	2	4	0.8
Yu et al. (2009) [96]		1	26	2.0
Zhang et al. (2018) [97]	T-Forest	1	4	1.0
Chen et al. (2015) [14]	RPF	1	3	0.4
Shen et al. (2016) [75]	EGiTree	1	2	0.3
Hariri et al. (2021) [32]	EIF	*1	*1	*1.0
Mensi et al. (2019) [66]		1	1	0.3
Gopalan et al. (2019) [29]	PIDForest	1	1	0.3
Liao et al. (2019) [50]	E-iForest	0	5	1.7
Chen et al. (2011) [15]	kpList	0	4	0.4
Aryal et al. (2021) [5]	usfAD	*0	*1	*1.0
Buschjager et al. (2020) [11]	GIF	*0	*1	*0.5
Park et al. (2021) [67]		*0	*0	*0.0
Holmer et al. (2019) [37]	HEIF	0	0	0.0
Ghaddar et al. (2019) [26]		0	0	0.0
Yao et al. (2019) [95]	dForest	0	0	0.0
Karczmarek et al. (2020) [42]	n-ary IF	*0	*0	*0.0
Sternby et al. (2020) [77]	ADF	*0	*0	*0.0
Xiang et al. (2020) [93]	OPHIForest	*0	*0	*0.0
Gao et al. (2019) [25]	CBIF	0	0	0.0
Leveni et al. (2021) [47]	PIF	*0	*0	*0.0
Lyu et al. (2020) [59]	RMSHForest	*0	*0	*0.0
Karczmarek et al. (2020) [41]	Fuzzy IF	*0	*0	*0.0
Qu et al. (2020) [70]		*0	*0	*0.0

MassAD gathers the citations from Ting et al. (2010) [84] and Ting et al. (2013) [83].

Table 6 Dynamic methods. The '*' highlights methods published in 2020 or 2021 for which the reported statistics at the time of the writing of this work (March 2021) is of course not reliable

Paper	Acronym	Citations in 2020	Total citations	Annual rate
Ding et al. (2013) [20]	iForestASD	25	69	7.7
Tan et al. (2011) [81]	Streaming HS	17	82	7.5
Guha et al. (2016) [30]	RRCF	9	23	3.8
Wu et al. (2014) [91]	RS-Forest	7	39	4.9
Ding et al. (2015) [21]	AHIForest	1	3	0.4
Sun et al. (2019) [79]	Streaming LSHiForest	0	2	0.7
Ma et al. (2020) [60]	Isolation Mondrian Forest	*0	*0	*0.0
Dickens et al. (2020) [18]	Mondrian Polya Forest	*0	*0	*0.0
Togbe et al. (2021) [85]		*0	*0	*0.0

unfair to compare citation of recently introduced methods versus established ones, the proposed list should be taken as a loose reference for listed methods importance.

5 Conclusion and Future Work

In this work we focused on anomaly detection, a practical problem that many times arises in industrial applications. Indeed, the detection of product defects or production instruments faults can be quickly addressed by this kind of techniques.

This review dealt with a particular type of algorithms based on tree structure. These have many advantages, like fast computations, low latency, low memory requirements, parallelism and high detection performances. Moreover, they can cope with the streaming data scenario where the model has to adapt to new incoming data. Moreover, recent efforts have been made by the scientific community to equip such methods with interpretable traits, making them particularly appealing in real-world contexts where root cause analysis is also of paramount importance.

The main procedural differences between the different methods have been discussed and the performances declared by their authors have been compared.

Use cases and a list of ready to use implementations has been made in order to provide practitioners an effective review. The methods performances over different datasets have been grouped together in a unique table.

This chapter has highlighted the many advantages of tree-based approaches over competing alternatives, and the different strategies proposed by the authors.

Some of them are very similar but others introduced very interesting novelties. Just to name a few, the authors found very promising the isolation principle, the anomaly score based on the mass in addition to tree depth, the weighted trees, the pattern anomalies, the continuous training made on data streams and the split criterions that try to accelerate the isolation.

On the other side, criteria that rely on distances other than $L1$, or that try to directly estimate the density, risk to quickly lose the advantage over more traditional methods.

The present study has some limitations, mainly due to the fact that many methods are recent or do not have public implementations provided by the proposing authors, nevertheless this work is intended to be a stating point for future investigations. The most important one lies in the performance comparison; moreover the provided tables have been assembled using results declared in the reviewed papers, so caution must be taken when looking at this comparison and the time complexities. To solve these issues the authors created a public repository at https://github.com/fdallac/treebasedAD where researchers and practitioners can find the codes implemented by us and quantitatively compare the methods. We also invite developers to share the implementations of their approaches in such repository to foster research in the field.

References

1. Ahmed S et al (2019) Unsupervised machine learning-based detection of covert data integrity assault in smart grid networks utilizing isolation forest. IEEE Trans Inf Forensics Secur 14(10):2765–2777
2. Alsini R et al (2021) Improving the outlier detection method in concrete mix design by combining the isolation forest and local outlier factor. Constr Build Mater 270:121396
3. Angiulli F, Pizzuti C (2002) Fast outlier detection in high dimensional spaces. In: European conference on principles of data mining and knowledge discovery. Springer, pp 15–27
4. Antonini M et al (2018) Smart audio sensors in the internet of things edge for anomaly detection. IEEE Access 6:67594–67610
5. Aryal S, Santosh KC, Dazeley R (2020) usfAD: a robust anomaly detector based on unsupervised stochastic forest. Int J Mach Learn Cybern 12(4):1137–1150
6. Aryal S, et al (2014) Improving iForest with relative mass. In: Pacific-Asia conference on knowledge discovery and data mining. Springer, pp 510–521
7. Bandaragoda TR et al (2018) Isolation-based anomaly detection using nearest-neighbor ensembles. Comput Intell 34(4):968–998
8. Barbariol T, Feltresi E, Susto GA (2020) Self- diagnosis of multiphase flow meters through machine learning-based anomaly detection. Energies 13(12):3136
9. Breiman L (2001) Random forests. Mach Learn 45(1):5–32
10. Brito LC, et al (2021) An explainable artificial intelligence approach for unsupervised fault detection and diagnosis in rotating machinery. arXiv preprint arXiv:2102.11848
11. Buschjager, S., Honysz, PJ, Morik, K (2020) Randomized outlier detection with trees. Int J Data Sci Anal 1–14
12. Carletti M, Terzi M, Susto GA (2020) Interpretable anomaly detection with DIFFI: depth-based feature importance for the isolation forest. arXiv preprint arXiv:2007.11117

13. Carletti M, et al (2019) Explainable machine learning in industry 4.0: evaluating feature importance in anomaly detection to enable root cause analysis. In: 2019 IEEE international conference on systems, man and cybernetics (SMC). IEEE, pp 21–26
14. Chen F, Liu Z, Sun M (2015) Anomaly detection by using random projection forest. In: 2015 IEEE international conference on image processing (ICIP). IEEE, pp 1210–1214
15. Chen G, Cai YL, Shi J (2011) Ordinal isolation: an efficient and effective intelligent outlier detection algorithm. In: 2011 IEEE international conference on cyber technology in automation, control, and intelligent systems. IEEE, pp 21–26
16. Das M, Parthasarathy S (2009) Anomaly detection and spatio-temporal analysis of global climate system. In: Proceedings of the 3rd international workshop on knowledge discovery from sensor data, pp 142–150
17. Désir C et al (2013) One class random forests. Pattern Recogn 46(12):3490–3506
18. Dickens C et al (2020) Interpretable anomaly detection with Mondrian Polya forests on data streams. arXiv preprint arXiv:2008.01505
19. Ding Z-G, Da-Jun D, Fei M-R (2015) An isolation principle based distributed anomaly detection method in wireless sensor networks. Int J Autom Comput 12(4):402–412
20. Ding Z, Fei M (2013) An anomaly detection approach based on isolation forest algorithm for streaming data using sliding window. IFAC Proc Vol 46(20):12–17
21. Ding Z, Fei M, Dajun D (2015) An online anomaly detection method for stream data using isolation principle and statistic histogram. Int J Model Simul Sci Comput 6(2):1550017
22. Du J et al (2020) ITrust: an anomaly-resilient trust model based on isolation forest for underwater acoustic sensor networks. IEEE Trans Mob Comput
23. Dua D, Graff C (2017) UCI Machine Learning Repository. http://archive.ics.uci.edu/ml
24. Flach PA, Kull M (2015) Precision-recall-gain curves: PR analysis done right. NIPS, vol. 15
25. Gao R et al (2019) Research and improvement of isolation forest in detection of local anomaly points. J Phys Conf Ser 1237(5):052023
26. Ghaddar A, Darwish L, Yamout F (2019) Identifying mass-based local anomalies using binary space partitioning. In: 2019 International conference on wireless and mobile computing, networking and communications (WiMob). IEEE, pp 183–190
27. Goix N, et al (2017) One class splitting criteria for random forests. In: Asian conference on machine learning. PMLR, pp 343–358
28. Goldstein M, Dengel A (2012) Histogram-based outlier score (HBOS): a fast unsupervised anomaly detection algorithm. In: KI-2012: poster and demo track, pp 59–63
29. Gopalan P, Sharan V, Wieder U (2019) Pidforest: anomaly detection via partial identification. arXiv preprint arXiv:1912.03582
30. Guha S et al (2016) Robust random cut forest based anomaly detection on streams. In: International conference on machine learning. PMLR, pp 2712–2721
31. Hara Y, et al (2020) Fault detection of hydroelectric generators using isolation forest. In: 2020 59th annual conference of the society of instrument and control engineers of Japan (SICE). IEEE, pp 864–869
32. Hariri S, Kind MC, Brunner RJ (2021) Extended isolation forest. IEEE Trans Knowl Data Eng 33(4):1479–1489 (2021). https://doi.org/10.1109/TKDE.2019.2947676. https://www.scopus.com/inward/record.uri?eid=2-s2.0-85102315664&doi=10.1109%2fTKDE.2019.2947676&partnerID=40&md5=2b9a150220b5e76da6945c12c631f6ff
33. Hariri S, Kind MC, Brunner RJ (2018) Extended isolation forest. arXiv preprint arXiv:1811.02141
34. Hawkins DM (1980) Identification of outliers, vol 11. Springer
35. Hill DJ, Minsker BS (2010) Anomaly detection in streaming environmental sensor data: a data-driven modeling approach. Environ Model Softw 25(9):1014–1022
36. Hofmockel J, Sax E (2018) Isolation forest for anomaly detection in raw vehicle sensor data. In: VEHITS 2018, pp 411–416
37. Holmér V (2019) Hybrid extended isolation forest: anomaly detection for bird alarm
38. Iglewicz B, Hoaglin DC (1993) How to detect and handle outliers, vol. 16. ASQ press

39. Jiang S, An Q (2008) Clustering-based outlier detection method. In: 2008 5th international conference on fuzzy systems and knowledge discovery, vol 2. IEEE, pp 429–433
40. John H, Naaz S (2019) Credit card fraud detection using local outlier factor and isolation forest. Int J Comput Sci Eng 7(4):1060–1064
41. Karczmarek P, Kiersztyn A, Pedrycz W (2020) Fuzzy set-based isolation forest. In: 2020 IEEE international conference on fuzzy systems (FUZZ-IEEE). IEEE, pp 1–6
42. Karczmarek, P, Kiersztyn A, Pedrycz W (2020) n-ary isolation forest: an experimental comparative analysis. In: International conference on artificial intelligence and soft computing. Springer, pp 188– 198
43. Karczmarek P, et al (2020) K-means-based isolation forest. In: Knowledge-based systems, vol 195, p 105659
44. Kim D et al (2018) Squeezed convolutional variational autoencoder for unsupervised anomaly detection in edge device industrial internet of things. In: 2018 international conference on information and computer technologies (ICICT). IEEE, pp 67–71
45. Kim J et al (2017) Applications of clustering and isolation forest techniques in real-time building energy-consumption data: application to LEED certified buildings. J Energy Eng 143(5):04017052
46. Kopp M, Pevny T, Holena M (2020) Anomaly explanation with random forests. Exp Syst Appl 149:113187
47. Leveni F et al (2020) PIF: anomaly detection via preference embedding
48. Li C et al (2021) Similarity-measured isolation forest: anomaly detection method for machine monitoring data. IEEE Trans Instrum Meas 70:1–12
49. Li S et al (2019) Hyperspectral anomaly detection with kernel isolation forest. IEEE Trans Geosci Remote Sens 58(1):319–329
50. Liao L, Luo B (2018) Entropy isolation forest based on dimension entropy for anomaly detection. In: International symposium on intelligence computation and applications. Springer, pp 365–376
51. Lin Z, Liu X, Collu M (2020) Wind power prediction based on high-frequency SCADA data along with isolation forest and deep learning neural networks. Int J Electr Power Energy Syst 118:105835
52. Liu FT, Ting KM, Zhou Z-H (2012) Isolation-based anomaly detection. ACM Trans Knowl Disc Data (TKDD) 6(1):1–39
53. Liu FT, Ting KM, Zhou Z-H (2008) Isolation forest. In: 2008 8th IEEE international conference on data mining. IEEE, pp 413–422
54. Liu FT, Ting KM, Zhou Z-H (2010) On detecting clustered anomalies using SCiForest. In: Joint european conference on machine learning and knowledge discovery in databases. Springer, pp 274–290
55. Liu J et al (2018) Anomaly detection in manufacturing systems using structured neural networks. In: 2018 13th world congress on intelligent control and automation (WCICA). IEEE, pp 175–180
56. Liu W et al (2019) A method for the detection of fake reviews based on temporal features of reviews and comments. IEEE Eng Manage Rev 47(4):67–79
57. Liu Z et al (2018) An optimized computational framework for isolation forest. In: Mathematical problems in engineering 2018
58. Luo S et al (2019) An attribute associated isolation forest algorithm for detecting anomalous electro-data. In: 2019 chinese control conference (CCC). IEEE, pp 3788–3792
59. Lyu Y et al (2020) RMHSForest: relative mass and half-space tree based forest for anomaly detection. Chin J Electr 29(6):1093–1101
60. Ma H et al (2020) Isolation Mondrian forest for batch and online anomaly detection. In: 2020 IEEE international conference on systems, man, and cybernetics (SMC). IEEE, pp 3051–3058
61. Maggipinto M, Beghi A, Susto GA (2019) A deep learning-based approach to anomaly detection with 2-dimensional data in manufacturing. In: 2019 IEEE 17th international conference on industrial informatics (INDIN), vol 1. IEEE, pp 187–192

62. Malanchev KL et al (2019) Use of machine learning for anomaly detection problem in large astronomical databases. In: DAMDID/RCDL, pp 205–216
63. Mao W et al (2018) Anomaly detection for power consumption data based on isolated forest. In: 2018 international conference on power system technology (POWERCON). IEEE, pp 4169–4174
64. Marteau P-F, Soheily-Khah S, Béchet N (2017) Hybrid isolation forest-application to intrusion detection. arXiv preprint arXiv:1705.03800
65. Meneghetti L et al (2018) Data-driven anomaly recognition for unsupervised model-free fault detection in artificial pancreas. IEEE Trans Control Syst Technol 28(1):33–47
66. Mensi A, Bicego M (2019) A novel anomaly score for isolation forests. In: International conference on image analysis and processing. Springer, pp 152–163
67. Park CH, Kim J (2021) An explainable outlier detection method using region-partition trees. J Supercomput 77(3):3062–3076
68. Pevny T (2016) Loda: lightweight on-line detector of anomalies. Mach Learn 102(2):275–304
69. Puggini L, McLoone S (2018) An enhanced variable selection and Isolation Forest based methodology for anomaly detection with OES data. Eng Appl Artif Intell 67:126–135
70. Qu H, Li Z, Wu J (2020) Integrated learning method for anomaly detection combining KLSH and isolation principles. In: 2020 IEEE congress on evolutionary computation (CEC). IEEE, pp 1–6
71. Rao GM, Ramesh D (2021) A hybrid and improved isolation forest algorithm for anomaly detection. In: Proceedings of international conference on recent trends in machine learning, IoT, smart cities and applications. Springer, pp 589–598
72. Riazi M, et al.: Detecting the onset of machine failure using anomaly detection methods. In: International conference on big data analytics and knowledge discovery. Springer, pp 3–12
73. Saito T, Rehmsmeier M (2015) The precision-recall plot is more informative than the ROC plot when evaluating binary classifiers on imbalanced datasets. PLOS ONE 10(3):e0118432
74. de Santis RB, Costa MA (2020) Extended isolation forests for fault detection in small hydroelectric plants. Sustainability 12(16):6421
75. Shen Y et al (2016) A novel isolation-based outlier detection method. In: Pacific rim international conference on artificial intelligence. Springer, pp 446–456
76. Staerman G et al (2019) Functional isolation forest. In: Asian conference on machine learning. PMLR, pp 332–347
77. Sternby J, Thormarker E, Liljenstam M (2020) Anomaly detection forest
78. Stojanovic L et al (2016) Big-data-driven anomaly detection in industry (4.0): an approach and a case study. In: 2016 IEEE international conference on big data (big data). IEEE, pp 1647–1652
79. Sun H, et al (2019) Fast anomaly detection in multiple multi-dimensional data streams. In: 2019 IEEE international conference on big data (Big Data). IEEE, pp 1218–1223
80. Susto GA, Beghi A, McLoone S (2017) Anomaly detection through on-line isolation forest: an application to plasma etching. In: 2017 28th annual SEMI advanced semiconductor manufacturing conference (ASMC). IEEE, pp 89–94
81. Tan SC, Ting KM, Liu TF (2011) Fast anomaly detection for streaming data. In: 22nd international joint conference on artificial intelligence
82. Tan Y, et al (2020) Decay detection of a marine gas turbine with contaminated data based on isolation forest approach. In: Ships and offshore structures, pp 1–11
83. Ting KM, et al (2013) Mass estimation. In: Machine learning, vol 90, no 1, pp 127–160
84. Ting KM et al (2010) Mass estimation and its applications. In: Proceedings of the 16th ACM SIGKDD international conference on knowledge discovery and data mining, pp 989–998
85. Togbe MU et al (2021) Anomalies detection using isolation in concept-drifting data streams. Computers 10(1):13
86. Tran PH, Heuchenne C, Thomassey S (2020) An anomaly detection approach based on the combination of LSTM autoencoder and isolation forest for multivariate time series data. In: FLINS 2020: proceedings of the 14th international FLINS conference on robotics and artificial intelligence. World Scientific, pp 18–21

87. Tsou Y-L, et al (2018) Robust distributed anomaly detection using optimal weighted one-class random forests. In: 2018 IEEE international conference on data mining (ICDM). IEEE, pp 1272–1277
88. Wang Y-B et al (2019) Separating multi-source partial discharge signals using linear prediction analysis and isolation forest algorithm. IEEE Trans Instrum Meas 69(6):2734–2742
89. Weber M, et al (2018) Embedded hybrid anomaly detection for automotive CAN communication. In: ERTS 2018: 9th european congress on embedded real time software and systems
90. Wetzig R, Gulenko A, Schmidt F (2019) Unsupervised anomaly alerting for iot-gateway monitoring using adaptive thresholds and half- space trees. In: 2019 6th international conference on internet of things: systems, management and security (IOTSMS). IEEE, pp 161–168
91. Wu K, et al (2014) RS-forest: a rapid density estimator for streaming anomaly detection. In: 2014 IEEE international conference on data mining. IEEE, pp 600–609
92. Wu T, Zhang Y-JA, Tang X (2018) Isolation forest based method for low-quality synchrophasor measurements and early events detection. In: 2018 IEEE international conference on communications, control, and computing technologies for smart grids (SmartGridComm). IEEE, pp 1–7
93. Xiang H et al (2020) OPHiForest: order preserving hashing based isolation forest for robust and scalable anomaly detection. In: Proceedings of the 29th ACM international conference on information & knowledge management, pp 1655–1664
94. Yang Q, Singh J, Lee J (2019) Isolation-based feature selection for unsupervised outlier detection. In: Annual conference of the PHM society, vol 11
95. Yao C et al (2019) Distribution forest: an anomaly detection method based on isolation forest. In: International symposium on advanced parallel processing technologies. Springer, pp 135–147
96. Yu X, Tang LA, Han J (2009) Filtering and refinement: a two stage approach for efficient and effective anomaly detection. In: 2009 9th IEEE international conference on data mining. IEEE, pp 617–626
97. Zhang C et al (2018) A novel anomaly detection algorithm based on trident tree. In: International conference on cloud computing. Springer, pp 295–306
98. Zhang X et al (2017) LSHiForest: a generic framework for fast tree isolation based ensemble anomaly analysis. In: 2017 IEEE 33rd international conference on data engineering (ICDE). IEEE, pp 983–994
99. Zhang Y et al (2019) Anomaly detection for industry product quality inspection based on Gaussian restricted Boltzmann machine. In: 2019 IEEE international conference on systems, man and cybernetics (SMC). IEEE, pp 1–6
100. Zhong S et al (2019) A novel unsupervised anomaly detection for gas turbine using isolation forest. In: 2019 IEEE international conference on prognostics and health management (ICPHM). IEEE, pp 1–6

Joint Use of Skip Connections and Synthetic Corruption for Anomaly Detection with Autoencoders

Anne-Sophie Collin and Christophe De Vleeschouwer

Abstract In industrial vision, the anomaly detection problem can be addressed with an autoencoder trained to map an arbitrary image, i.e. with or without any defect, to a clean image, i.e. without any defect. In this approach, anomaly detection relies conventionally on the reconstruction residual or, alternatively, on the reconstruction uncertainty. To improve the sharpness of the reconstruction, we consider an autoencoder architecture with skip connections. In the common scenario where only clean images are available for training, we propose to corrupt them with a synthetic noise model to prevent the convergence of the network towards the identity mapping, and introduce an original Stain noise model for that purpose. We show that this model favors the reconstruction of clean images from arbitrary real-world images, regardless of the actual defects appearance. In addition to demonstrating the relevance of our approach, our validation provides the first consistent assessment of reconstruction-based methods, by comparing their performance over the MVTec AD dataset [1], both for pixel- and image-wise anomaly detection. Our implementation is available at https://github.com/anncollin/AnomalyDetection-Keras.

1 Introduction

Anomaly detection can be defined as the task of identifying all diverging samples that does not belong to the distribution of regular, also named clean, data. This task could be formulated as a supervised learning problem. Such an approach uses both clean and defective examples to learn how to distinguish these two classes or even to refine the classification of defective samples into a variety of subclasses. However, the scarcity and variability of the defective samples make the data collection challenging and frequently produce unbalanced datasets [2]. To circumvent the above-mentioned issues, anomaly detection is often formulated as an unsupervised learning task. This

A.-S. Collin (✉) · C. De Vleeschouwer
UCLouvain, Louvain-La-Neuve, Belgium
e-mail: anne-sophie.collin@uclouvain.be

C. De Vleeschouwer
e-mail: christophe.devleeschouwer@uclouvain.be

K. P. Tran (ed.), *Control Charts and Machine Learning for Anomaly Detection in Manufacturing*, Springer Series in Reliability Engineering,
https://doi.org/10.1007/978-3-030-83819-5_8

formulation makes it possible to either solve the detection problem itself or to ease the data collection process required by a supervised approach.

Anomaly detection has a broad scope of application, which implies that processed data can differ in nature. It typically corresponds to speech sequences, time series, images or video sequences [3]. Here, we consider the automated monitoring of production lines through visual inspection to detect defective samples. More specifically, we are interested in identifying abnormal structures in a manufactured object based solely on the analysis of one image of the considered item. Computer vision sensors offer the opportunity to be easily integrated in a production line without disturbing the production scenarios [4]. To handle such high-dimensional data, Convolutional Neural Networks (CCNs) provide a solution of choice, due to their capacity to extract rich and versatile representations.

The unsupervised anomaly detection framework considered in this work is depicted in Fig. 1. It builds on the training of an autoencoder to project an arbitrary image onto the clean distribution of images (blue block). The training set is constituted exclusively of clean images. Then, defective structures can be inferred from the reconstruction (red block), following a traditional approach based on the residual [2], or even from an estimation of the prediction uncertainty [5].

During training, the autoencoder is constrained to minimize the reconstruction error of clean structures in the images. Several loss functions, presented later in Sect. 2, can be considered to quantify this reconstruction error. With the objective of building our method on the Mean Squared Error (MSE) loss for its simplicity and widespread usage, we propose a new non-parametric approach that addresses the standard issues related to the use this loss function. To enhance the sharpness of the reconstruction, we consider an autoencoder equipped with skip connections, which allow the information to bypass the bottleneck. In order to prevent systematic transmission of the image structures through these links, the network is trained to reconstruct a clean image out of a corrupted version of the input, instead of an unmodified version of it. As discussed later, the methodology used to corrupt the training images has a huge impact on the overall performances. We introduce a new synthetic model, named Stain, that adds an irregular elliptic structure of variable color and size to the input image. Despite its simplicity, the Stain model is by far the best performing compared to the scene-specific corruption investigated in a previous study [6]. Our Stain model has the double advantage of performing consistently better, while being independent of the image content. We demonstrate that adding skip connections to the autoencoder architecture when simultaneously corrupting the training clean images with our Stain noise model addresses the blurry reconstruction issue related to the use of the MSE loss.

The present work extends our conference paper [8] by providing an extensive study of the internal statistics of the network. This analysis reveals the natural trend of the network to distinguish between the representations of regular and irregular samples, which is even more explicit in the bottleneck layer. Moreover, we show that, when applied simultaneously, the two modifications of the autoencoder-based reconstruction framework studied in this work, namely:

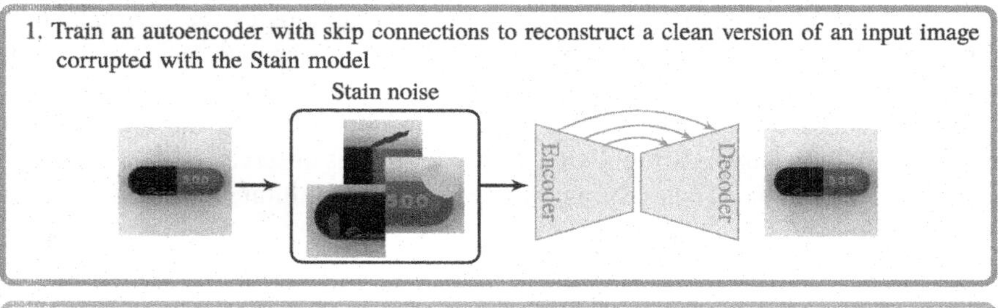

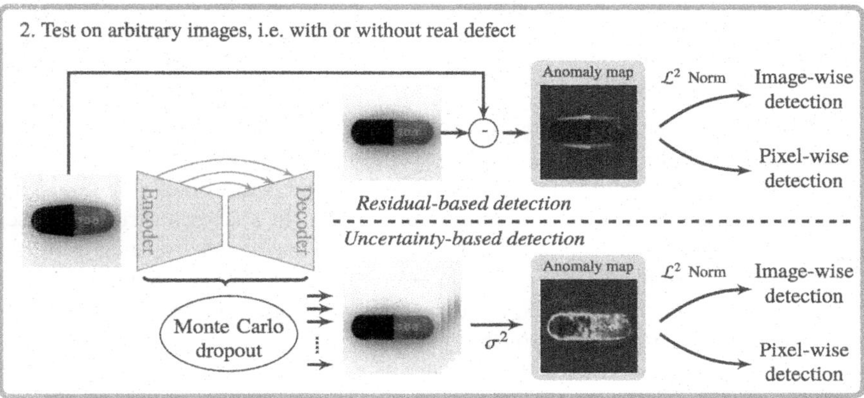

Fig. 1 We improve the quality of the reconstructed images by training an autoencoder with skip connections on corrupted images. *1. Blue block.* Corrupting the training images with our Stain noise model avoids the convergence of the network towards an unwanted identity operator. *2. Red block.* The two anomaly detection strategies. In the upper part, the anomaly map is generated by subtracting the input image from its reconstruction. In the lower part, the anomaly map is estimated by the variance between 30 reconstructions inferred with Monte Carlo dropout (MCDropout) [7]. It relies on the hypothesis that structures that are not seen during training (defective areas) correlate with higher reconstruction uncertainty

1. the corruption of the training images with our Stain-noise model, and
2. the addition of skip connections to the autoencoder architecture,

lead to a better separation of the internal representations associated to initial and reconstructed versions of irregular samples than regular ones. This phenomenon is observed even though the network has not been explicitly trained to separate these two classes.

In Sect. 2, we provide an overview of previous reconstruction-based methods addressing the anomaly detection problem, and motivate the use of the MSE loss to train our auto-encoder. Details of our method, including network architecture and the Stain noise model description, are provided in Sect. 3. In Sect. 4, we provide a comparative study of residual- and uncertainty-based anomaly detection strategies, both at the image and pixel level. This extensive comparative study demonstrates the benefit of our proposed framework, combining skip connections and our original corruption model. Section 5 further investigates and compares the internal representations of clean and defective images in an autoencoder with and without skip connections. This analysis provides insight into the variety of performance observed in our comparative study, and allows to develop intuition regarding the obtained results. Moreover, this study highlights the discrepancies observed across image categories of the MVTec AD dataset, and raises questions for future research. Section 6 concludes.

2 Related Work

Anomaly detection is a long-standing problem that has been considered in a variety of fields [2, 3] and the reconstruction-based approach is one popular way to address the issue. In comparison to other methods for which the detection of abnormal samples is performed in another domain than the image [9–13], reconstruction-based approaches offer the opportunity to identify the pixels that lead to the rejection of the image from the normal class. This section presents a literature review organized into three subsections, each one focusing on the main issues encountered when working with a reconstruction-based approach.

Low Contrast Defects Detection. Conventional reconstruction-based methods infer anomaly based on the reconstruction error between an arbitrary input and its reconstructed version. It assumes that clean structures are perfectly conserved while defective ones are replaced by clean content. However, when a defect contrasts poorly with its surroundings, replacing abnormal structures with clean content does not lead to a sufficiently high reconstruction error. In such cases, this methodology reaches the limit of its underlying assumptions. A previous study [5] detected anomalies by quantifying the prediction uncertainty with MCDropout [7] instead of the reconstruction residual.

Reconstruct Sharp Structures. To obtain a clean reconstruction out of an arbitrary image, an autoencoder is trained on clean images to perform an image-to-image identity mapping under the minimization of a loss function. The bottleneck forces the network to learn a compressed representation of the training data that is expected to regularize the reconstruction towards the normal class. In the literature, the use of the MSE loss to train an hourglass CNN, without skip connections, has been criticized for its trend to produce blurry output images [14, 15]. Since anomaly detection is based on the reconstruction residual, this behavior is detrimental because it alters the clean structures of an image as well as the defective ones.

A lot of effort has been made to improve the quality of the reconstructed images by the introduction of new loss functions. In this spirit, unsupervised methods based on Generative Adversarial Networks (GANs) have emerged [16–20]. If GANs are known for their ability to produce realistic high-quality synthetic images [21], they have major drawbacks. Usually, GANs are difficult to train due to their trend to converge towards mode collapse [22]. Moreover, in the context of anomaly detection, some GAN-based solutions fail to exclude defective samples from the generative distribution [19] and require an extra optimization step in the latent space during inference [17]. This process ensures that the defective structures of the input image are replaced by clean content. Performances of AnoGAN [17] over the MVTec AD dataset have been reported by Bergmann et al. [1]. Those are significantly lower than the method proposed in this work.

To improve the sharpness of the reconstruction, Bergmann et al. proposed a loss derived from the Structural SIMilarity (SSIM) index [15]. The use of the SSIM loss has been motivated by its ability to produce well looking images from a human perceptual perspective [14, 23]. The SSIM have shown some improvement over the MSE loss for the training of an autoencoder in the context of anomaly detection. However, the SSIM loss formulation does not generalize to color images and is parametric. Traditionally, these hyper-parameters are tuned based on a validation set. However, in a real-life scenarios of anomaly detection, samples with real defects are usually not available. For this reason, our chapter focuses on the MSE rather than on the parametric SSIM.

Prevent the Reconstruction of Defective Structures. It is usually expected that the compression induced by the bottleneck is sufficient to regularize the reconstruction towards the clean distribution of images. In practice, the autoencoder is not explicitly constrained to not reproduce abnormal content and often reconstructs defective structures. A recent method proposed to mitigate this issue by iteratively projecting the arbitrary input towards the clean distribution of images. The projection is constrained to be similar, in the sense of the $\mathcal{L}^1$ norm, to the initial input [24]. Instead of performing this optimization in the latent space as made with AnoGAN [17], they propose to find an optimal clean input image. If this practice enhances the sharpness of the reconstruction, the optimization step is resource consuming.

Also, the reconstruction task can be formulated as an image completion problem [25, 26]. To make the inference and training phases consistent, it is assumed that the defects are entirely contained in the mask during inference, which limits the practical usage of the method. Random inpainting masks have been considered to deal with this issue [27]. Mei et al. [28] also proposed to use a denoising autoencoder to reconstruct training images corrupted with salt-and-pepper noise. However they did not discuss the gain brought by this modification, and only considered it for an hourglass CNN, without skip connections.

The reconstruction of clean structures has also been promoted by constraining the latent representation. Practically, Gong et al. [29] learned a dictionary to describe the latent representation of clean samples based on a sparse linear combination of dictionary elements. In contrast, defective samples are assumed to require more complex combinations of the dictionary items. Based on this hypothesis, enforcing sparsity during inference prevents the reconstruction of defective structures. In a similar spirit, Wang et al. [30] considered the use of a VQ-VAE [31]. Their method incorporates a specific autoregressive model in the latent space which can be used to enforce similarity between encoded vectors of the training and the test sets, thereby preventing the reconstruction of defective structures.

The methodology proposed in this work presents a simple approach to enhance the sharpness of the reconstructed images. The skip connections allow the preservation of high frequency information by bypassing the bottleneck. However, we show that this practice penalizes anomaly detection when the model is trained to perform identity mapping on uncorrupted clean images. Nevertheless, the introduction of an original noise model allows to significantly improve the anomaly detection accuracy for the skipped architecture, which eventually outperforms the conventional one in many real-life cases. Also, we compare anomaly detection based on the reconstruction residual or uncertainty estimation. This second option appears to be of particular interest for the detection of low contrast defective structures.

3 Our Anomaly Detection Framework

Our method addressed anomaly detection based on the regularized reconstruction performed by an autoencoder. This section presents the different components of our approach, ranging from the training of the autoencoder to the strategies considered to detect defects based on the reconstruction residual or the reconstruction uncertainty.

3.1 Model Configuration

The reconstruction of a clean version of any input image is based on a CNN. Our architecture, referred to as **Autoencoder with Skip connections (AESc)** and shown in Fig. 2, is a variant of U-Net [32]. AESc takes input images of size 256×256 and projects them onto a latent space of $4 \times 4 \times 512$ dimension. The projection towards the lower dimensional space is performed by six consecutive convolutional layers strided by a factor two. The back projection is performed by six layers of convolution followed by an upsampling operator of factor two. All convolutions have a 5×5 kernel. Unlike the original U-Net version, our skip connections perform an addition, not a concatenation, of feature maps of the encoder to the decoder.

For the sake of comparison, we also consider the **Autoencoder (AE)** network which follows the same architecture but from which we removed the skip connections.

All models have been trained during 250 epochs to minimize the MSE loss over batches constituted of 16 images. We used the Adam optimizer [33] with an initial learning rate of 0.01. This learning rate is decreased by a factor of 2 when the PSNR over a validation set, constituted by 20% of the training images, reaches a plateau for at least 30 epochs. No additional data augmentation than the synthetic corruption model presented in Sect. 3.2 is applied on the training set.

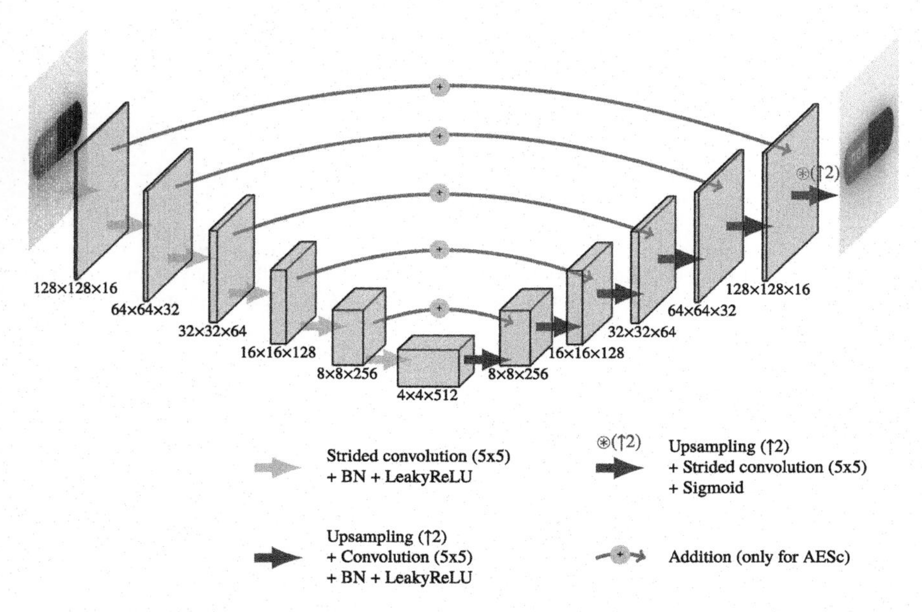

Fig. 2 AESc architecture performing the projection of an arbitrary 256 × 256 image towards the distribution of clean images with the same dimension. Note that the AE architecture shares the same specifications with the exception of the skip connections that have been removed

3.2 *Corruption Model*

Ideally, the autoencoder should preserve clean structures while modifying those that are not. To this end, it is wanted that defective structures are excluded from the generative model, eventhough the training task is an identity mapping. Due to the impossibility of collecting pairs of clean and defective versions of the same sample, we propose to introduce synthetic corruption during training to explicitly constrain the autoencoder to remove this additive noise. Our **Stain** noise model, illustrated in Fig. 1 and explained in Fig. 3, corrupts images by adding a structure whose color is randomly selected in the grayscale range and whose shape is an ellipse with irregular edges.

The intuition behind the definition of this noise model is that occluding large area of the training images is a data augmentation procedure that helps to improve the network training [34, 35]. Due to the skip connections in our network architecture, this form of data augmentation is essential to avoid the convergence of the model towards the identity operator. However, we noticed that the use of regular shapes, like a conventional ellipse, leads to overfitting to this additive noise structure, as also pointed out in a context of inpainting with rectangular holes [36].

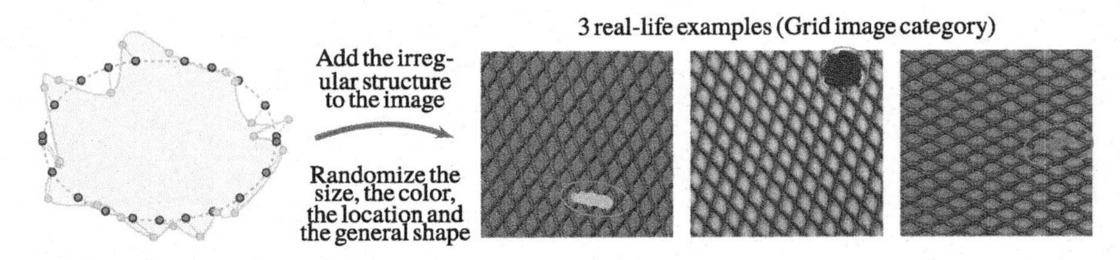

Fig. 3 The Stain noise model is a cubic interpolation between 20 points (orange dots), arranged in ascending order of polar coordinates, located around the border of an ellipse of variable size (blue line). The axes of the ellipse are comprised between 1 and 12% of the smallest image dimension and its eccentricity is randomly initialized

3.3 Anomaly Detection Strategies

We compare two approaches to obtain the anomaly map representing the likelihood that a pixel is abnormal. On the one hand, the **residual-based** approach evaluates the abnormality by measuring the absolute difference between the input image $\mathbf{x}$ and its reconstruction $\hat{\mathbf{x}}$. On the other hand, the **uncertainty-based** approach relies on the intuition that structures that are not seen during training, i.e. the anomalies, will correlate with higher uncertainties. This is estimated by the variance between 30 output images inferred with the MCDropout technique. Our experiments revealed that more accurate detection is obtained by applying an increasing level of dropout for deepest layers. More specifically, the dropout levels are [0, 0, 10, 20, 30, 40] percent for layers ranging from the highest spatial resolution to the lowest.

Out of the anomaly map, it is either possible to classify the entire image as clean/defective or to classify each pixel as belonging to a clean/defective structure. In the first case, referred to as **image-wise detection**, it is common to compute the $\mathcal{L}^p$ norm of the anomaly map given by

$$\mathcal{L}^p(\mathbf{x}, \hat{\mathbf{x}}) = \left(\sum_{i=0}^{m} \sum_{j=0}^{n} |\mathbf{x}_{i,j} - \hat{\mathbf{x}}_{i,j}|^p \right)^{1/p} \tag{1}$$

with $\mathbf{x}_{i,j}$ denoting the pixel belonging to the ith row and the jth column of the image $\mathbf{x}$ of size $m \times n$. Based on our experiments, we present results obtained for $p = 2$ since they achieve the most stable accuracy values across the experiments. Hence, all images for which the $\mathcal{L}^2$ norm of the abnormality map exceeds a chosen threshold are considered as defective. In the second case, referred to as **pixel-wise detection,** the threshold is applied directly on each pixel value of the anomaly map.

To perform image-wise or pixel-wise anomaly detection, a threshold has to be determined. Since this threshold value is highly dependent on the application, we present the performances in terms of Area Under the receiver operating characteristic Curve (AUC), obtained by varying over the full range of threshold values.

4 Anomaly Detection on the MVTec AD Dataset

Experiments have been conducted on grayscale images of the MVTec AD dataset [1], containing five categories of textures and ten categories of objects. In this dataset, defects are real and have various appearance. Their location is defined with a binary segmentation mask. All images have been scaled to a 256×256 size. Anomaly detection is performed at this resolution.

4.1 AESc + Stain: Qualitative and Quantitative Analysis

In this section, we compare qualitatively and quantitatively the results obtained with our AESc + Stain model for both image- and pixel-wise detection. This analysis focuses on the residual-based detection approach to emphasize the benefits of adding skip connections to the AE architecture. The comparison of the results obtained with residual- versus uncertainty-based strategies is discussed later in Sect. 4.2.

Qualitatively, Fig. 4 reveals the trends of the AE and AESc models trained with and without the Stain noise corruption. On the one hand, the AE network produces blurry reconstructions as depicted by the overall higher residual intensities. If the global structure of the object (Cable and Toothbrush) are properly reconstructed, the AE network struggles to infer the finer details of the texture images (Carpet sample). On the other hand, the AESc model shows finer reconstruction of the image details depicted by a nearly zero residual over the clean areas of the images. However, when ASEc is trained without corruption, the model converges towards an identity operator, as revealed by the close-to-zero residuals of defective structures. The corruption of the training images with the Stain model alleviates this unwanted behavior by leading to high reconstruction residuals in defective areas while simultaneously keeping low reconstruction residuals in clean structures.

Quantitatively, Table 1 presents the image-wise detection performances obtained with the AESc and AE networks trained with and without our Stain noise model. The last column provides a comparison with the ITEA method, introduced by Huang et al. [6]. ITAE is also a reconstruction-based approach which relies on an autoencoder with skip connections trained with images corrupted by random rotations and a graying operator (averaging of pixel value along the channel dimension) selected based on prior knowledge about the task.

This table highlights the superiority of our AESc + Stain noise model to solve the image-wise anomaly detection. The improvement brought by adding skip connections to an autoencoder trained with corrupted images is even more important for texture images than for object images. We also observe that, if the highest accuracy is consistently obtained with the residual-based approach, the uncertainty-based decision derived from the AESc + Stain model generally provides the second best (underlined in Table 1) performances among tested networks, attesting the quality of the AESc + Stain model for image-based decision.

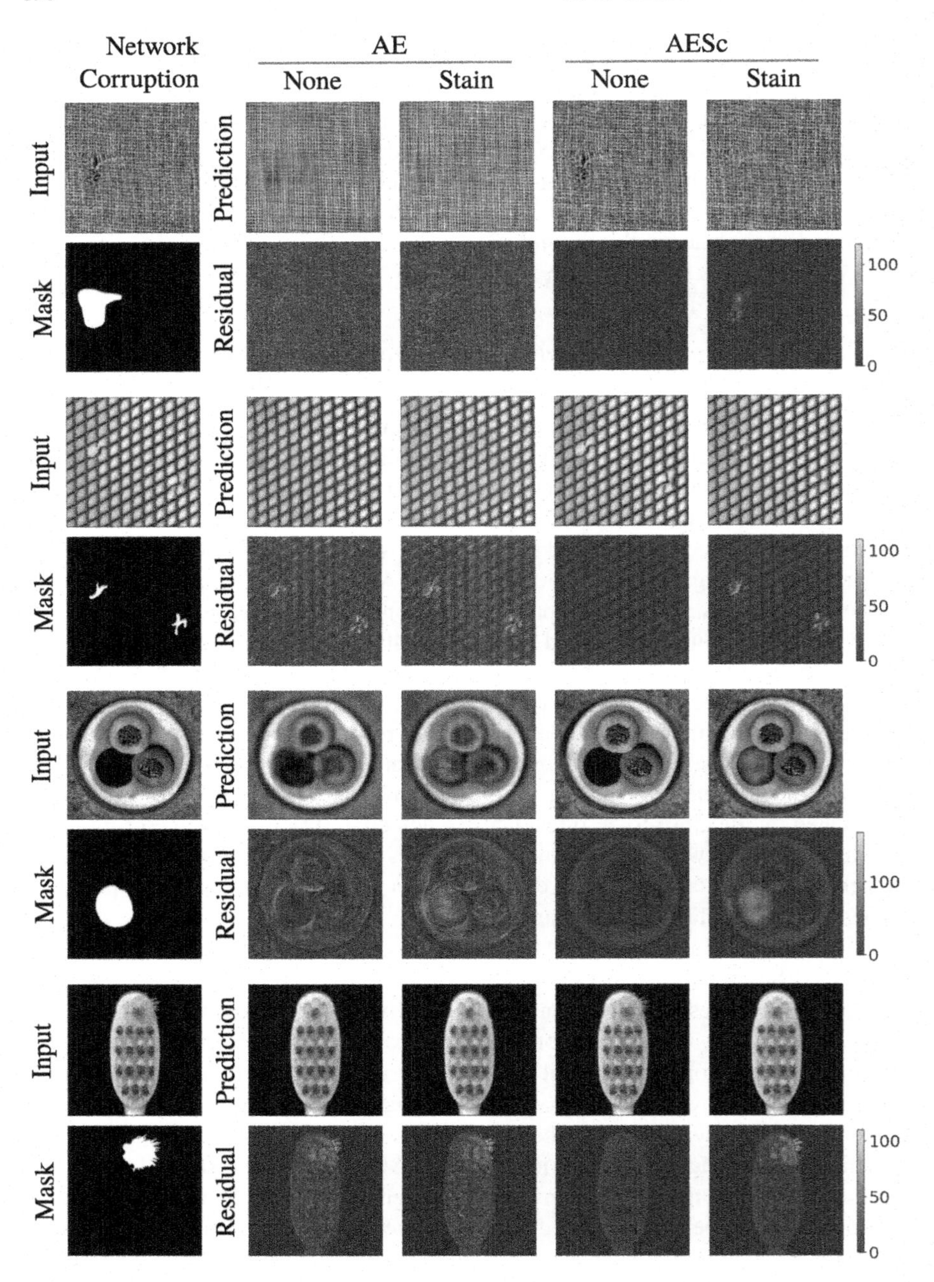

Fig. 4 Predictions obtained with the AESc and AE networks trained with and without our Stain noise model. Two defective textures are considered, namely a Carpet (first sample) and a Grid (second sample), as well as two defective objects, namely a Cable (third sample) and a Toothbrush (fourth sample). First column show the image fed in the networks and the mask locating the defect. Odd rows show the reconstructed images and even rows show the anomaly maps obtained with the residual-based strategy

Table 1 Image-wise detection AUC obtained with the residual- and uncertainty-based detection methods[a]

		Uncertainty				Residual				
Network		AE		AESc		AE		AESc		ITAE [6]
Corruption		None	Stain	None	Stain	None	Stain	None	Stain	
Textures	Carpet	0.41	0.30	0.44	0.80	0.43	0.43	0.48	**0.89**	0.71
	Grid	0.69	0.66	0.12	**0.97**	0.80	0.84	0.52	**0.97**	0.88
	Leather	0.86	0.57	0.88	0.72	0.45	0.54	0.56	**0.89**	0.87
	Tile	0.73	0.50	0.72	0.95	0.49	0.57	0.88	**0.99**	0.74
	Wood	0.87	0.86	0.78	0.78	0.92	0.94	0.92	**0.95**	0.92
	Mean[b]	0.71	0.58	0.59	0.84	0.62	0.66	0.67	**0.94**	0.82
Objects	Bottle	0.72	0.41	0.71	0.82	**0.98**	0.97	0.77	**0.98**	0.94
	Cable	0.64	0.48	0.52	0.87	0.70	0.77	0.55	**0.89**	0.83
	Capsule	0.55	0.49	0.44	0.71	**0.74**	0.64	0.60	**0.74**	0.68
	Hazelnut	0.83	0.60	0.68	0.90	0.90	0.88	0.85	**0.94**	0.86
	Metal nut	0.38	0.33	0.41	0.62	0.57	0.59	0.24	**0.73**	0.67
	Pill	0.63	0.48	0.55	0.62	0.76	0.76	0.70	**0.84**	0.79
	Screw	0.45	0.77	0.13	0.80	0.68	0.60	0.30	0.74	**1.00**
	Toothbrush	0.36	0.44	0.51	0.99	0.93	0.96	0.78	**1.00**	**1.00**
	Transistor	0.67	0.59	0.55	0.90	0.84	0.85	0.46	**0.91**	0.84
	Zipper	0.44	0.41	0.70	0.93	0.90	0.88	0.72	**0.94**	0.80
	Mean[c]	0.57	0.50	0.52	0.82	0.80	0.79	0.60	**0.87**	0.84
	Global mean[d]	0.62	0.53	0.54	0.83	0.74	0.75	0.62	**0.89**	0.84

[a]For each row, the best performing approach is highlighted in boldface and the second best is underlined
[b]Mean AUC obtained over the classes of images belonging to the texture categories
[c]Mean AUC obtained over the classes of images belonging to the object categories
[d]Mean AUC obtained over the entire dataset

Table 2 presents the pixel-wise detection performances obtained with our approaches and compares them with the method reported in [1], referred to as AE_{L2}. This residual-based method relies on an autoencoder without skip connections, and provides state of the art performance in the pixel-wise detection scenario. Similarly to our AE model, AE_{L2} is trained to minimize the MSE of the reconstruction of images that are not corrupted with synthetic noise. AE_{L2} however differs from our AE model in several aspects, including a different network architecture, data augmentation, patch-based inference for the texture images, and anomaly map post-processing with mathematical morphology. Despite our efforts, in absence of public code, we have been unable to reproduce the results presented in [1]. Hence, our table just copy the results from [1]. For fair comparison between AE and AESc + Stain, the table also provides the results obtained with our AE, since our AE and AESc + Stain models adopt the same architecture (up to the skip connections) and the same training procedure.

In the residual-based detection strategy, our AESc + Stain method obtains similar performances as the AE_{L2} approach when averaged over all the image categories of the MVTec AD dataset. However, as already pointed in the image-wise detection scenario, AESc + Stain performs better with texture images and worse with object images. Regarding the decision strategy, we observe an opposite trend than the one encountered for image-wise detection: the uncertainty-based approach performs a bit better than the residual-based strategy when it comes to pixel-wise decisions. This difference is further investigated in the next section.

4.2 Residual- Vs. Uncertainty-Based Detection Strategies

Figure 5 provides a visual comparison between residual- and uncertainty-based strategies. Globally, we observe that the reconstruction residual mostly correlates with the uncertainty. However, the uncertainty indicator is usually more widespread. This behavior can sometimes lead to a better coverage of the defective structures (Bottle and Pill) or to an increase of the number of false positive pixels that are detected (Carpet and Cable).

One important observation concerns the relationship between the detection of a defective structure and its contrast with its surroundings. In the residual-based approach, regions of an image are considered as defective if their reconstruction error exceeds a threshold. In the proposed formulation, the network is explicitly constrained to replace synthetic defective structures with clean content. No constraint is introduced regarding the contrast of the reconstructed structure and its surroundings. Hence, defects that are poorly contrasted lead to small residual intensities. On the contrary, the intensity of the uncertainty indicator does not depend on the contrast between a structure with the surroundings. For low contrast defects, it enhances their detection as illustrated (Bottle and Pill). On the contrary, it can deteriorate the location of high contrast defects for which the residual map is an appropriate anomaly indicator (Carpet and Cable). In theses cases, the sharp prediction obtained with the residual-based approach is preferred over the uncertainty-based one.

Table 2 Pixel-wise detection AUC obtained with the residual- and uncertainty-based detection methods[a]

		Uncertainty				Residual				
Network		AE		AESc		AE		AESc		AE_{L2} [1]
Corruption		None	Stain	None	Stain	None	Stain	None	Stain	
Textures	Carpet	0.55	0.54	0.43	**0.91**	0.57	0.62	0.52	0.79	0.59
	Grid	0.52	0.49	0.50	**0.95**	0.81	0.82	0.57	0.89	0.90
	Leather	0.86	0.52	0.58	0.87	0.79	0.82	0.71	**0.95**	0.75
	Tile	0.54	0.50	0.53	**0.79**	0.45	0.54	0.62	0.74	0.51
	Wood	0.61	0.48	0.51	**0.84**	0.64	0.71	0.65	**0.84**	0.73
	Mean[b]	0.62	0.51	0.51	**0.87**	0.65	0.70	0.61	0.84	0.70
Objects	Bottle	0.68	0.63	0.64	**0.88**	0.85	**0.88**	0.47	0.84	0.86
	Cable	0.54	0.70	0.66	0.84	0.62	0.83	0.72	0.85	**0.86**
	Capsule	0.92	0.89	0.65	**0.93**	0.87	0.87	0.63	0.83	0.88
	Hazelnut	**0.95**	0.91	0.60	0.89	0.92	0.93	0.79	0.88	**0.95**
	Metal nut	0.79	0.73	0.50	0.62	0.82	0.84	0.52	0.57	**0.86**
	Pill	0.82	0.82	0.61	**0.85**	0.81	0.81	0.64	0.74	**0.85**
	Screw	0.94	0.94	0.61	0.95	0.93	0.93	0.72	0.86	**0.96**
	Toothbrush	0.84	0.83	0.79	**0.93**	0.92	**0.93**	0.73	**0.93**	**0.93**
	Transistor	0.79	0.64	0.51	0.78	0.79	0.82	0.56	0.80	**0.86**
	Zipper	0.78	0.77	0.60	**0.90**	0.73	0.75	0.60	0.78	0.77
	Mean[c]	0.81	0.79	0.62	0.86	0.83	0.86	0.64	0.81	**0.88**
	Global mean[d]	0.74	0.69	0.58	**0.86**	0.77	0.81	0.63	0.82	0.82

[a] For each row, the best performing approach is highlighted in boldface and the second best is underlined
[b] Mean AUC obtained over the classes of images belonging to the texture categories
[c] Mean AUC obtained over the classes of images belonging to the object categories
[d] Mean AUC obtained over the entire dataset

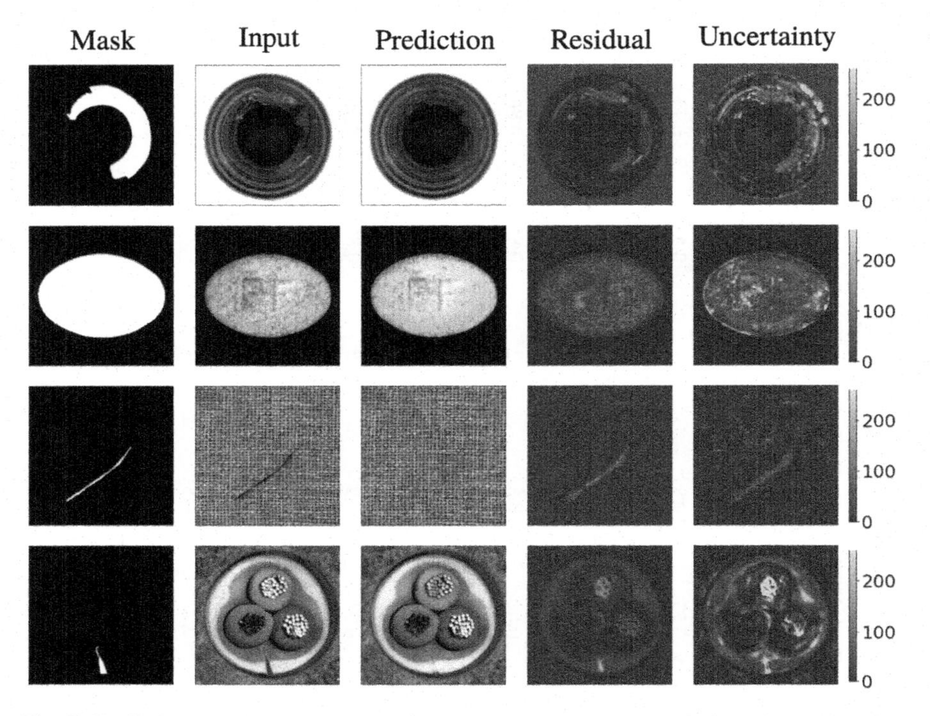

Fig. 5 Predictions obtained with the AESc network trained with our Stain noise model. One defective texture is considered, namely a Carpet (third row) as well as three defective objects, namely a Bottle (first row), a Pill (second row) and a Cable (fourth row). From left to right, columns represent the ground-truth, the image fed to the network, the prediction (without MCDropout), the reconstruction residual and the reconstruction uncertainty

As reported in Sect. 4.1, we observe that the uncertainty-based detection perform generally worse than the residual-based approach for image-wise detection. We explain this drop of performance by an increase of the intensities of the uncertainty maps inferred from the clean images belonging to the test set. As the image-wise detection is based on the $\mathcal{L}^2$ norm of the anomaly map, the lowest the anomaly maps of clean images, the better the detection of defective images. For image-wise detection, the performances are less sensitive to the optimal coverage of the defective area as long as the overall intensity of the clean anomaly maps is low.

On the contrary, the uncertainty-based strategy improves the pixel-wise detection of the AESc + Stain model. For this use case, a better coverage of the defective structure is crucial. As previously mentioned, AESc + Stain model used usually leads to reconstruction residual constituted of sporadic spots and misses low contrast defects. The uncertainty-based strategy compensates these two issues.

4.3 Comparative Study of Corruption Models

Up to now, we considered only the Stain noise model to corrupt training data. In this comparative study we consider other noise models to confirm the relevance of our previous approach over other types of corruption that could have been considered. We provide here a comparison with three other synthetic noise models represented in Fig. 6:

a- Gaussian noise. Corrupt by adding white noise applied uniformly over the entire image. For normalized intensities between 0 and 1, a corrupted pixel value x', corresponding to an initial pixel value x, is the realization of a random variable given by a normal distribution of mean x and variance σ^2 in the set: $[0.1, 0.2, 0.4, 0.8]$.
b- Scratch. Corrupt by adding one curve connecting two points whose coordinates are randomly chosen in the image and whose color is randomly selected in the gray scale range. The curve can follow a straight line, a sinusoidal wave or the path of a square root function.
c- Drops. Corrupt by adding 10 droplets whose color are randomly selected in the gray scale range and whose shape are circular with a random diameter (chosen between 1 and 2% of the smallest image dimension). The droplets partially overlap.

In addition, we have also considered the possibility to corrupt the training images with a combination of several models. We propose two hybrid models:

d- Mix1. This configuration corrupts training images with a combination of the Stain, Scratch and Drops models. We fix that 60% of the training images are corrupted with the Stain model while the remaining 40% are corrupted with the Scratch and Drops models in equal proportions.
e- Mix2. This configuration corrupts training images with a combination of the Stain and the Gaussian noise models. We fix that 60% of the training images are corrupted with the Stain model while the remaining 40% are corrupted with the Gaussian noise model.

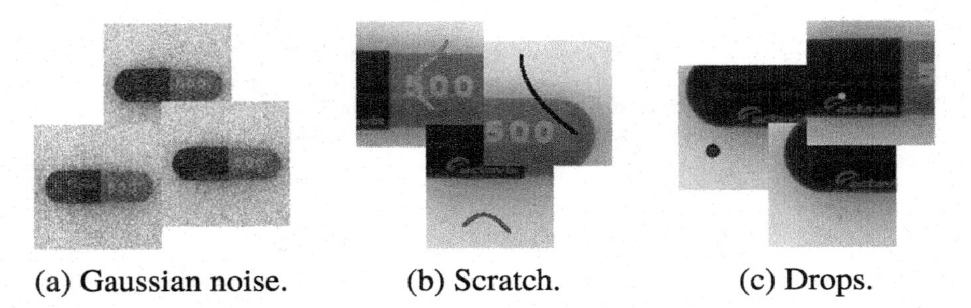

(a) Gaussian noise. (b) Scratch. (c) Drops.

Fig. 6 Illustration of the Gaussian noise, scratch and drops models. The original clean image is the one presented in Fig. 1

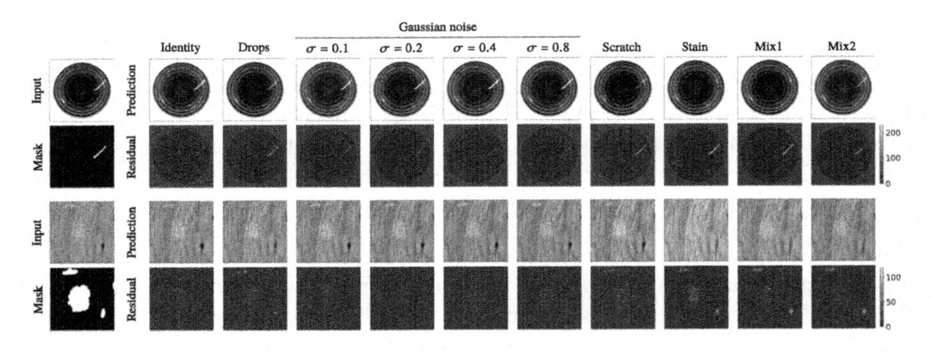

Fig. 7 Reconstructions obtained with the AESc network trained with different noise models. We consider here one defective object, namely a Bottle (first sample) and a defective texture, namely a Wood (second sample). Rows and columns are defined as in Fig. 4

Table 3 Image-wise AUC obtained with the AESc network trained with different noise models with the residual-based problem formulation[a]

	Corruption	None	Drops	Gaussian noise (σ)				Scratch	Stain	Mix1	Mix2
				0.1	0.2	0.4	0.8				
Textures	Carpet	0.48	0.87	0.52	0.46	0.51	0.53	0.63	**0.89**	0.84	0.84
	Grid	0.52	0.94	0.55	0.69	0.59	0.72	0.79	**0.97**	0.91	0.96
	Leather	0.56	0.87	0.72	0.71	0.74	0.71	0.77	**0.89**	0.88	**0.89**
	Tile	0.88	0.94	0.94	0.92	0.90	0.92	0.95	**0.99**	0.98	0.96
	Wood	0.92	**0.99**	0.89	0.90	0.91	0.85	0.96	0.95	0.94	0.79
	Mean[b]	0.67	0.92	0.72	0.74	0.73	0.75	0.82	**0.94**	0.91	0.89
Objects	Bottle	0.77	**0.99**	0.82	0.85	0.81	0.75	0.91	0.98	0.98	0.97
	Cable	0.55	0.60	0.58	0.53	0.49	0.46	0.60	0.89	0.87	**0.90**
	Capsule	0.60	0.71	0.58	0.68	0.57	0.59	0.66	**0.74**	**0.74**	0.53
	Hazelnut	0.85	**0.98**	0.75	0.73	0.92	0.73	0.96	0.94	0.93	0.81
	Metal Nut	0.24	0.54	0.32	0.27	0.28	0.24	0.44	0.73	0.71	**0.86**
	Pill	0.70	0.79	0.69	0.71	0.73	0.68	0.78	**0.84**	0.77	0.78
	Screw	0.30	0.46	0.91	**0.99**	0.78	0.65	0.71	0.74	0.22	0.72
	Toothbrush	0.78	**1.00**	0.99	0.98	0.79	0.82	0.87	**1.00**	**1.00**	**1.00**
	Transistor	0.46	0.83	0.55	0.49	0.48	0.50	0.68	0.91	**0.92**	**0.92**
	Zipper	0.72	0.93	0.66	0.63	0.69	0.58	0.79	0.94	0.90	**0.98**
	Mean[c]	0.60	0.78	0.68	0.69	0.65	0.60	0.74	**0.87**	0.80	0.85
	Global mean[d]	0.62	0.83	0.70	0.70	0.68	0.65	0.77	**0.89**	0.84	0.86

[a]For each row, the best performing approach is highlighted in boldface and the second best is underlined
[b]Mean AUC obtained over the classes of images belonging to the texture categories
[c]Mean AUC obtained over the classes of images belonging to the object categories
[d]Mean AUC obtained over the entire dataset

Figure 7 allows to compare reconstructions obtained when the AESc network is trained over images corrupted with the newly introduced noise models with respect to the Stain noise model. First, these examples illustrate the convergence of the model towards the identity mapping when the Gaussian noise model is used as synthetic corruption. An analysis of the results obtained over the entire dataset reveals that the

AESc + Gaussian noise model does almost not differ from the AESc network trained with unaltered images.

Compared to the Gaussian noise, other models introduced before improve the identification of defective areas in the images. This is reflected by higher intensities of the reconstruction residual in the defective areas and close-to-zero reconstruction residual in the clean areas. With the exception of the Gaussian noise model, the Scratch model is the most conservative, among those considered, in the sense that most of the structures of the input images tend to be reconstructed identically. This practice increases the number of false negative. Also, the Drops model restricts the structures detected as defective to sporadic spots. Finally, the three models based on the Stain noise (Stain, Mix1 and Mix2) provide the residuals that correlate the most with the segmentation mask.

Generally, models based on the Stain noise (Stain, Mix1 and Mix2) lead to the most relevant reconstruction for anomaly detection, i.e. lower residual intensities in clean areas and higher residual intensities in defective areas. More surprisingly, this statement remains true even if the actual defect looks more similar to the Scratch model than the Stain noise (Bottle sample in Fig. 7). We recall that defects contained in the MVTec AD dataset are real observations of an anomaly. This reflects that models trained with synthetic corruption models that look similar to real ones do not necessarily generalize well to real defects.

Table 3 quantifies the impact of the synthetic noise model on the performances of the ASEc network to solve the image-wise detection task with a residual-based approach. The AESc + Stain configuration is the best performing in all use cases when considering the mean performances that are obtained over the entire dataset, as revealed by the previous qualitative study. The two hybrid models (Mix1 and Mix2) lead usually to slightly lower performances than those obtained with the Stain model. Those observations attest that the Stain model is superior to others and justify the choice of the Stain noise as our newly introduced approach to corrupt the training images with synthetic noise.

5 Internal Representation of Defective Images

The choice of an autoencoder network architecture, with or without skip connections, has been motivated by the willingness to compress the input image representation in order to regularize the reconstruction towards clean images only.

This section analyzes the latent representation of clean and defective images for both the AE + Stain and the AESc + Stain methods. The purpose of this study is to supplement the results presented in Sect. 4 by introducing a new problem formulation. A strong relation will be highlighted between our previous approach and this new study. In comparison to the initial *reconstruction-based* approach, this discussion addresses the anomaly detection task as the identification of *out-of-distribution* samples. In this formulation, each input image is represented in a another domain. This step has for purpose to produce tensors whose definition is particularly suited for this task. Then, a new metric quantifies the likelihood of a tensor to be sampled from the clean data distribution or not.

5.1 Formulating Anomaly Detection as an Out-of-Distribution Sample Detection Problem

In this section, two fundamental notions are introduced in order to formulate the anomaly detection task as the identification of out-of-distribution samples. First, each sample is represented in every layer, by the layer feature map tensors. The originality of our formulation lies in the use of the output image of the network to represent each image as a tensor. As a reminder, this output is assumed to be a clean reconstruction of the arbitrary input image. Second, a measure allows to evaluate the distance of the input to the clean distribution, in each tensor space.

Definition of the Tensor Space. Let's denote an input image by $\mathbf{x}$ and the output of the network by $\hat{\mathbf{x}}$, with both $\mathbf{x}, \hat{\mathbf{x}} \in \mathbb{R}^{256\times256}$. Note that we fixed the image resolution to 256×256 for the MVTev AD dataset, but the equations can be adapted to handle any other image spatial dimension. We also define the operator $l_i(\cdot) : \mathbb{R}^{256\times256} \rightarrow \mathbb{R}^{n_i \times n_i \times c_i}$ which projects an input image to its activation tensor in the ith layer. In this notation, n_i and c_i respectively denote the spatial dimension and the number of channels in layer i. As depicted in Fig. 8, $l_1(\mathbf{x})$ corresponds to the activation tensor obtained after one convolutional layer while $l_6(\mathbf{x})$ is the activation tensor in the bottleneck.

For each layer of the encoder, a Δl_i tensor is computed for any input image $\mathbf{x}$ by subtracting the activation maps of $\hat{\mathbf{x}}$ from those of $\mathbf{x}$:

$$\Delta l_i(\mathbf{x}) = l_i(\mathbf{x}) - l_i(\hat{\mathbf{x}}), \tag{2}$$

with $\hat{\mathbf{x}}$ being the closest clean reconstruction of $\mathbf{x}$, obtained by inferring $\mathbf{x}$ through the AE or AESc autoencoder. If $\mathbf{x}$ is sampled from the clean distribution, the images $\mathbf{x}$ and $\hat{\mathbf{x}}$ should be almost identical, as their activation tensors. This implies that the resulting Δl_i tensor of $\mathbf{x}$ should contains close-to-zero values. On the contrary, if $\mathbf{x}$ is sampled from the defective distribution, the corresponding Δl_i tensor should deviate from the zero tensor.

Definition of the Distance to the Clean Distribution. It is expected that all the Δl_i tensors of a clean image contain close-to-zero values. However, it has been observed that, for a set of clean images, the components diverge more or less from this zero value depending on the input image. In order to take this phenomenon into account, we propose to represent each component of the Δl_i tensors, denoted by the index j, by the mean ($\mu_{i,j}$) and variance ($\sigma_{i,j}$) observed across the training set as described

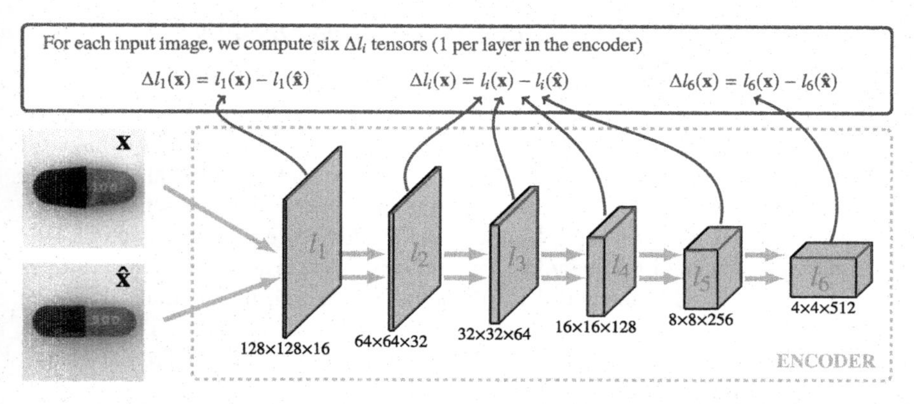

Fig. 8 For each input image $\mathbf{x}$, we infer both this image and its closest clean version $\hat{\mathbf{x}}$ obtained by passing $\mathbf{x}$ through AE or AESc. Hence, we obtain two activation tensors per layer i denoted by the operator $l_i(\cdot)$, one corresponding to $\mathbf{x}$ and the other to $\hat{\mathbf{x}}$. The six Δl_i tensors of an input image are obtained by computing the difference between the activation tensor of the input image and the ones of its reconstructed clean version

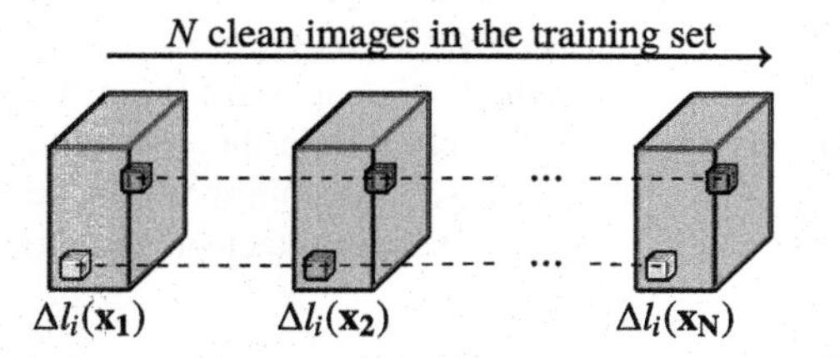

For each component j in the Δl_i tensor with dimension $n_i \times n_i \times c_i$:

- $\mu_{i,j} = \sum_{n=1}^{N} \frac{\Delta l_{i,j}(\mathbf{x_n})}{N}$
- $\sigma_{i,j} = \sqrt{\frac{\sum_{n=1}^{N} (\Delta l_{i,j}(\mathbf{x_n}) - \mu_{i,j})^2}{N}}$

Fig. 9 For each component j of the activation tensors, we compute the mean and the standard deviation of the neuron values across all clean images in the training set. Then, μ_j and σ_j represent the normal clean distribution of an activation value for a given component

in Fig. 9. The activation of the jth component in each Δl_i is considered as abnormal (or out-of-distribution) if its value is not contained in the range $\mu_{i,j} \pm 3\sigma_{i,j}$.

Then, we quantify the level of anomaly affecting an image, by the number of out-of-distribution activations divided by the tensor dimension. This value is referred as the Out-of-Distribution Ratio (OoDR):

$$\text{OoDR}_i(\mathbf{x}) = \sum_{j=1}^{n_i \times n_i \times c_i} \frac{[\Delta l_{i,j}(\mathbf{x})]_{\text{OoD}}}{n_i \times n_i \times c_i} \tag{3}$$

$$\text{with} \begin{cases} [\Delta l_{i,j}(\mathbf{x})]_{\text{OoD}} = 0, & \text{if } \mu_{i,j} - 3\sigma_{i,j} \leq \Delta l_{i,j}(\mathbf{x}) \leq \mu_{i,j} + 3\sigma_{i,j} \\ [\Delta l_{i,j}(\mathbf{x})]_{\text{OoD}} = 1, & \text{else.} \end{cases}$$

The principal motivation behind this definition lies in a metric which scales well when dealing with high dimensional tensors.

5.2 *Out-of-Distribution Ratio for Clean and Defective Samples*

As introduced in Sect. 3, our networks are trained to minimize the reconstruction error over clean structures in the image. Despite training does not explicitly encourage the separation of clean and defective image tensors, we observe that this separation naturally emerges. This is revealed by the proximity/distance to zero of the OoDR_i distributions associated to clean/defective images. Moreover, the OoDR_i distributions appear to provide insightful justifications of the various anomaly detection performance obtained for different categories of images in Table 1. This analysis summarizes the results obtained over the image categories of the MVTec AD dataset by presenting three illustrative cases.

Case 1: Image category for which both AESc + Stain and AE + Stain achieve high performance. This first case focuses on the Bottle image category, for which both networks achieve close to 0.98 image-wise detection AUC (see Table 1). The OoDR_i distributions in both AE and AESc networks over the Bottle category are shown in Fig. 10. As expected, we observe that the OoDR_i of clean images (blue) lead to smaller values than those of the corrupted/defective images (red). In the training set, the clean OoDR_i distributions are peaky in zero, while the corrupted ones are spread over a larger range of values. In the test set, the sharpness of the clean distributions is reduced which increases the overlap between the clean and the defective distributions.

With respect to the layer index, we observe that the OoDR_i distribution separation increases as we go deeper into the network, which is even more evident with the AESc architecture. Particularly, in the bottleneck ($i = 6$) of the AESc network, the two image distributions are almost perfectly separable.

Case 2: Image categories for which high performance is achieved with AESc + Stain but not with AE + Stain. This second case focuses on two texture image categories, Tile and Carpet. The AESc + Stain method achieves accurate detection of defective images (0.99 and 0.89 on Tile and Carpet, respectively) while image-wise detection AUC obtained with the AE + Stain methods are poor (0.57 and 0.43 on Tile and Carpet, respectively).

As shown in Fig. 10, the clean (blue) and corrupted/defective (red) OoDR_i distributions of the AE + Stain method obtained over the Tile category follow the same trend. Moreover, the OoDR_i distributions of the clean images sampled from the test set are shifted to the right in comparison to those of the training set. Such trend is not visible with the AESc + Stain method, for which the OoDR_i distributions follow a similar behavior to the one observed for the Bottle dataset.

Regarding the Carpet category, the poor anomaly detection performance achieved by the AE + Stain model is also reflected by overlapping OoDR_i distributions. The OoDR_i distributions on test samples appear to be bi-modal and widely spread

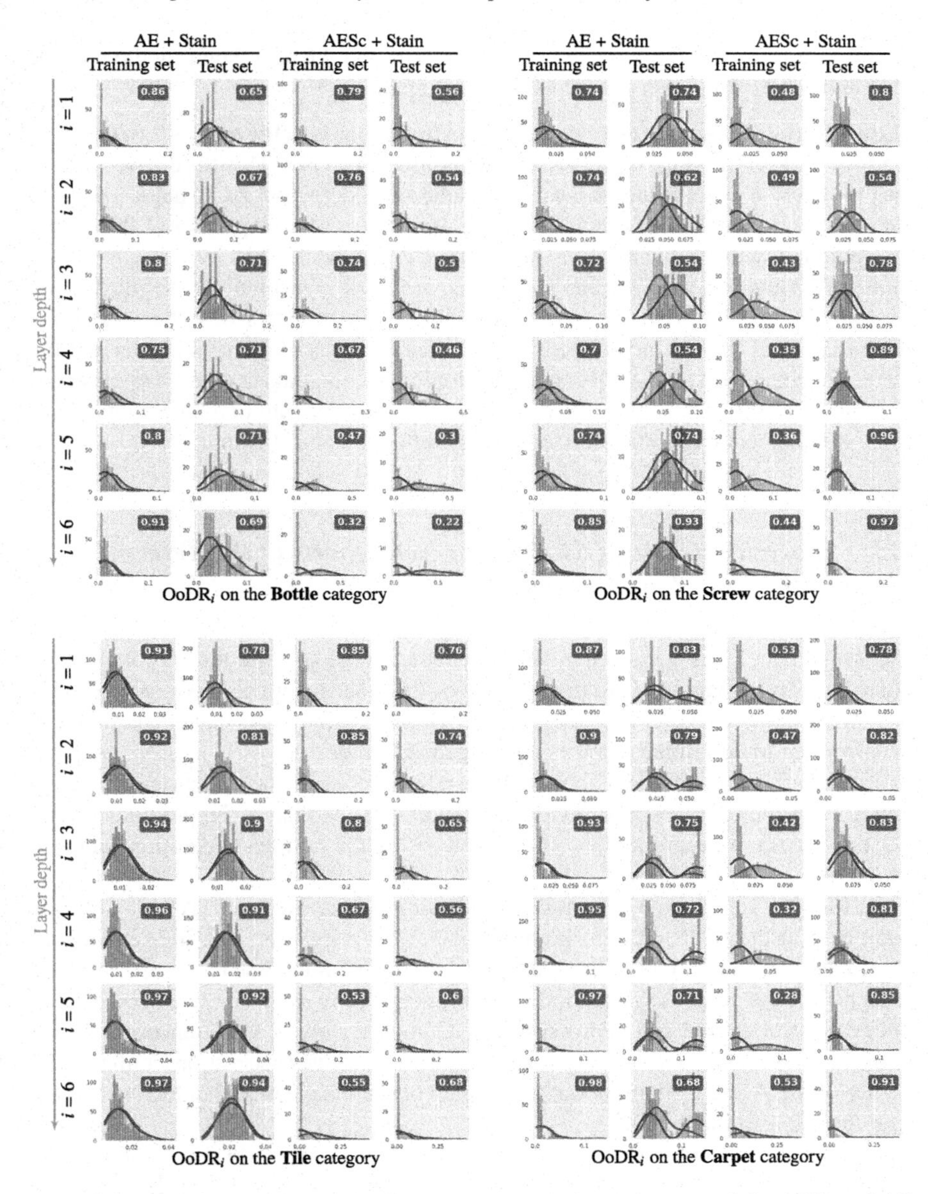

Fig. 10 Distribution of the OoDR$_i$ in each layer (only the encoder part) of the AESc and AE networks trained over images of the Bottle (up left), Screw (up right), Tile (down left) and Carpet (down right) categories and corrupted with the Stain noise model. For both architectures, the distribution for the training set (odd columns) and those for the test set (even columns) are presented apart. In each graph, OoDR$_i$ distributions related to clean images are represented in blue while OoDR$_i$ distributions related to corrupted (training)/defective (test) images are represented in red. The ratio of overlapping area between the two curves is provided in the gray rectangles (0 for non overlapping curves, 1 for identical curves)

patterns. Again, those trends are not present with the AESc + Stain method where blue and red distributions appear to be more separable.

Case 3: Image category for which poor performance is achieved both with AESc + Stain and AE + Stain compared to the state of the art. This third case focuses on the Screw category due to the poor image-wise detection AUC achieved (0.74 for AESc + Stain and 0.6 for AE + Stain) with respect to the perfect (1.00) detection obtained with the state of the art method. As shown in Fig. 10, we observe an unexpected inversion of the clean and defective OoDR_i distributions over the test set for both AE and AESc networks. One possible explanation for this phenomenon is the variation of the luminosity between the clean images of the training set and the clean images of the test set. By carefully looking at the dataset, we have observed a dimming of the lightning in the clean image test set leading to a global change of the screws appearance characterized by less reflection and shading.

5.3 Analyzing Out-of-Distribution Activations at the Tensor Component Level

The objective of this section is to study whether some specific tensor components contribute more than others to the increase of the OoDR_i for defective images. This is of interest to potentially identify a subset of tensor components that are particularly discriminant to determine whether an image is clean or not.

In Sect. 5.2, we observed that the separation between the clean and defective OoDR_i distributions was more evident in the bottleneck than in the previous layers. For this reason, we focus the rest of our analysis on the samples separation achieved exclusively in the bottleneck.

In Fig. 11, we provide a partial visualization of the Δl_6 and $[\Delta l_6]_{\text{OoD}}$ tensors computed in the bottleneck for both AE and AESc networks. The image category chosen for this experiment is the Bottle one, for which our method performs well. The tendency of the AESc + Stain model to produce near-to-zero Δl_6 tensors for clean images is strongly marked in this situation. Comparatively, the Δl_6 tensors obtained with AE + Stain for clean images are more noisy. This observation confirms the ability of the AESc + Stain model to separate better the clean and defective images than the AE + Stain model.

A more detailed analysis of this figure does not allow to state that some specific tensor components are particularly relevant to distinguish clean and defective images. If this were the case, vertical stripes would have been visible on the "unit blocks" obtained from the $[\Delta l_6]_{\text{OoD}}$ tensors.

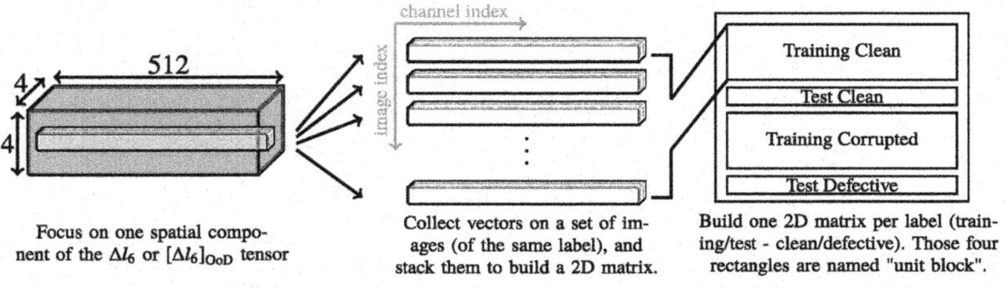

(a) Legend describing how the dataset is summarized in one "unit block". Out of a Δl_6 or $[\Delta l_6]_{\mathrm{OoD}}$ tensor of dimension $4 \times 4 \times 512$, only one one spatial resolution in considered by keeping only one $1 \times 1 \times 512$ vector per input image. This procedure is repeated for each image in the training clean/corrupted and the test clean/defective sets. All the vectors obtained from images belonging to the same class label are stacked together. Finally, one dataset is summarized by one "unit block" which is the ensemble of the training clean, test clean, training corrupted and test defective stacked vectors.

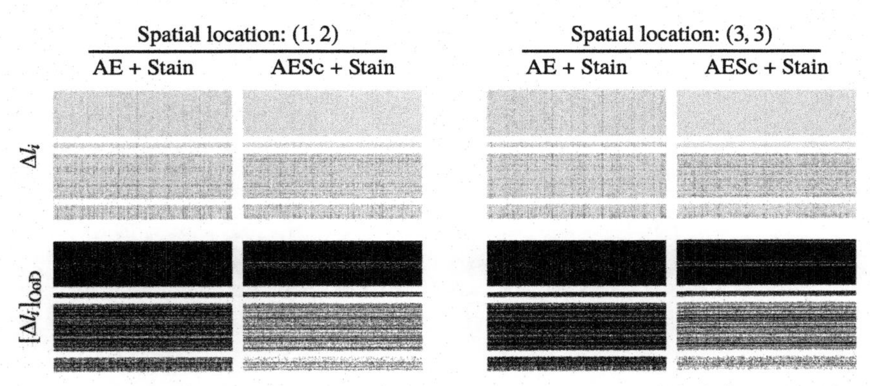

(b) According to the legend described above, the subfigure presents the "unit blocks" corresponding to the Bottle category. In the first row, the "unit blocks" are constructed from the Δl_6 tensors. The color map varies from the blue (negative activation values) to the red (positive activation values). In the second row, the "unit blocks" are constructed from the $[\Delta l_6]_{\mathrm{OoD}}$ tensors. A black value indicates that the corresponding component j of the Δl_6 ranges between $\mu_{6,j} \pm 3\,\sigma_{6,j}$, while a white value stands for an out-of-distribution component.

Fig. 11 Visualization of the "unit blocks" obtained both from the Δl_6 and the $[\Delta l_6]_{\mathrm{OoD}}$ tensors at two spatial locations in the bottleneck: (1, 2) and (3, 3). Those results are generated from the images of the Bottle category

5.4 Using the Out-of-Distribution Ratio as a Measure Quantifying the Level of Abnormality

Despite the fact that our networks are not explicitly trained to address the anomaly detection task as an out-of-distribution problem, it is possible to study the detection performances obtained when using the OoDR_i as a criterion to distinguish clean from defective images. Since we have observed previously that the best OoDR_i separation is achieved in the bottleneck, we carry out this analysis exclusively on the OoDR_6

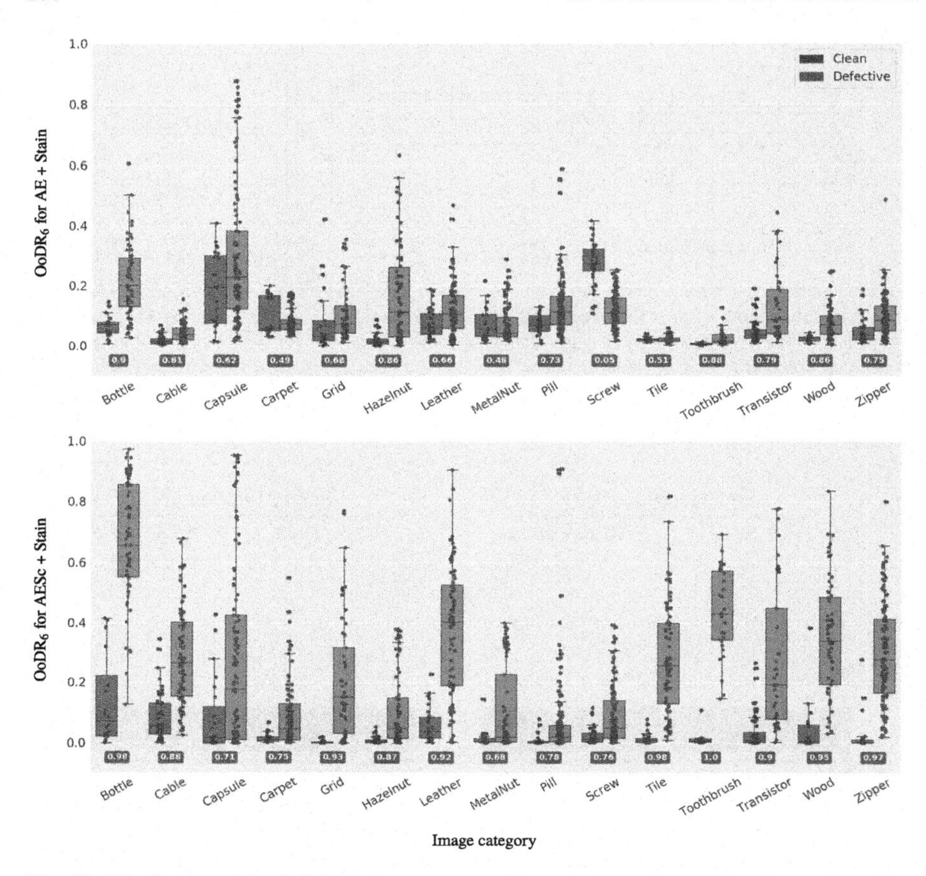

Fig. 12 Distribution of the OoDR_6 values obtained on the test clean (blue) and test defective (red) images on all image categories. Values below each box plot provide the image-wise detection AUC obtained when using the OoDR_6 as a decision rule. Upper graph provides the results obtained with the AE + Stain method and lower graph for the AESc + Stain method

values. Results for all image categories are represented in Fig. 12. In general, AESc + Stain achieves better image-wise detection AUC values than AE + Stain, which is consistent with previous observations. In comparison with the reconstruction-based approach (involving the computation of the $\mathcal{L}^2$-norm between an input image and its reconstruction), the out-of-distribution formulation achieves lower detection rates in general.

Intuitively, the design of a detector that would combine both the out-of-distribution and the reconstruction-based approaches into account could be used to enhance the detection of defective samples. Nevertheless, an improvement of the detection performance will be obtained only if the two decision criteria are complementary. Figure 13 however reveals that, in the current training configuration, the two criteria are strongly correlated. In this figure, the $\mathcal{L}^2$-norm is plotted as a function of the

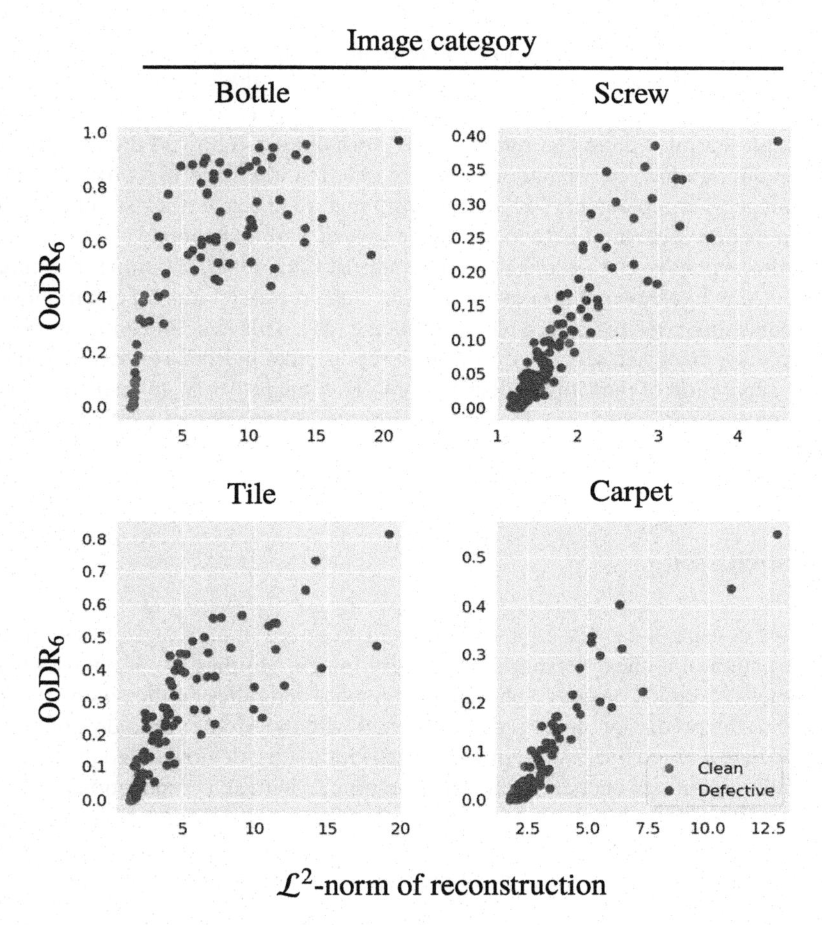

Fig. 13 For the test clean images (blue) and test defective images (red), we show the $OoDR_6$ value depending on the $\mathcal{L}^2$-norm of the reconstruction error. We performed this analysis for the four image categories addressed in Sect. 5.2 (Bottle, Screw, Tile and Capet)

$OoDR_6$ value. Generally, there is no decision boundary that performs significantly better than a vertical or an horizontal line. Hence, using a single criterion is sufficient, and there is no significant gain to expect from their combination.

5.5 Related Works and Perspectives

Multiple recent works have studied the internal representation of an image by a CNN to address anomaly detection [37, 38] or adversarial attacks detection [39, 40]. In the particular context of detecting adversarial samples, Granda et al. [40] revealed that only a limited subset of neurons is relevant to predict the class label. Hence, adversarial samples can be detected by observing how the state of those relevant neurons change compared to a class centroid. In comparison, since no class label is available during training in our case, our formulation relies on the comparison between the internal representations associated to the input and to its reconstructed clean image. It complements [40] by showing that, in absence of class supervision, all neurons of a layer are likely to change in presence of an anomaly.

Another recent work has considered the internal CNN representation for anomaly detection, and has shown that constraining the latent representation of clean samples to be sparse improves the detection performance [41]. This is in line with our observations, since increased sparsity of the clean representation is also expected to favor a better separation of the OoDR_i distributions. This suggests that an interesting path for future research could consist in promoting OoDR_i separation explicitly during training.

6 Conclusion

Our work considers the detection of abnormal structure in an images based on the reconstruction of a clean version of this query image. Similarly to previous works, our framework builds on convolutional autoencoder and relies on the reconstruction residual or the prediction uncertainty, estimated with the Monte Carlo dropout technique, to detect anomalies. As an original contribution, we demonstrated the benefits of considering an autoencoder architecture equipped with skip connections, as long as the training images are corrupted with our Stain noise model to avoid convergence towards an identity operator. This new approach performs significantly better than traditional autoencoders to detect real defects on texture images of the MVTec AD dataset.

Furthermore, we also provided a fair comparison between the residual- and uncertainty-based detection strategies relying on our AESc + Stain model. Unlike the reconstruction residual, the uncertainty indicator is independent of the contrast between the defect and its surroundings, which is particularly relevant for low contrast defects localization. However, in comparison to the residual-based detection strategy, the uncertainty-based approach increases the false positive rate in the clean structures.

To better understand our evaluation, we conducted a throughout analysis of the internal representations of clean and defective samples. For this purpose, the anomaly detection task has been formulated as the identification of out-of-distribution samples. In other words, it identifies the samples for which the internal representations of the input and reconstructed image significantly differ, compared to an estimated normal variation level observed among clean samples. Our study revealed that some correlation exists between the performance of our initial approach and the natural trend of the network to separate clean from defective samples in this new problem formulation. This suggests that controlling explicitly the separation of clean and corrupted image representations during training could help in improving the anomaly detection performance on datasets where our approach is now ineffective.

Acknowledgments Part of this work has been funded by the "Pôle Mecatech ADRIC" Walloon Region project, and by the Belgian National Fund for Scientific Research (FRS-FNRS).

References

1. Bergmann P, Fauser M, Sattlegger D, Steger C (2019) MVTec AD - a comprehensive real-world dataset for unsupervised anomaly detection. In: Proceedings of the IEEE/CVF conference on computer vision and pattern recognition (CVPR), pp 9592–9600
2. Pimentel MAF, Clifton DA, Clifton L, Tarassenko L (2014) A review of novelty detection. Signal Process 99:215–249
3. Chandola V, Banerjee A, Kumar V (2009) Anomaly detection: a survey. ACM Comput Surv (CSUR) 41:1–58
4. Rahmatov N, Paul A, Saeed F, Hong WH, Seo HC, Kim J (2019) Machine learning-based automated image processing for quality management in industrial Internet of Things. Int J Distrib Sens Netw (IJDSN), 15(10)
5. Seebock P (2019) Exploiting epistemic uncertainty of anatomy segmentation for anomaly detection in retinal OCT. IEEE Trans Med Imaging 39(1):87–98
6. Huang C, Cao J, Ye F, Li M, Zhang Y, Lu C (2019) Inverse-transform autoencoder for anomaly detection. arXiv preprint arXiv:1911.10676
7. Kendall A, Gal Y (2017) What uncertainties do we need in Bayesian deep learning for computer vision? In: Proceedings of the advances in neural information processing systems conference (NeurIPS), pp 5574–5584
8. Collin A-S, de Vleeschouwer C (2021) Improved anomaly detection by training an autoencoder with skip connections on images corrupted with stain-shaped noise. In: Proceedings of the international conference on pattern recognition (ICPR)
9. Carrera D, Boracchi G, Foi A, Wohlberg B (2015) Detecting anomalous structures by convolutional sparse models. In: Proceedings of the international joint conference on neural networks (IJCNN), pp 1–8
10. Napoletano P, Piccoli F, Schettini R (2018) Anomaly detection in nanofibrous materials by CNN-based self-similarity. Sensors 18(1):209
11. Staar B, Lütjen M, Freitag M (2019) Anomaly detection with convolutional neural networks for industrial surface inspection. Procedia CIRP 79:484–489
12. Zhou C, Paffenroth RC (2017) Anomaly detection with robust deep autoencoders. In: Proceedings of the ACM SIGKDD international conference on knowledge discovery and data mining (SIGKDD), pp 665–674
13. Leveni F, Deib M, Boracchi G, Alippi C (2021) PIF: anomaly detection via preference embedding. In: Proceedings of the international conference on pattern recognition (ICPR)
14. Zhao H, Gallo O, Frosio I, Kautz J (2016) Loss functions for image restoration with neural networks. IEEE Trans Comput Imaging 3(1):47–57
15. Bergmann P, Löwe S, Fauser M, Sattlegger D, Steger C (2019) Improving unsupervised defect segmentation by applying structural similarity to autoencoders. In: Proceedings of the international joint conference on computer vision, imaging and computer graphics theory and applications (VISIGRAPP), pp 372–380
16. Sabokrou M, Khalooei M, Fathy M, Adeli E (2018) Adversarially learned one-class classifier for novelty detection. In: Proceedings of the IEEE computer society conference on computer vision and pattern recognition (CVPR), pp 3379–3388

17. Schlegl T, Seeböck P, Waldstein SM, Schmidt-Erfurth U, Langs G (2017) Unsupervised anomaly detection with generative adversarial networks to guide marker discovery. In: Proceedings of the international conference on information processing in medical imaging (IPMI), pp 146–157
18. Schlegl T, Seeböck P, Waldstein SM, Langs G, Schmidt-Erfurth U (2019) f-AnoGAN: fast unsupervised anomaly detection with generative adversarial networks. Med Image Anal 54:30–44
19. Akçay S, Atapour-Abarghouei A, Breckon TP (2019) Skip-GANomaly: skip connected and adversarially trained encoder-decoder anomaly detection. In: Proceedings of the international joint conference on neural networks (IJCNN), pp 1–8
20. Baur C, Wiestler B, Albarqouni S, Navab N (2018) Deep autoencoding models for unsupervised anomaly segmentation in brain MR images. In: Proceedings of the international MICCAI Brainlesion workshop, pp 161–169
21. Radford A, Metz L, Chintala S (2016) Unsupervised representation learning with deep convolutional generative adversarial networks. In: Proceedings of the international conference on learning representations (ICLR), pp 1–16
22. Goodfellow I, et al (2014) Generative adversarial nets. In: Proceedings of the advances in neural information processing systems conference (NeurIPS), pp 2672–2680
23. Snell J, et al (2017) Learning to generate images with perceptual similarity metrics. In: Proceedings of the IEEE international conference on image processing (ICIP), pp 4277–4281
24. Dehaene D, Frigo O, Combrexelle S, Eline P (2020) Iterative energy-based projection on a normal data manifold for anomaly localization. In: Proceedings of the international conference on learning representations (ICLR)
25. Haselmann M, Gruber DP, Tabatabai P (2018) Anomaly detection using deep learning based image completion. In: Proceedings of the IEEE international conference on machine learning and applications (ICMLA), pp 1237–1242
26. Munawar A, Creusot C (2015) Structural inpainting of road patches for anomaly detection. In: Proceedings of the IAPR international conference on machine vision applications (MVA), pp 41–44
27. Zavrtanik V, Kristan M, Skočaj D (2021) Reconstruction by inpainting for visual anomaly detection. Pattern Recogn 112:107706
28. Mei S, Wang Y, Wen G (2018) Automatic fabric defect detection with a multi-scale convolutional denoising autoencoder network model. Sensors 18(4):1064
29. Gong D, et al (2019) Memorizing normality to detect anomaly: memory-augmented deep autoencoder for unsupervised anomaly detection. In: Proceedings of the IEEE/CVF international conference on computer vision (ICCV), pp 1705–1714
30. Wang L, Zhang D, Guo J, Han Y (2020) Image anomaly detection using normal data only by latent space resampling. Appl Sci 10(23):8660–8679
31. Van den Oord A, Vinyals O, Kavukcuoglu K (2017) Neural discrete representation learning. Proc Adv Neural Inf Process Syst Conf (NeurIPS) 30:6306–6315
32. Ronneberger O, Fischer P, Brox T (2015) U-Net: convolutional networks for biomedical image segmentation. In Proceedings of the international conference on medical image computing and computer-assisted intervention (MICCAI), pp 234–241
33. Kingma DP, Ba JL (2015) Adam: a method for stochastic optimization. In: Proceedings of the international conference on learning representations (ICLR)
34. Zhong Z, Zheng L, Kang G, Li S, Yang Y (2020) Random erasing data augmentation. In: Proceedings of the AAAI conference on artificial intelligence, pp 13001–13008
35. Fong R, Vedaldi A (2019) Occlusions for effective data augmentation in image classification. In: Proceedings of the international conference on computer vision workshop (ICCVW), pp 4158–4166
36. Liu G, Reda FA, Shih KJ, Wang TC, Tao A, Catanzaro B (2018) Image inpainting for irregular holes using partial convolutions. In: Proceedings of the European conference on computer vision (ECCV), pp 85–100

37. Rippel O, Mertens P, Merhof D (2020) Modeling the distribution of normal data in pre-trained deep features for anomaly detection. arXiv preprint arXiv:2005.14140
38. Massoli FV, Falchi F, Kantarci A, Akti Ş, Ekenel HK, Amato G (2020) MOCCA: multi-layer one-class classification for anomaly detection. arXiv preprint arXiv:2012.12111
39. Lee K, Lee K, Lee H, Shin J (2018) A simple unified framework for detecting out-of-distribution samples and adversarial attacks. In: Proceedings of the advances in neural information processing systems conference (NeurIPS), pp 7167–7177
40. Granda R, Tuytelaars T, Oramas J (2020) Can the state of relevant neurons in a deep neural networks serve as indicators for detecting adversarial attacks? arXiv preprint arXiv:2010.15974
41. Abati D, Porrello A, Calderara S, Cucchiara R (2019) Latent space autoregression for novelty detection. In: Proceedings of the IEEE/CVF conference on computer vision and pattern recognition (CVPR), pp 481–490

A Comparative Study of L_1 and L_2 Norms in Support Vector Data Descriptions

Edgard M. Maboudou-Tchao and Charles W. Harrison

Abstract The Support Vector Data Description (L_1 SVDD) is a non-parametric one-class classification algorithm that utilizes the L_1 norm in its objective function. An alternative formulation of SVDD, called L_2 SVDD, uses a L_2 norm in its objective function and has not been extensively studied. L_1 SVDD and L_2 SVDD are formulated as distinct quadratic programming (QP) problems and can be solved with a QP-solver. The L_2 SVDD and L_1 SVDD's ability to detect small and large shifts in data generated from multivariate normal, multivariate t, and multivariate Laplace distributions is evaluated. Similar comparisons are made using real-world datasets taken from various applications including oncology, activity recognition, marine biology, and agriculture. In both the simulated and real-world examples, L_2 SVDD and L_1 SVDD perform similarly, though, in some cases, one outperforms the other. We propose an extension of the SMO algorithm for L_2 SVDD, and we compare the runtimes of four algorithms: L_2 SVDD (SMO), L_2 SVDD (QP), L_1 SVDD (SMO), and L_1 SVDD (QP). The runtimes favor L_1 SVDD (QP) versus L_2 SVDD (QP), sometimes substantially; however using SMO reduces the difference in runtimes considerably, making L_2 SVDD (SMO) feasible for practical applications. We also present gradient descent and stochastic gradient descent algorithms for linear versions of both the L_1 SVDD and L_2 SVDD. Examples using simulated and real-world data show that both methods perform similarly. Finally, we apply the L_1 SVDD and L_2 SVDD to a real-world dataset that involves monitoring machine failures in a manufacturing process.

Keywords One-class classification · Support Vector Data Description · Sequential minimum optimization · L1-norm · L2-norm · SVDD · L2-SVDD · L1-SVDD · Quadratic programming · Gradient descent · Stochastic gradient descent · Monitoring · Manufacturing control chart · Machine failure

---

E. M. Maboudou-Tchao (✉) · C. W. Harrison
Department of Statistics and Data Science, University of Central Florida, Orlando, USA
e-mail: edgard.maboudou@ucf.edu

K. P. Tran (ed.), *Control Charts and Machine Learning for Anomaly Detection in Manufacturing*, Springer Series in Reliability Engineering,
https://doi.org/10.1007/978-3-030-83819-5_9

1 Introduction

One-class classification (OCC) refers to the problem of constructing a classifier using only data with a single label, called the target class, in order to distinguish between target class data and non-target class data. If the target class data are assumed to be, in some sense, typical, then the non-target class may be viewed as atypical and thus one-class classification can be viewed as a form of outlier detection.

Although there are a variety of methods that can be used for OCC [8], this chapter focuses on two OCC techniques called L_1 Norm Support Vector Data Description (L_1 SVDD) and L_2 Norm Support Vector Data Description (L_2 SVDD) [3, 21]. Both methods are based on constructing a description of the target class data using support vectors. The description is a set of observed data vectors that together form a hypersphere that encompasses the target class data. This description is then utilized in order to determine if unlabeled data should be classified as either the target class or non-target class. L_1 and L_2 SVDD have a number of advantages that make them useful in practical settings where important characteristics of the data, such as its distribution, are unknown. Crucially, both methods do not need to make assumptions about the underlying probability distribution of the data, can be used when the number of variables exceeds the number of observed data vectors, and can model nonlinear boundaries using a kernel function.

L_1 SVDD, typically referred to as just SVDD, is the most commonly used version of SVDD and has been utilized in numerous applications. Sun and Tsung [20] proposed a control chart based on the L_1 SVDD, called the k-chart, to monitor data in a Statistical Process Control (SPC) context and then applied the k-chart to monitor chemical process data. Maboudou-Tchao [15] suggested a L_1 SVDD control chart using a Mahalanobis kernel. L_1 SVDD was used by Duan, Liu, and Gao [7] to monitor structural health. Sanchez-Hernandez, Boyd, and Foody [18] applied L_1 SVDD in a remote sensing context for fenland classification whereas Chaki et al. [2] used L_1 SVDD for water well saturation classification. Camerini, Coppotelli, and Bendisch [1] used L_1 SVDD to detect faults in helicopter drivetrain components whereas Luo, Wang, and Cui [10] used L_1 SVDD for analog circuit fault detection. L_1 SVDD was also used as a component in a methodology to detect anomalies in network logs [9]. Modifications of L_1 SVDD have been successfully applied for one-class classification of tensor datasets. Maboudou-Tchao, Silva, and Diawara [13] proposed Support Matrix Data Description (SMDD) to monitor changes in matrix datasets. Maboudou-Tchao [12] used tensor methods to construct control charts to monitor high-dimensional data. Maboudou-Tchao [14] suggested support tensor data description (STDD) for change-point detection in second-order tensor datasets with an application to image data. Note that least-squares one-class classification methods are also available. Choi [4] proposed least-squares one-class support vector machines (LS-OCSVM) and Maboudou-Tchao [11] used LS-OCSVM to detect change-points in the mean vector of a process.

In practice, the number of data vectors available to train a one-class classifier may be large. Larger datasets, which can range from thousands of records to millions of records, can pose a challenge for obtaining solutions for the L_1 and L_2 SVDD depending on the algorithm used to compute the solution. Both the L_1 and L_2 SVDD require a numerical method to obtain a solution and a common algorithm for doing so is called quadratic programming (QP). Unfortunately, QP becomes infeasible as the number of data vectors increases, so Platt [17] proposed a fast method called sequential minimum optimization (SMO) as an alternative to obtaining solutions using QP. Originally the SMO was proposed to obtain solutions for a support vector machine (SVM) and then SMO was later extended to L_1 SVDD. To our knowledge, SMO has not been extended to L_2 SVDD, and thus we propose an extension to the SMO algorithm for L_2 SVDD.

In cases where the data are linearly separable, the use of a non-linear kernel function is unnecessary. The linear L_1 SVDD and linear L_2 SVDD may be solved via unconstrained optimization of the primal problem using gradient descent and stochastic gradient descent. Gradient descent is an iterative optimization method for finding the minimum of a differentiable function. Gradient descent uses the entire training dataset in each epoch to accomplish its goal, but a disadvantage is that sometimes it can be trapped in local minima. Stochastic gradient descent attempts to circumvent this problem by instead using a randomly drawn observation in each epoch to update the gradient.

As implied above, the L_2 SVDD has not been as well studied as L_1 SVDD, and this fact motivates an evaluation of the L_2 SVDD in comparison to its more popular counterpart. This chapter makes the following contributions. We provide a comparison of the Type II error rate of the L_1 and L_2 SVDD for detecting small and large shifts in an underlying statistical process using data generated from the multivariate normal, multivariate t, and multivariate Laplace distributions. We propose a SMO algorithm for fitting the L_2 SVDD and compare the runtimes of four algorithms including L_1 SVDD (QP), L_1 SVDD (SMO), L_2 SVDD (QP), and L_2 SVDD (SMO) using simulated and real-world datasets.

The remainder of this chapter is structured as follows. Section 2 and Sect. 3 review the L_1 SVDD and L_2 SVDD, respectively. Section 4 presents a simulation study to evaluate the performance of the L_1 and L_2 SVDD. Section 5 provides a review of the SMO algorithm for L_1 SVDD and presents the SMO algorithm for L_2 SVDD. Section 5 also compares the SMO algorithms to their corresponding quadratic programming counterparts. Section 6 presents both gradient descent and stochastic gradient descent algorithms for solving the unconstrained (linear) L_1 SVDD as well as the unconstrained (linear) L_2 SVDD; their performance is compared on both simulated and real-world datasets. Section 7 presents a real-world application of L_1 and L_2 SVDD for monitoring machine failures in a manufacturing process. Section 8 provides a brief summary of this chapter.

2 L_1 Norm Support Vector Data Description (L_1 SVDD)

Let $\mathbf{x}_i, i = 1, 2, \ldots, N$ be a sequence of $p-$variate training (or target) observations. The Support Vector Data Description with a L_1 norm [21] tries to find a sphere with minimum volume containing all (or most of) of the observations and can be formulated as an optimization problem:

$$\min_{r,\mathbf{a},\xi_i} r^2 + C\sum_{i=1}^{N} \xi_i,$$

$$\text{subject to } ||\phi(\mathbf{x}_i) - \mathbf{a}||^2 = (\phi(\mathbf{x}_i) - \mathbf{a})'(\phi(\mathbf{x}_i) - \mathbf{a}) \leq r^2 + \xi_i \tag{2.1}$$

$$\xi_i \geq 0, i = 1, 2, \ldots, N$$

where $\mathbf{a}$ is the center of the sphere, r is the radius of the sphere, ξ_i are the slack variables, $C > 0$ is a parameter introduced to control the influence of the slack variables, and $\phi(\cdot)$ is a mapping that takes an input $\mathbf{x} \in \mathcal{X} \subseteq \mathbb{R}^p$ and maps it to a feature space $\mathcal{F}$; that is, $\phi : \mathcal{X} \rightarrow \mathcal{F}$ where $\mathcal{F}$ is a Hilbert space. The corresponding dual problem after applying the kernel trick $k(\mathbf{x}_i, \mathbf{x}_j) = \phi(\mathbf{x}_i)'\phi(\mathbf{x}_j)$ is given by

$$\max_{\boldsymbol{\beta}} \sum_{i=1}^{N} \beta_i k(\mathbf{x}_i, \mathbf{x}_i) - \sum_{i=1}^{N}\sum_{j=1}^{N} \beta_i\beta_j k(\mathbf{x}_i, \mathbf{x}_j),$$

$$\text{subject to} \sum_{i=1}^{N} \beta_i = 1 \tag{2.2}$$

$$0 \leq \beta_i \leq C, i = 1, 2, \ldots, N.$$

Note that the dual problem (2.2) is a convex QP problem and can be solved using a quadratic programming solver. The solution $\boldsymbol{\beta}$ is used to compute the optimal center $\mathbf{a}$, and the observed data vectors $\mathbf{x}_i$ where $\beta_i > 0$ for $i = 1, 2, \ldots, N$ are referred to as support vectors. The support vectors $\mathbf{x}_t$ that correspond to vectors located on the sphere boundary have $0 < \beta_t < C$ and are referred to as boundary support vectors whereas support vectors $\mathbf{x}_k$ such that $\beta_k = C$ are called non-boundary support vectors. Let N_1 be the number of boundary support vectors, then the squared radius r^2 is the squared distance from the center of the hypersphere $\mathbf{a}$ to the boundary support vectors $\mathbf{x}_t$:

$$r^2 = \frac{1}{N_1}\sum_{t=1}^{N_1}\left(k(\mathbf{x}_t, \mathbf{x}_t) - 2\sum_{j=1}^{N_1} \beta_j k(\mathbf{x}_t, \mathbf{x}_j) + \sum_{j=1}^{N_1}\sum_{l=1}^{N_1} \beta_j\beta_l k(\mathbf{x}_j, \mathbf{x}_l)\right) \tag{2.3}$$

For classification, a test vector **u** is in the target class if the following condition is true

$$d_{\mathbf{u}} = k(\mathbf{u}, \mathbf{u}) - 2\sum_{j=1}^{N} \beta_j k(\mathbf{u}, \mathbf{x}_j) + \sum_{j=1}^{N}\sum_{l=1}^{N} \beta_j \beta_l k(\mathbf{x}_j, \mathbf{x}_l) \leq r^2 \tag{2.4}$$

If condition (2.4) is false, then the test vector **u** is declared to be in the non-target class.

3 L2 Norm Support Vector Data Description (L2 SVDD)

Let $\mathbf{x}_j$, $j = 1, 2, \ldots, N$ be a sequence of $p-$variate training (or target) observations. The Support Vector Data Description with a L_2 norm, referred to here as L_2 SVDD, is given by the following optimization problem:

$$\min_{R^2, \mathbf{c}, \xi} R^2 + \frac{C}{2}\sum_{j=1}^{N} \xi_j^2,$$

$$\text{subject to } ||\phi(\mathbf{x}_j) - \mathbf{c}||^2 = (\phi(\mathbf{x}_j) - \mathbf{c})'(\phi(\mathbf{x}_j) - \mathbf{c}) \leq R^2 + \xi_j,\ j = 1, 2, \ldots, N \tag{3.1}$$

where **c** is the center of the sphere, R is the radius of the sphere, ξ_j are the slack variables, $C > 0$ is a parameter that controls influence of the slack variables, and $\phi(\cdot)$ is a mapping that takes an input $\mathbf{x} \in \mathcal{X} \subseteq \mathbb{R}^p$ and maps it to a feature space $\mathcal{F}$; that is, $\phi : \mathcal{X} \rightarrow \mathcal{F}$ where $\mathcal{F}$ is a Hilbert space. Introducing Lagrange multipliers α_j for $j = 1, 2, \ldots, N$ yields:

$$L = R^2 + C\sum_{j=1}^{N} \xi_j^2 - \sum_{j=1}^{N} \alpha_j \left(R^2 + \xi_j - (\phi(\mathbf{x}_j) - \mathbf{c})'(\phi(\mathbf{x}_j) - \mathbf{c})\right)$$

$$= R^2\left(1 - \sum_{j=1}^{N} \alpha_j\right) + C\sum_{j=1}^{N} \xi_j^2 - \sum_{j=1}^{N} \alpha_j \xi_j + \sum_{j=1}^{N} \alpha_j \left(\phi(\mathbf{x}_j) - \mathbf{c})'(\phi(\mathbf{x}_j) - \mathbf{c}\right) \tag{3.2}$$

Differentiating Eq. (3.2) with respect to R^2, **c**, ξ_j yields the following three equations. Each is set equal to zero and we obtain the following expressions:

$$\frac{\partial L}{\partial R^2} = 0 \Longrightarrow \sum_{j=1}^{N} \alpha_j = 1 \tag{3.3}$$

$$\frac{\partial L}{\partial \mathbf{c}} = \mathbf{0} \implies \mathbf{c} = \sum_{j=1}^{N} \alpha_j \phi(\mathbf{x}_j) \tag{3.4}$$

$$\frac{\partial L}{\partial \xi_j} = 0 \implies \xi_j = \frac{\alpha_j}{2C} \tag{3.5}$$

Substituting Eqs. (3.3), (3.4), and (3.5) into Eq. (3.2) and applying the kernel trick $k(\mathbf{x}_i, \mathbf{x}_j) = \phi(\mathbf{x}_i)'\phi(\mathbf{x}_j)$ yields the dual problem.

$$\sum_{j=1}^{N} \alpha_j k(\mathbf{x}_j, \mathbf{x}_j) - \sum_{i=1}^{N}\sum_{j=1}^{N} \alpha_i \alpha_j \left(k(\mathbf{x}_i, \mathbf{x}_j) + \frac{1}{2C}\delta_{ij} \right),$$

$$\text{subject to} \sum_{j=1}^{N} \alpha_j = 1$$

$$0 \leq \alpha_j < \infty \tag{3.6}$$

where δ_{ij} is the Kronecker Delta function.

$$\delta_{ij} = \begin{cases} 0 & \text{if } i \neq j \\ 1 & \text{if } i = j \end{cases}$$

Similarly to L_1 SVDD, the dual problem (3.6) can be solved using a QP-solver, and the solution $\boldsymbol{\alpha}$ is used to compute the optimal center $\mathbf{c}$. The Karush-Kuhn-Tucker (KKT) conditions are necessary and sufficient conditions for any optimal solution. For the L_2 SVDD, these conditions are

(i) Stationarity

1. $1 - \sum_{j=1}^{N} \alpha_j = 0$
2. $\mathbf{c} - \sum_{i=1}^{N} \alpha_j \phi(\mathbf{x}_j) = \mathbf{0}$
3. $2C\xi_j - \alpha_j = 0$

(ii) Primal feasibility

1. $||\phi(\mathbf{x}_j) - \mathbf{c}||^2 \leq R^2 + \xi_j$

(iii) Dual feasibility

1. $\alpha_j \geq 0$

(iv) Complementary Slackness

1. $\alpha_j \left(||\phi(\mathbf{x}_j) - \mathbf{c}||^2 - R^2 - \xi_j \right) = 0$

From the complementary slackness of the KKT conditions, the observed data vectors $\mathbf{x}_j$ where $\alpha_j > 0$ for $j = 1, 2, \ldots, N$ are the support vectors. Let N_2 be the number of support vectors for the L_2 SVDD, then R^2 is the squared kernel distance from the center of the hypersphere $\mathbf{c}$ to the support vectors $\mathbf{x}_s$:

$$R^2 = \frac{1}{N_2}\sum_{s=1}^{N_2}\left(k(\mathbf{x}_s, \mathbf{x}_s) - 2\sum_{j=1}^{N_2}\alpha_j k(\mathbf{x}_s, \mathbf{x}_j) + \sum_{j=1}^{N_2}\sum_{l=1}^{N_2}\alpha_j\alpha_l k(\mathbf{x}_j, \mathbf{x}_l)\right) \tag{3.7}$$

For classification, an unseen vector $\mathbf{z}$ is in the target class if the following condition is true

$$d_{\mathbf{z}} = k(\mathbf{z}, \mathbf{z}) - 2\sum_{j=1}^{N}\alpha_j k(\mathbf{z}, \mathbf{x}_j) + \sum_{j=1}^{N}\sum_{l=1}^{N}\alpha_j\alpha_l k(\mathbf{x}_j, \mathbf{x}_l) \leq R^2 \tag{3.8}$$

If condition (3.8) is false, then the unseen vector $\mathbf{z}$ is declared to be in the non-target class.

4 Simulation Study

In this section, the ability of L_1 SVDD and L_2 SVDD to detect small and large changes is assessed. Quadratic programming is used to obtain the solution in the simulations, and each simulation proceeds as follows. The data used to train the model are generated from a specified distribution with sample size 100. A test vector $\mathbf{z}_i$ is generated from the specified distribution but the first component of the distribution's mean vector is shifted by δ. The kernel distance for $\mathbf{z}_i$ is computed using the left-hand side of either condition (2.4) or condition (3.8), respectively. This process is repeated 25,000 times and yields a set of 25, 000 kernel distance values. Using a control limit h, the percentage of kernel distance values falling below h is computed to yield the false positive rate for the specified distribution for a shift in the first component of the mean vector of size δ. We choose $\delta \in [0.0, 0.1, 0.2, \ldots, 1.0, 1.5, 2.0, 2.5, 3.0, 3.5, 4.0, 5.0]$ to account for both small and large shifts. The control limit h is determined via the same procedure except that $\delta = 0$; h is computed as the $1 - \alpha$ percentile of the distributions of test kernel distances where $\alpha = 0.05$. Three distributions are evaluated in this simulation: multivariate normal, multivariate t, and multivariate Laplace. The training data used in each simulation are generated as a random draw of $N = 100$ vectors $\mathbf{x}_i \in \mathbb{R}^p$ for $i = 1, 2, \ldots, N$ where p is the number of predictor variables. The training data for the multivariate normal and multivariate Laplace have the mean vector equal to $\mathbf{0}$ and the covariance matrix equal to the identity matrix $\mathbf{I}$. The training data for the multivariate t-distribution has the mean vector equal to $\mathbf{0}$ and the covariance matrix equal to $\frac{\nu-2}{\nu}\mathbf{I}$ with $\nu = 3$ degrees of freedom. The multivariate Laplace simulations use $p = 5, 10$; the multivariate normal and multivariate t simulations use $p = 5, 10,$ and 100. The kernel function used in the simulations is the Gaussian kernel function:

$$k(\mathbf{x}_i, \mathbf{x}_j) = \frac{1}{\sqrt{2\pi}} \exp\left(-\frac{(\mathbf{x}_i - \mathbf{x}_j)'(\mathbf{x}_i - \mathbf{x}_j)}{2\sigma^2}\right) \tag{4.1}$$

4.1 Simulation Results

The results for $p = 5$ are provided in Fig. 1 and show that the L_1 SVDD and L_2 SVDD perform similarly for all values of δ for both the multivariate normal and multivariate Laplace distributions. For the multivariate t-distribution, L_2 SVDD outperforms L_1 SVDD when $\delta \in \{1.0, 1.5, 2.0, 2.5, 3.0\}$. The results for $p = 10$ are shown in Fig. 2. The L_2 SVDD marginally outperforms the L_1 SVDD for the multivariate normal data when $\delta \in \{1, 1.5, 2.0, 2.5, 3.0, 3.5, 4.0, 5.0\}$ and both models perform similarly for the multivariate t-distribution. For the multivariate Laplace distribution, the L_1 SVDD outperforms L_2 SVDD for $\delta \in \{1, 1.5, 2.0, 2.5, 3.0, 3.5, 4.0, 5.0\}$. The results for $p = 100$ are shown in Fig. 3. Both models have a similar performance for the multivariate normal distribution whereas the L_2 SVDD marginally outperforms the L_1 SVDD for the multivariate t-distribution.

5 Sequential Minimum Optimization (SMO)

Quadratic programming (QP) does not scale well as the number of observations N increases. Consequently, sequential minimum optimization (SMO) was proposed by Platt [17] as a fast alternative for computing the solutions for a support vector machine (SVM). The key insight of SMO is that the smallest optimization problem for a SVM contains two Lagrange multipliers that adhere to a linear equality constraint, so the resulting solution of this small problem has a closed-form. SMO proceeds by iteratively solving a series of the smallest possible optimization at each step.

5.1 SMO for L_1 SVDD

Consider two Lagrange multipliers β_1 and β_2 corresponding to two observations while treating the remaining $N - 3$ Lagrange multipliers as constants. At this step in the SMO algorithm, the dual problem can be expressed as the following:

$$\max_{\beta_i, \beta_j} \sum_{i=1}^{2} \beta_i k(\mathbf{x}_i, \mathbf{x}_i) - \sum_{i=1}^{2}\sum_{j=1}^{2} \beta_i \beta_j k(\mathbf{x}_i, \mathbf{x}_j) - \sum_{i=1}^{2} \beta_i \mu_i + \sum_{i=3}^{N} \beta_i k(\mathbf{x}_i, \mathbf{x}_i) - \sum_{i=3}^{N}\sum_{j=3}^{N} \beta_i \beta_j k(\mathbf{x}_i, \mathbf{x}_j),$$

$$\text{subject to} \sum_{i=1}^{2} \beta_i = \Delta \tag{5.1}$$

$$0 \leq \beta_1, \beta_2 \leq C$$

where $\mu_i = \sum_{j=3}^{N} \beta_j k(\mathbf{x}_i, \mathbf{x}_j)$ and $\Delta = 1 - \sum_{i=3}^{N} \beta_i$. Let $k_{ij} = k(\mathbf{x}_i, \mathbf{x}_j)$. Using the fact $\beta_1 = \Delta - \beta_2$ and focusing only on terms that involve β_1 and β_2, the following optimization problem is obtained:

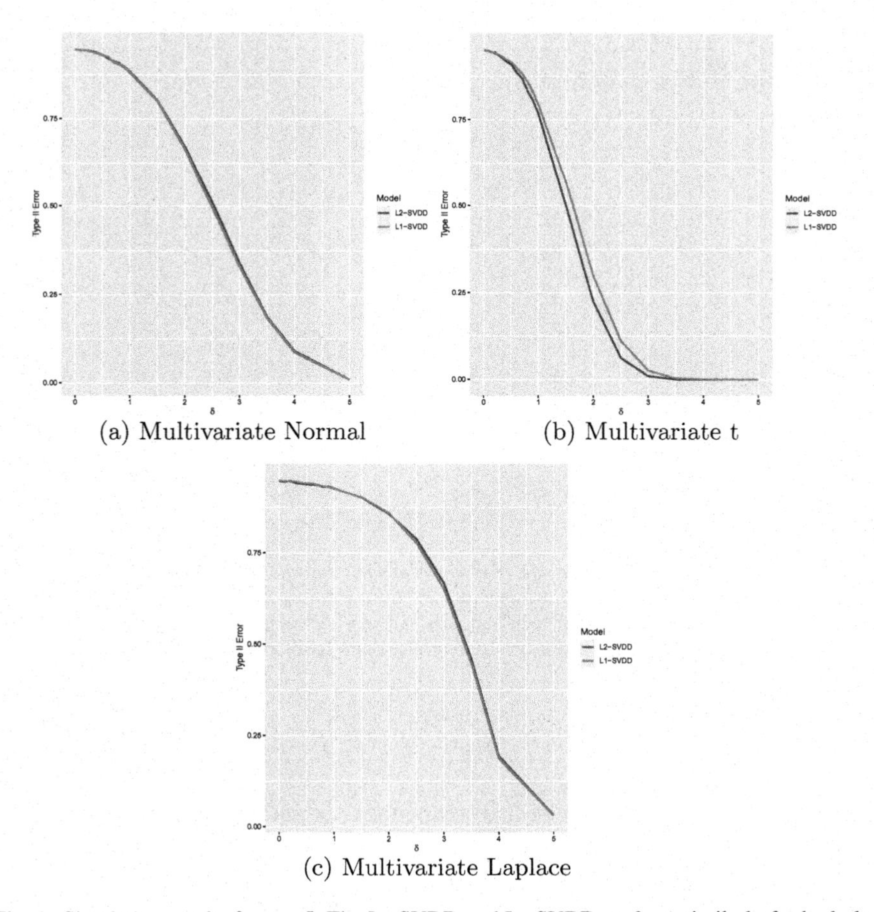

(a) Multivariate Normal (b) Multivariate t

(c) Multivariate Laplace

Fig. 1 Simulation results for p = 5. The L_1 SVDD and L_2 SVDD perform similarly for both the multivariate normal and the multivariate Laplace distributions, though the L_2 SVDD outperforms the L_1 SVDD when $\delta \in \{1.0, 1.5, 2.0, 2.5, 3.0\}$ for the multivariate t-distribution

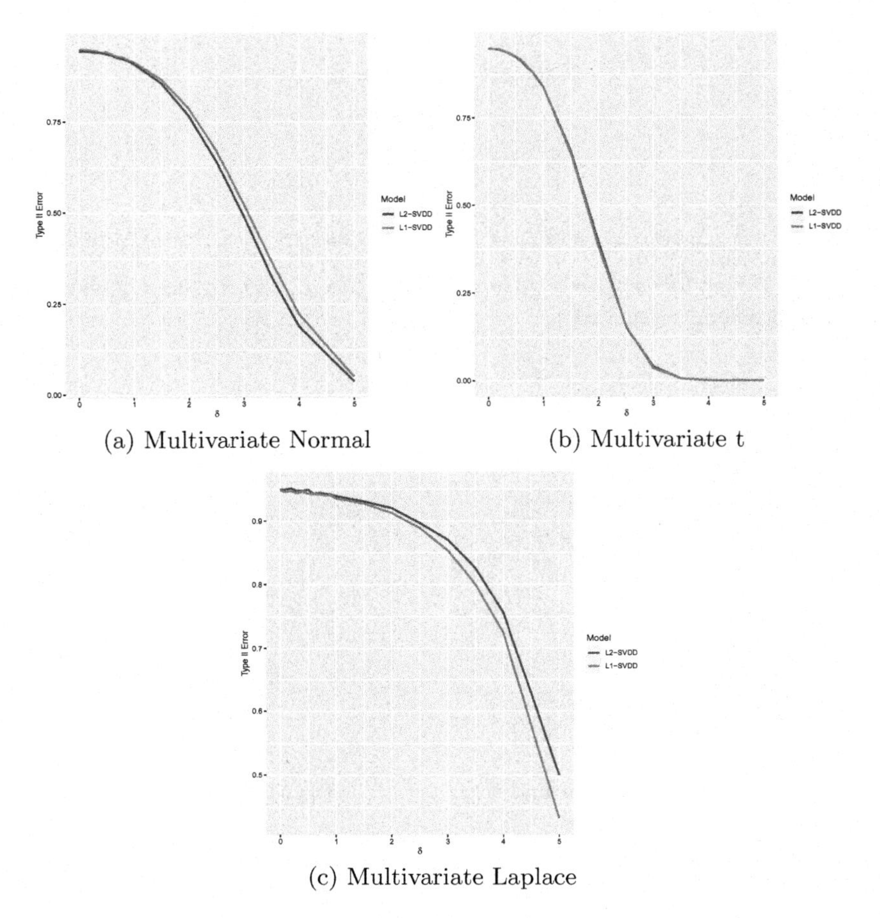

(a) Multivariate Normal (b) Multivariate t

(c) Multivariate Laplace

Fig. 2 Simulation results for p = 10. The L_2 SVDD marginally outperforms the L_1 SVDD for the multivariate normal data when $\delta \in \{1, 1.5, 2.0, 2.5, 3.0, 3.5, 4.0, 5.0\}$, both perform similarly for the multivariate t-distribution, and L_1 SVDD outperforms L_2 SVDD for $\delta \in \{1, 1.5, 2.0, 2.5, 3.0, 3.5, 4.0, 5.0\}$ for the multivariate Laplace distribution

$$\max_{\beta_2} \left((\Delta - \beta_2) - (\Delta - \beta_2)^2\right) k_{11} - 2(\Delta - \beta_2)\beta_2 k_{12} + (\beta_2 - \beta_2^2)k_{22} \tag{5.2}$$

Differentiating with respect to β_2 and setting the expression equal to zero yields

$$-k_{11} + k_{22} - (-2(\Delta - \beta_2)k_{11}) + \mu_1 - \mu_2 - 2(\Delta - 2\beta_2)k_{12} - 2\beta_2 k_{22} = 0$$

Solving for β_2 yields:

$$\beta_2 = \frac{2\Delta(k_{11} - k_{12}) - k_{11} + k_{22} + \mu_1 - \mu_2}{2(k_{11} + k_{12}) - 4k_{12}} \tag{5.3}$$

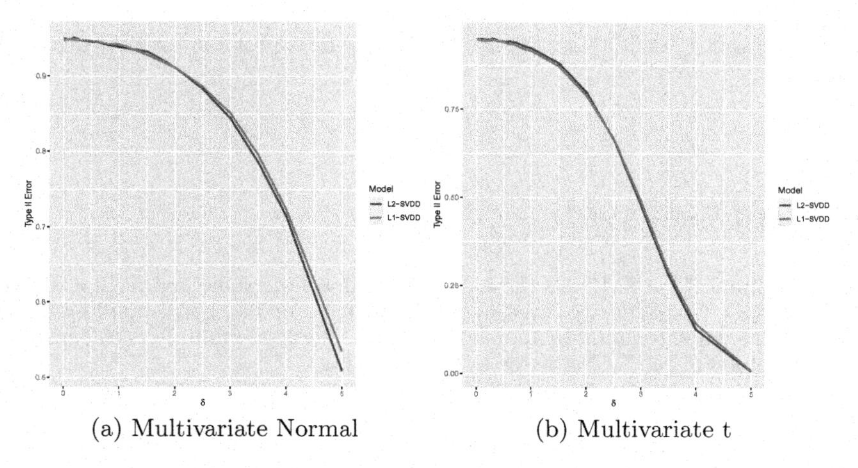

Fig. 3 Simulation results for $p = 100$. The L_2 SVDD and L_1 SVDD both have a similar performance for both the multivariate t-distribution whereas L_2 SVDD marginally outperforms the L_1 SVDD for the multivariate normal distribution when $\delta \in \{3, 3.5, 4, 5\}$

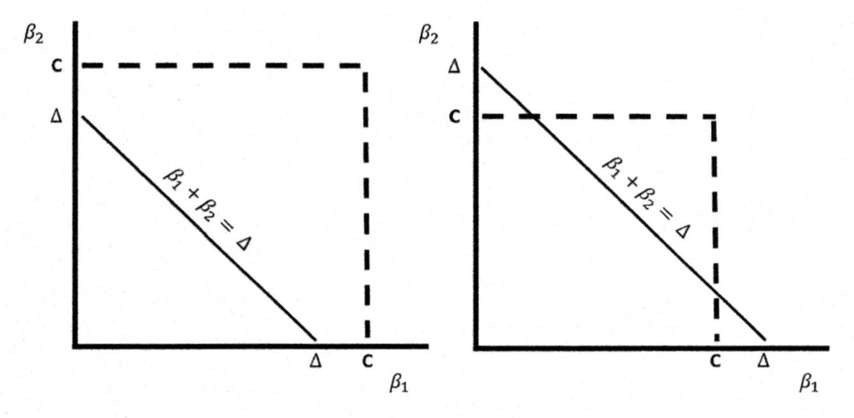

Fig. 4 The constraints for β_2 in SMO for L_1 SVDD

where $\mu_i = \sum_{j=3}^{N} \beta_j k_{ij}$. Since $0 \leq \beta_i \leq C$ for $i = 1, 2, \ldots, N$ and $\beta_1 + \beta_2 = \Delta$, there are additional constraints on β_2. If $C > \Delta$, then $0 \leq \beta_2 \leq \Delta$ whereas if $C \leq \Delta$, then $\Delta - C \leq \beta_2 \leq C$. Equivalently, the lower and upper bound for β_2 is given by

$$L_{\beta_2} = \max(0, \beta_1 + \beta_2 - C) \tag{5.4}$$

$$H_{\beta_2} = \min(C, \beta_1 + \beta_2). \tag{5.5}$$

A graphical depiction of the constraints, similar to that of Platt [17], is provided in Fig. 4.

Using Eq. (5.3) and the fact that $\beta_1 + \beta_2 = \Delta$, the expression for β_1 is given by

$$\beta_1 = \Delta - \beta_2 \tag{5.6}$$

5.2 SMO for L_2 SVDD

The derivation of the SMO for the L_2 SVDD follows the same approach as that of the L_1 SVDD. Consider two Lagrange multipliers α_1 and α_2 corresponding to two observations while treating the remaining $N - 3$ Lagrange multipliers as constants. The dual problem can be expressed as the following:

$$\max_{\alpha_i,\alpha_j} \sum_{i=1}^{2} \alpha_i k(\mathbf{x}_i, \mathbf{x}_i) - \sum_{i=1}^{2}\sum_{j=1}^{2} \alpha_i \alpha_j \left(k(\mathbf{x}_i, \mathbf{x}_i) + \delta_{ij}\right) - \sum_{i=1}^{2} \alpha_i \mathrm{H}_i + \sum_{i=3}^{N} \alpha_i k(\mathbf{x}_i, \mathbf{x}_i)$$
$$- \sum_{i=3}^{N}\sum_{j=3}^{N} \alpha_i \alpha_j \left(k(\mathbf{x}_i, \mathbf{x}_i) + \delta_{ij}\right),$$

$$\text{subject to} \sum_{i=1}^{2} \alpha_i = \Delta \tag{5.7}$$

$$0 \leq \alpha_1, \alpha_2 < \infty$$

where $\mathrm{H}_i = \sum_{j=3}^{N} \alpha_j k(\mathbf{x}_i, \mathbf{x}_j)$ and $\Delta = 1 - \sum_{i=3}^{N} \alpha_i$. Let $\mathrm{H}_{ij} = k(\mathbf{x}_i, \mathbf{x}_j) + \delta_{ij}$ and $k_{ii} = k(\mathbf{x}_i, \mathbf{x}_i)$. Using the fact $\alpha_1 = \Delta - \alpha_2$ and focusing only on terms that involve α_1 and α_2 yields the following optimization problem:

$$\max_{\alpha_2} (\Delta - \alpha_2)k_{11} + \alpha_2 k_{22} - (\Delta - \alpha_2)^2 \mathrm{H}_{11} - 2(\Delta - \alpha_2)\alpha_2 \mathrm{H}_{12} - \alpha_2^2 \mathrm{H}_{22} \tag{5.8}$$
$$- (\Delta - \alpha_2)\mathrm{H}_1 - \alpha_2 \mathrm{H}_2$$

Differentiating with respect to α_2 and setting the expression equal to zero yields

$$-k_{11} + k_{22} + \mathrm{H}_1 - \mathrm{H}_2 + 2\Delta \mathrm{H}_{11} - 2\Delta \mathrm{H}_{12} - 2\alpha_2 \mathrm{H}_{11} + 4\alpha_2 \mathrm{H}_{12} - 2\alpha_2 \mathrm{H}_{22} = 0$$

Solving for α_2 yields:

$$\alpha_2 = \frac{2\Delta(\mathrm{H}_{11} - \mathrm{H}_{12}) + \mathrm{H}_1 - \mathrm{H}_2 - k_{11} + k_{22}}{2\mathrm{H}_{11} - 4\mathrm{H}_{12} + 2\mathrm{H}_{22}} \tag{5.9}$$

Since $0 \leq \alpha_i < \infty$ for $i = 1, 2, \ldots, N$ and $\alpha_1 + \alpha_2 = \Delta$, the lower and upper bound for α_2 is given by:

$$L_{\alpha_2} = 0 \tag{5.10}$$

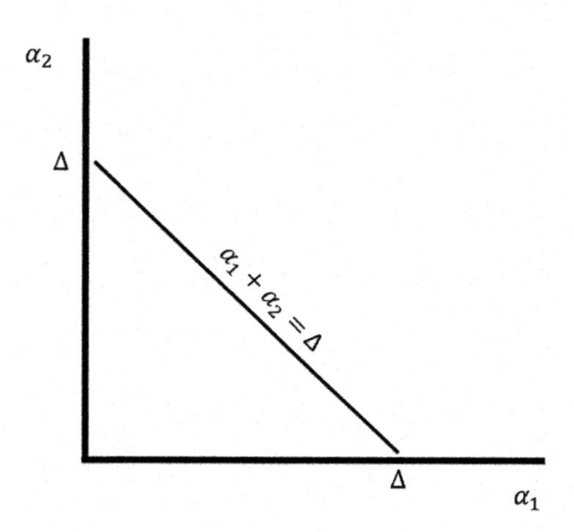

Fig. 5 The constraints for α_2 in SMO for L_2 SVDD

$$H_{\alpha_2} = \alpha_1 + \alpha_2 \tag{5.11}$$

A graphical depiction of the constraints is provided in Fig. 5.

Using Eq. (5.9) and the fact that $\alpha_1 + \alpha_2 = \Delta$, the expression for α_1 is given by

$$\alpha_1 = \Delta - \alpha_2 \tag{5.12}$$

5.3 Performance Study

In this section a comparison of four algorithms is presented: L_1 SVDD (QP), L_1 SVDD (SMO), L_2 SVDD (QP), and L_2 SVDD (SMO). The timing of any individual model depends primarily on the training sample size N and the set of hyperparmaeters such as the SVDD parameter C and parameters used in the kernel function. We present the training times corresponding to the optimal parameters as determined by the model's accuracy on unseen test data. The datasets used include data simulated from a multivariate normal distribution and four real world dataset concerning activity recognition [16], breast cancer [19], Abalone sea snails [22], and rice varieties [5]; all datasets were accessed through the UCI machine learning repository [6]. The kernel function here is the Gaussian kernel (4.1), so there are two hyperparmaeters: the SVDD parameter C and the Gaussian kernel function parameter σ. The timing results are displayed in Table 1. For a fixed optimization algorithm (i.e. QP or SMO), L_1 SVDD tends to be faster than L_2 SVDD, especially for QP when the sample size is larger. However, the difference in runtimes becomes smaller with the use of SMO. For example, L_2 SVDD's runtime for the Rice dataset is almost twice that of L_1 SVDD (approximately 53 s longer), but using the SMO reduces the difference in runtime to approximately 30% (just over 8 s).

Table 1 The model runtimes (seconds) of the L_2 SVDD (QP), L_2 SVDD (SMO), L_1 SVDD (QP), and L_1 SVDD (SMO), respectively, for various simulated and real-world datasets

	Dataset	n	p	L_2 (QP)	L_2 (SMO)	L_1 (QP)	L_1 (SMO)
1	MVN	1,000	10	23.56	18.26	21.07	13.79
2	MVN	750	10	12.24	10.61	11.01	7.80
3	MVN	500	10	5.24	4.33	4.36	3.48
4	MVN	250	10	1.23	1.10	0.91	0.91
5	MVN	100	10	0.24	0.18	0.16	0.14
6	Activity Recognition	1110	7	120.56	64.33	113.73	58.66
7	Breast Cancer	347	30	2.35	2.28	2.19	1.77
8	Abalone	1297	8	74.70	25.10	49.80	21.03
9	Rice	1620	7	118.38	39.47	65.42	31.29

Next, we evaluate the accuracy of the four methods. For the multivariate normal datasets, the test set consists of 20 observations where the first 10 observations are generated from the in-control distribution $N(\mathbf{0}, \mathbf{I})$ whereas the next 10 observations are generated from the out-of-control distribution $N(\boldsymbol{\mu}_1, \mathbf{I})$ where the first component of the mean vector $\boldsymbol{\mu}_1$ is 5 and the other components are 0. In the real world examples the test dataset is comprised of 10 observations from the in-control class and 10 from the out-of-control class. The breast cancer dataset consists of two classes of interest: malignant and benign tumors; the in-control class (target class) is taken to be the benign tumors. In the activity recognition dataset, people are monitored via sensors on their bodies while they perform a variety of physical, emotional, mental, or neurtral activities; the in-control class (target class) is taken to be a physical activity. In the Abalone sea snail dataset, the in-control class is taken to be female Abalones whereas the out-of-control class are the male Abalones. In the rice dataset, two types of rice, Osmancik and Cammeo, are photographed and then seven features are extracted; the target class is taken to be Cammeo. The results are contained in Table 2. For the simulated multivariate normal data, all methods perform similarly and classify all or most of the testing observations correctly. For the activity recognition dataset, all methods classify only half of the records correctly whereas in the breast cancer dataset, all four methods perform similarly. For the abalone dataset, L_2 SVDD (SMO) has the best performance followed by L_1 SVDD (SMO), L_2 SVDD (QP), and L_1 SVDD (QP). For the rice dataset the L_2 SVDD (SMO) has the best performance by a small margin (1 additional record classified correctly) followed by a tie between L_1 SVDD (SMO) and L_2 SVDD (QP) which both outperform L_1 SVDD (QP).

Table 2 The model accuracy of the L_2 SVDD (QP), L_2 SVDD (SMO), L_1 SVDD (QP), and L_1 SVDD (SMO), respectively, for various simulated and real-world datasets

	Dataset	n	p	L_2 (QP)	L_2 (SMO)	L_1 (QP)	L_1 (SMO)
1	MVN	1,000	10	100%	100%	100%	100%
2	MVN	750	10	100%	95%	100%	100%
3	MVN	500	10	100%	100%	100%	100%
4	MVN	250	10	100%	100%	100%	100%
5	MVN	100	10	100%	100%	100%	100%
6	Activity Recognition	1110	7	50%	50%	50%	50%
7	Breast Cancer	347	30	100%	90%	95%	100%
8	Abalone	1297	7	85%	95%	75%	90%
9	Rice	1620	7	75%	80%	65%	75%

6 Stochastic Sub-gradient Descent Solutions

SVDD models often use a non-linear kernel function to map the training vectors into some high-dimensional space in order to solve the dual problem. In some cases, data in the original input space contain a lot of information so that fitting a SVDD without a non-linear mapping can obtain a similar performance to a SVDD that uses a non-linear kernel function. In the case where data are not mapped to some high-dimensional space, we call such situations linear SVDD. One can still solve the dual problem for linear SVDD, but it will be more beneficial if we can solve the SVDD primal problem directly. The objective function of L_1 SVDD is nondifferentiable, so typical optimization methods cannot be directly applied. However, L_2 SVDD is a piecewise quadratic and strongly convex function, which is differentiable.

The first approach discussed for solving SVDD problem was to use quadratic optimization with linear constraints. However, the memory requirements of quadratic programming methods renders a direct use of quadratic programming methods for SVDD very difficult when the training sample consists of many observations.

The second approach presented to solve SVDD problem was SMO. SMO is an approach to overcome the quadratic memory requirement of quadratic programming. SMO solves the dual problem by using an active set of constraints and hence, works on a subset of dual variables. Since SMO finds a feasible dual solution and its goal is to maximize the dual objective function, it often results in a rather slow convergence rate to the optimum of the primal objective function.

6.1 L_1 SVDD

Let $\mathbf{x}_i$, $i = 1, 2, \ldots, N$ be a sequence of $p-$variate training (or target) observations. L_1 SVDD tries to find a sphere with minimum volume containing all (or most of) vectors. Learning an SVDD has been formulated as a constrained optimization problem over r^2 and $\mathbf{a}$.

$$\begin{aligned} \underset{r,\mathbf{a},\xi_i}{\text{minimize}} \quad & r^2 + C\sum_{i=1}^{N} \xi_i, \\ \text{subject to} \quad & ||\mathbf{x}_i - \mathbf{a}||^2 \le r^2 + \xi_i, \quad i = 1, 2, \ldots, N \\ & \xi_i \ge 0, \quad i = 1, 2, \ldots, N \end{aligned} \tag{6.1}$$

The learning problem (6.1) can be replaced by the unconstrained optimization problem

$$\min_{r,\mathbf{a}} \frac{\lambda}{2} r^2 + \frac{1}{2} \sum_{i=1}^{N} \max\left(0, ||\mathbf{x}_i - \mathbf{a}||^2 - r^2\right) \tag{6.2}$$

where $\lambda = 1/C$ and $\max\left(0, ||\mathbf{x}_i - \mathbf{a}||^2 - r^2\right)$ is the 'hinge' loss. Define J_1 as

$$J_1(r, \mathbf{a}) = \frac{\lambda}{2} r^2 + \frac{1}{2} \sum_{i=1}^{N} \max\left(0, ||\mathbf{x}_i - \mathbf{a}||^2 - r^2\right) \tag{6.3}$$

It immediately follows that an observation $\mathbf{x}_i$ inside the sphere is not penalized while there is a penalty for an observation $\mathbf{x}_i$ outside of the sphere. Since the hinge loss is convex, then the objective function J_1 is convex and a locally optimal point is globally optimal (provided the optimization is over a convex set, which it is in our case).

6.1.1 Gradient Descent Algorithm for L_1 SVDD

To minimize the objective J_1, we will use the gradient descent algorithm. Since the hinge loss is not differentiable, a sub-gradient is computed, and the sub-gradients are

$$\frac{\partial J_1}{\partial r} = \begin{cases} \lambda r, & \text{if } ||\mathbf{x}_i - \mathbf{a}||^2 - r^2 \le 0 \\ r(\lambda - N), & \text{if } ||\mathbf{x}_i - \mathbf{a}||^2 - r^2 > 0 \end{cases} \tag{6.4}$$

and

$$\frac{\partial J_1}{\partial \mathbf{a}} = \begin{cases} \mathbf{0}, & \text{if } ||\mathbf{x}_i - \mathbf{a}||^2 - r^2 \le 0 \\ -\sum_{i=1}^{N} (\mathbf{x}_i - \mathbf{a}), & \text{if } ||\mathbf{x}_i - \mathbf{a}||^2 - r^2 > 0 \end{cases} \tag{6.5}$$

Let $\boldsymbol{\omega}_1$ be a 2×1 vector holding the values of r and $\mathbf{a}$, that is $\boldsymbol{\omega}_1 = (r, \mathbf{a})'$, and let $\boldsymbol{\omega}_1^{(t)}$ be the values of $\boldsymbol{\omega}$ at iteration t, then the simultaneous sub-gradient descent update for r and $\mathbf{a}$ is given by:

$$\boldsymbol{\omega}_1^{(t+1)} = \boldsymbol{\omega}_1^{(t)} - \eta \nabla J_1(\boldsymbol{\omega}_1^{(t)}) \tag{6.6}$$

where η is a hyperparameter that represents the step size.

6.1.2 Stochastic Gradient Descent Algorithm for L_1 SVDD

With stochastic gradient descent, at each iteration, we randomly select a single observation $\mathbf{x}_i$ from the training set and use that observation to obtain an approximation of the gradient of the objective function. A step with pre-determined step size is taken in the opposite direction to guarantee minimization. Define J_{s_1} as

$$J_{s_1}(r, \mathbf{a}) = \frac{\lambda}{2} r^2 + \frac{1}{2} \max\left(0, ||\mathbf{x}_i - \mathbf{a}||^2 - r^2\right) \tag{6.7}$$

Consequently, it follows that the sub-gradients are

$$\frac{\partial J_{s_1}}{\partial r} = \begin{cases} \lambda r, & \text{if } ||\mathbf{x}_i - \mathbf{a}||^2 - r^2 \leq 0 \\ r(\lambda - 1), & \text{if } ||\mathbf{x}_i - \mathbf{a}||^2 - r^2 > 0 \end{cases} \tag{6.8}$$

and

$$\frac{\partial J_{s_1}}{\partial \mathbf{a}} = \begin{cases} \mathbf{0}, & \text{if } ||\mathbf{x}_i - \mathbf{a}||^2 - r^2 \leq 0 \\ -(\mathbf{x}_i - \mathbf{a}), & \text{if } ||\mathbf{x}_i - \mathbf{a}||^2 - r^2 > 0 \end{cases} \tag{6.9}$$

Consequently, the simultaneous stochastic sub-gradient descent update for r and $\mathbf{a}$ is given by:

$$\boldsymbol{\omega}_1^{(t+1)} = \boldsymbol{\omega}_1^{(t)} - \eta \nabla J_{s_1}(\boldsymbol{\omega}_1^{(t)}) \tag{6.10}$$

where η is a hyperparameter that represents the step size or learning rate.

6.2 *L_2 SVDD*

Let $\mathbf{x}_j$, $j = 1, 2, \ldots, N$ be a sequence of $p-$variate training (or target) observations. L_2 SVDD tries to find a sphere with minimum volume containing all (or most) of the training observations. Learning a L_2 SVDD can be formulated as a constrained optimization problem over R^2 and $\mathbf{c}$:

$$\begin{aligned} \underset{R,\mathbf{c},\xi_j}{\text{minimize}} \quad & R^2 + C\sum_{i=j}^{N}\xi_j^2, \\ \text{subject to} \quad & ||\mathbf{x}_j - \mathbf{c}||^2 \leq R^2 + \xi_j, \quad j = 1, 2, \ldots, N \end{aligned} \tag{6.11}$$

The learning problem (6.11) can be replaced by the unconstrained optimization problem

$$\min_{R,\mathbf{c}} \frac{\lambda}{2}R^2 + \frac{1}{2}\sum_{j=1}^{N} \max\left\{\left(0, ||\mathbf{x}_j - \mathbf{c}||^2 - R^2\right)\right\}^2 \tag{6.12}$$

where $\lambda = 1/C$.

Define J_2 as

$$J_2(R, \mathbf{c}) = \frac{\lambda}{2}R^2 + \frac{1}{2}\sum_{j=1}^{N} \max\left\{\left(0, ||\mathbf{x}_j - \mathbf{c}||^2 - R^2\right)\right\}^2 \tag{6.13}$$

Similarly to the L_1 SVDD, an observation $\mathbf{x}_j$ inside the sphere is not penalized while there is a cost for an observation $\mathbf{x}_j$ outside the sphere. Since the objective function J_2 is strongly convex, a locally optimal point is globally optimal.

6.2.1 Gradient Descent Algorithm for L_2 SVDD

To minimize the objective J_2, we will use gradient descent algorithm. The sub-gradients are

$$\frac{\partial J_2}{\partial R} = \begin{cases} \lambda R, & \text{if } ||\mathbf{x}_j - \mathbf{c}||^2 - R^2 \leq 0 \\ R\left\{\lambda - 2\sum_{j=1}^{N}\left(||\mathbf{x}_j - \mathbf{c}||^2 - R^2\right)\right\}, & \text{if } ||\mathbf{x}_j - \mathbf{c}||^2 - R^2 > 0 \end{cases} \tag{6.14}$$

and

$$\frac{\partial J_2}{\partial \mathbf{c}} = \begin{cases} \mathbf{0}, & \text{if } ||\mathbf{x}_j - \mathbf{c}||^2 - R^2 \leq 0 \\ -2\sum_{j=1}^{N}(\mathbf{x}_j - \mathbf{c})(||\mathbf{x}_j - \mathbf{c}||^2 - R^2), & \text{if } ||\mathbf{x}_j - \mathbf{c}||^2 - R^2 > 0 \end{cases} \tag{6.15}$$

Let $\boldsymbol{\omega}_2$ be a 2×1 vector holding the values of R and $\mathbf{c}$, that is $\boldsymbol{\omega}_2 = (R, \mathbf{c})'$, and let $\boldsymbol{\omega}_2^{(t)}$ be the values of R and $\mathbf{c}$ at iteration t. It follows that the simultaneous sub-gradient descent update for R and $\mathbf{c}$ is given by:

$$\boldsymbol{\omega}_2^{(t+1)} = \boldsymbol{\omega}_2^{(t)} - \eta\nabla J_2(\boldsymbol{\omega}_2^{(t)}) \tag{6.16}$$

where η is the step size.

6.2.2 Stochastic Gradient Descent Algorithm for L$_2$ SVDD

In stochastic gradient descent, we randomly select one observation $\mathbf{x}_j$ from the training set and use that observation to approximate the gradient of the objective function. Define J_{s_2} as

$$J_{s_2}(R, \mathbf{c}) = \frac{\lambda}{2}R^2 + \frac{1}{2}\left\{\max\left(0, ||\mathbf{x}_j - \mathbf{c}||^2 - R^2\right)\right\}^2 \tag{6.17}$$

$$\frac{\partial J_{s_2}}{\partial R} = \begin{cases} \lambda R, & \text{if } ||\mathbf{x}_j - \mathbf{c}||^2 - R^2 \leq 0 \\ R(\lambda - 2(||\mathbf{x}_j - \mathbf{c}||^2 - R^2)), & \text{if } ||\mathbf{x}_j - \mathbf{c}||^2 - R^2 > 0 \end{cases} \tag{6.18}$$

and

$$\frac{\partial J_{s_2}}{\partial \mathbf{c}} = \begin{cases} \mathbf{0}, & \text{if } ||\mathbf{x}_j - \mathbf{c}||^2 - R^2 \leq 0 \\ -2(\mathbf{x}_j - \mathbf{c})\left(||\mathbf{x}_j - \mathbf{c}||^2 - R^2\right), & \text{if } ||\mathbf{x}_j - \mathbf{c}||^2 - R^2 > 0 \end{cases} \tag{6.19}$$

then the simultaneous stochastic sub-gradient descent update for R and $\mathbf{c}$ is given by:

$$\boldsymbol{\omega}_2^{(t+1)} = \boldsymbol{\omega}_2^{(t)} - \eta \nabla J_{s_2}(\boldsymbol{\omega}_2^{(t)}) \tag{6.20}$$

where η is the learning rate.

6.3 *Performance Evaluation*

In this section we apply linear L$_1$ SVDD and linear L$_2$ SVDD to various datasets and assess their relative computation time as well as their accuracy. The datasets used in this section correspond to those used in Sect. 5.3. It is important to note that there are 3 hyperparameters for fitting the four linear SVDD models considered in this section: the step size, number of epochs, and C. The models were all optimized for accuracy and then the corresponding time was recorded. The timing results are presented in Table 3. We note that the model runtimes depend on the three aforementioned hyperparameters including $C = \frac{1}{\lambda}$ which is a component of the gradient update equations and controls, at least partially, the extent to which the radius and center are updated. As expected, stochastic gradient descent results in much faster runtimes even when the sample size is larger.

For both the linear L$_1$ SVDD and linear L$_2$ SVDD, stochastic gradient descent provides a faster model runtime as well as similar, in some cases superior, accuracy on the test data. Both linear L$_1$ SVDD and linear L$_2$ SVDD perform well on the multivariate normal data irrespective of the underlying optimization algorithm. For the Abalone dataset, both the linear L$_1$ SVDD and linear L$_2$ SVDD perform similarly:

Table 3 The model runtimes (seconds) of the linear L_2 SVDD using gradient descent (GD), linear L_2 SVDD using stochastic gradient descent (SGD), linear L_1 SVDD using gradient descent (GD), and linear L_2 SVDD using stochastic gradient descent (SGD) for various simulated and real-world datasets

	Dataset	n	p	L_2 (GD)	L_2 (SGD)	L_1 (GD)	L_1 (SGG)
1	MVN	1,000	10	46.319	0.17	29.98	0.43
2	MVN	750	10	9.61	0.19	17.98	0.26
3	MVN	500	10	6.52	0.14	12.17	0.21
4	MVN	250	10	3.63	0.04	2.74	0.15
5	MVN	100	10	0.12	0.12	3.56	0.26
6	Activity Recognition	1110	7	0.99	0.11	3.02	0.22
7	Breast Cancer	347	30	10.31	2.55	8.08	3.75
8	Abalone	1297	8	0.09	0.09	3.48	0.09
9	Rice	1620	7	9.33	0.25	0.779	0.25

linear L_2 SVDD (GD) has an accuracy of 65%, linear L_2 SVDD (SGD) 70%, linear L_1 SVDD (GD) 70%, and linear L_1 SVDD (SGD) 70%. All four algorithms predict only 50% on the other three real-world datasets. This perhaps is caused by the four models' inability to model non-linearly separable data, and note that from Table 2 the L_1 SVDD and L_2 SVDD models (SMO or QP) that utilize the Gaussian kernel function perform well on the three datasets which implies that the decision boundary is non-linear (Table 4).

7 Case Study

In this section we apply L_1 SVDD and L_2 SVDD to monitor a machine in a manufacturing process [23]. The dataset contains 7,905 anonymized observations where each observation is represented as a $p = 17$ dimensional vector of measurements including temperature, humidity, and an additional 15 measurements whose names are anoymized. In addition to the predictor information, each observation is associated with a timestamp so that one observation associated with the machine corresponds to a specific date and hour of operation; also, each observation has a label indicating if the machine failed during the specified hour of operation. Of the 7,905 hours of operation, there are 75 failures and 7,830 non-failures. We define the target class as the observations where the machine did not fail during the hour and take the non-target class (i.e. outlier) to be the observations where the machine failed during the hour. Since the data are collected on an hourly basis, we train the model using target class data from hour 150 (i.e. January 7, 2016 at 6 A.M.) to hour 499 so that the training sample size is 350 observations (350 h; 1 observation per hour). The monitoring dataset corresponds to 19 observations collected at hours 500 to 518 where the

Table 4 The model accuracy of the linear L_2 SVDD using gradient descent (GD), linear L_2 SVDD using stochastic gradient descent (SGD), linear L_1 SVDD using gradient descent (GD), and linear L_2 SVDD using stochastic gradient descent (SGD) for various simulated and real-world datasets

	Dataset	n	p	L_2 (GD)%	L_2 (SGD)%	L_1 (GD)%	L_1 (SGG)%
1	MVN	1,000	10	95	95	100	100
2	MVN	750	10	100	95	100	95
3	MVN	500	10	95	95	100	95
4	MVN	250	10	95	100	100	100
5	MVN	100	10	100	100	95	100
6	Activity Recognition	1110	7	50	50	50	50
7	Breast Cancer	347	30	50	50	50	50
8	Abalone	1297	8	65	70	70	70
9	Rice	1620	7	50	50	50	50

first 16 observations correspond to the target class and the next 3 observations correspond to the non-target class. The monitoring is terminated once a non-target class prediction is produced by the model, so the ideal result occurs when the monitoring is terminated at hour 516.

7.1 L_1 SVDD

The L_1 SVDD is trained using quadratic programming to solve the dual problem (2.2). Once the solution to the dual problem (2.2) is determined, the radius is computed using equation (2.3). Next, we compute the kernel distance for the monitoring observations using the equation on the left hand side of condition (2.4) and then determine if condition (2.4) is true or false. The L_1 SVDD correctly classifies the data at hours 500 to 515 correctly as the target class and correctly classifies the data at hour 516 to be in the non-target class, so the monitoring terminates at hour 516. The control chart for the L_1 SVDD is shown in Fig. 6 and plots the kernel distance against the hour (i.e. one observation corresponds to one hour) in the testing dataset with a dotted horizontal line corresponding to the radius. From the control chart, we can see that the kernel distance values corresponding to the observations at hours 500 to 515 is less than the radius and only the kernel distance for the observation at hour 516 is greater than the radius which yields a non-target classification and terminates the monitoring.

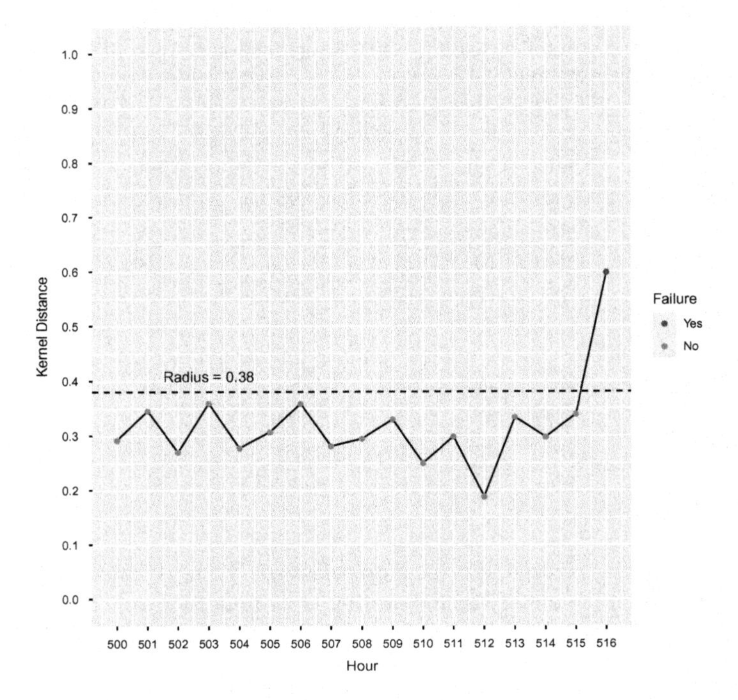

Fig. 6 The control chart for the L_1 SVDD for the machine failure dataset

7.2 L_2 SVDD

The L_2 SVDD is trained using quadratic programming to solve the dual problem (3.6). Using the solution to dual problem (3.6), the radius is computed using equation (3.7). We then compute the kernel distance for the monitoring observations using the equation on the left hand side of condition (3.8) to determine if condition (3.8) is true or false. The L_2 SVDD correctly classifies the data corresponding to hours 500 to 515 as the target class and correctly classifies the data corresponding to hour 516 to be in the non-target class, so the monitoring terminates at hour 516. The control chart for the L_2 SVDD is shown in Fig. 7 and plots the kernel distance against the hour (i.e. one observation corresponds to one hour) in the monitoring dataset with a dotted horizontal line corresponding to the radius. From the control chart, we can see that the kernel distance for the observations at hours 500 to 515 is less than the radius and only the kernel distance for the observation at hour 516 is greater than the radius which yields a non-target classification and terminates the monitoring.

We note that though the two control charts are similar, there are some subtle differences. For example, the kernel distances corresponding to the observations at hour 503 and 506, respectively, are closer to the radius for the L_2 SVDD as compared

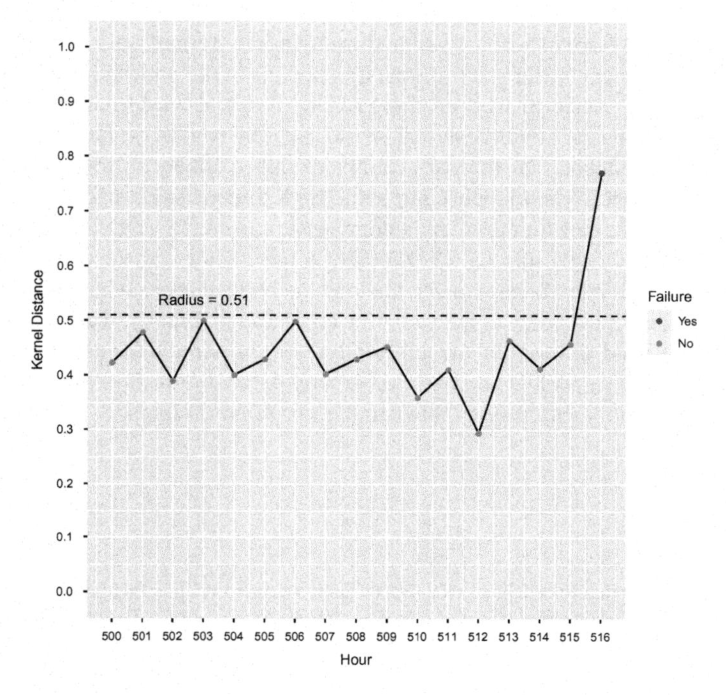

Fig. 7 The control chart for the L_2 SVDD for the machine failure dataset

to the L_1 SVDD. Also, the scale of the kernel distance values and the radius for the L_2 SVDD is larger than that of the L_1 SVDD.

8 Conclusion

L_1 SVDD, commonly referred to as just SVDD, is a commonly used one-class classification algorithm that utilizes a L_1 norm whereas the L_2 SVDD is an alternative formulation of SVDD that uses a L_2 norm. Performance simulations show that L_2 SVDD and L_1 SVDD tend to have a similar performance, though one algorithm may sometimes outperform the other. We propose an extension of the sequential minimum optimization (SMO) algorithm for L_2 SVDD. Timing results show that the L_2 SVDD is generally slower than the L_1 SVDD for a fixed optimization algorithm (i.e. QP or SMO), but using the SMO tends to reduce the difference in runtime between the L_1 and L_2 SVDD, making the use of L_2 SVDD (SMO) more feasible in practice. We presented update equations used in gradient descent and stochastic gradient descent algorithms that are used to solve both the unconstrained (i.e. linear) L_1 SVDD and the unconstrained (i.e. linear) L_2 SVDD. Our simulations for the unconstrained formulations show that both linear L_1 SVDD and linear L_2 SVDD have a similar performance for simulated multivariate normal data as well as real-world data. The linear SVDD algorithms, irrespective of the underlying optimization algorithm, did not perform well on three of the four real-world datasets. This can be explained by the fact that these real-world datasets cannot be modeled as linear problems which means that the data cannot fit into a sphere in the original input space and will need to be mapped into some high dimensional space by the use of some non-linear kernel function. As a comparison, the L_1 SVDD (QP), L_1 SVDD (SMO), L_2 SVDD (QP), and L_2 SVDD (SMO) use a Gaussian kernel and all have superior accuracy on the breast cancer, abalone, and rice datasets whereas the activity recognition dataset accuracy is the same for all methods. Finally, the L_1 SVDD and L_2 SVDD are used to monitor machine failures in a manufacturing process. Both the L_1 SVDD and L_2 SVDD are able to correctly detect the shift in the process, and there are some minor differences between the two control charts.

References

1. Camerini V, Coppotelli G, Bendisch S (2018) Fault detection in operating helicopter drivetrain components based on support vector data description. Aerospace Sci Technol 73:48–60
2. Chaki S, et al (2014) A one-class classification framework using SVDD: application to an imbalanced geological dataset. In: Proceedings of the 2014 IEEE Students' Technology Symposium. IEEE, pp 76–81
3. Chang W-C, Lee C-P, Lin, C-J (2013) A revisit to support vector data description. In: Dept. Comput. Sci., Nat. Taiwan Univ., Taipei, Taiwan, Tech. Rep (2013)
4. Choi Y-S (2009) Least squares one-class support vector machine. Pattern Recogn Lett 30(13):1236–1240
5. Cinar I, Koklu M (2019) Classification of rice varieties using artificial intelligence methods. Int J Intell Syst Appl Eng 7(3):188–194
6. Dua D, Graff C (2017) UCI Machine Learning Repository. http://archive.ics.uci.edu/ml
7. Duan CD, Liu YY, Gao Q (2012) Structure health monitoring using support vector data description and wavelet packet energy distributions. Appl Mech Mater 135:930–937 (Trans Tech Publ. 2012)
8. Khan SS, Madden MG (2014) One-class classification: taxonomy of study and review of techniques. Knowl Eng Rev 29(3):345–374
9. Liu S, et al (2019) Network log anomaly detection based on GRU and SVDD. In: 2019 IEEE international conference on parallel & distributed processing with applications, big data & cloud computing, sustainable computing & communications, social computing & networking (ISPA/BDCloud/SocialCom/SustainCom). IEEE. 2019, pp 1244–1249
10. Luo H, Wang Y, Cui J (2011) A SVDD approach of fuzzy classification for analog circuit fault diagnosis with FWT as preprocessor. Expert Syst Appl 38(8):10554–10561
11. Maboudou-Tchao EM (2020) Change detection using least squares one-class classification control chart. In: Quality Technology & Quantitative Management (2020), pp 1–18
12. Maboudou-Tchao EM (2019) High-dimensional data monitoring using support machines. In: Communications in statistics-simulation and computation, pp 1–16
13. Maboudou-Tchao EM (2018) Kernel methods for changes detection in covariance matrices. Commun Stat-Simul Comput 47(6):1704–1721
14. Maboudou-Tchao EM (2021) Support tensor data description. J Qual Technol 53(2):109–134
15. Maboudou-Tchao EM, Silva IR, Diawara N (2018) Monitoring the mean vector with Mahalanobis kernels. Quality Technol Quantit Manage 15(4):459–474

16. Mohino-Herranz I, et al (2019) Activity recognition using wearable physiological measurements: selection of features from a comprehensive literature study. Sensors 19(24):5524
17. Platt J (1998) Sequential minimal optimization: a fast algorithm for training support vector machines
18. Sanchez-Hernandez C, Boyd DS, Foody GM (2007) One-class classification for mapping a specific land-cover class: SVDD classification of fenland. IEEE Trans Geosci Remote Sens 45(4):1061–1073
19. Street WN, Wolberg WH, Mangasarian OL (1993) Nuclear feature extraction for breast tumor diagnosis. In: Biomedical image processing and biomedical visualization, vol 1905. International Society for Optics and Photonics, pp 861–870 (1993)
20. Sun R, Tsung F (2003) A kernel-distance-based multivariate control chart using support vector methods. Int J Prod Res 41(13):2975–2989
21. Tax DMJ, Duin RPW (2004) Support vector data description. Mach Learn 54(1):45–66
22. Waugh SG (1995) Extending and benchmarking cascade-correlation: extensions to the Cascade-Correlation architecture and benchmarking of feed-forward supervised artificial neural networks. PhD thesis. University of Tasmania
23. Zuriaga C (2017) Machine failures. https://bigml.com/user/czuriaga/gallery/dataset/587d062d49c4a16936000810. Accessed 05 Sept 2021

Feature Engineering and Health Indicator Construction for Fault Detection and Diagnostic

Khanh T. P. Nguyen

Abstract Nowadays, the rapid growth of modern technologies in Internet of Things (IoT) and sensing platforms is enabling the development of autonomous health management systems. This can be done, in the first step, by using intelligent sensors, which provide reliable solutions for systems monitoring in real-time. Then, the monitoring data will be treated and analyzed in the second step to extract health indicators (HIs) for maintenance and operation decisions. This procedure called feature engineering (FE) and HI construction is the key step that decides the performance of condition monitoring systems. Hence, in this chapter we present a comprehensive review and new advances of FE techniques and HI construction methods for fault detection and diagnostic (FDD) of engineering systems. This chapter would also serve as an instructive guideline for industrial practitioners and researchers with different levels of experience to broaden their skills about system condition monitoring procedure.

1 Condition Monitoring Data Acquisition for Fault Detection and Diagnostic

Reliable condition monitoring data (CM) is a key factor for an efficient deployment of fault detection and diagnostic solutions. This data can measure the system's behavior characteristics and capture its slight changes in real-time. For example, oil analysis can help detecting machine oxidation, dilution, moisture while pressure sensors allows finding leakages or faulty connectors. Besides, thermal imaging sensors allows monitoring temperature of interacting machine parts, and then warning any abnormalities. Similarly, vibration sensors can allow track deviations from nominal

K. T. P. Nguyen (✉)
Production Engineering Laboratory, INP Toulouse, The National Engineering School of Tarbes, 65000 Tarbes, France
e-mail: thi-phuong-khanh.nguyen@enit.fr

K. P. Tran (ed.), *Control Charts and Machine Learning for Anomaly Detection in Manufacturing*, Springer Series in Reliability Engineering,
https://doi.org/10.1007/978-3-030-83819-5_10

vibration frequencies of system components for early detection of their stress and unexpected faults.

1.1 Choice of Condition Monitoring Parameters

Although condition monitoring data has enable the capacity of tracking system states in real-time, it is not trivial to identify the appropriate physical parameters to monitor the system degradation phenomena. A recommended scenario requires both domain experts and data scientists for planning the condition monitoring process. In detail, while the data scientists investigate the operating and maintenance historical data to identify critical components and important failure types, the experts can provide their valuable knowledge about the system's characteristics and also verify the useful information extracted by the data scientists. Besides, the architectural, structural, and functional analyses of the system should be performed to isolate the failure mechanisms. This action, also known as critical component identification, can be realized using qualitative analysis approaches, such as experience feedback, failure tree, event tree, cause, and effect tree, or through operator knowledge in case of insufficient information about the system [1].

Once the critical components are localized, depending on their characteristics and on the information available, a proper definition of the monitoring process, such as monitoring parameters and the appropriate sensors, can be defined, see Fig. 1. The critical components can be electronic, mechanical, hydraulic, pneumatic, etc. Their heterogeneity poses a big challenge in the choice of monitoring sensors that requires expert knowledge in multiple domains. In [2], the authors review different parameters that can be used to monitor the state of health of systems and in Table 1, the appropriate sensors are also proposed to capture degradation mechanisms according to every category: thermal, electrical, chemical, etc. For example, numerous studies propose to use vibration sensors to monitor the bearing condition because their defects generally produce fault signatures in the machine vibration. Besides, current sensor is

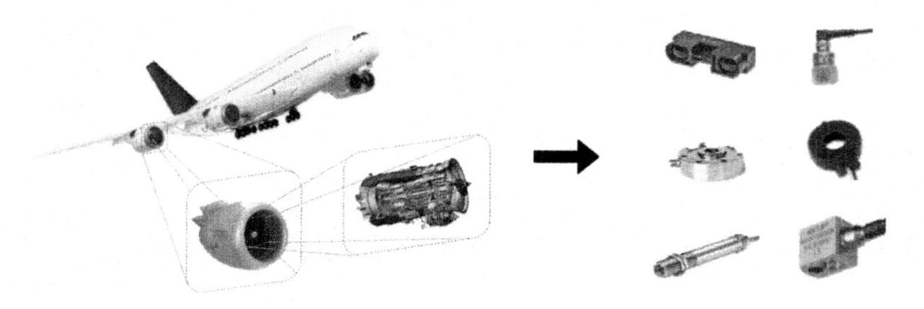

Fig. 1 From system analysis to the selection of appropriate sensors.

Table 1 Examples of parameters and sensors for condition monitoring [2]

Category	Parameter	Sensor
Thermal	Temperature, heat flux, heat dissipation	Negative temperature coefficient thermistors, resistance temperature detectors, thermocouples, semiconductor-based sensors
Electrical	Voltage, current, resistance, inductance, capacitance, dielectric constant, charge, polarization, electric field, frequency power, noise level, impedance	Open-loop circuit, closed-loop circuit, Rogowski coil, current clamp meters
Chemical	Species concentration, gradient, re-activity, mess, molecular weight	Ion sensor, humidity sensor, gas sensor, biosensor
Mechanical	Length, area, volume, velocity or acceleration, mass flow, force, torque, stress, strain, density, stiffness, strength, angular, direction, pressure, acoustic intensity, power, acoustic spectral distribution	Bourdon tube, manometer, diaphragms, pressure gauge, strain gauge, load cell, tachometer, encoder
Optical	Intensity, phase, wavelength, polarization, reflection, transmittance, refraction, index, distance, vibration, amplitude and frequency	Photoconductive devices, photovoltaic cell, photodiodes
Magnetic	Magnetic field, flux density, magnetic moment, permeability, direction, distance, position, flow	Coils, reed switch, hall elements, magnetoresistive element, semiconductor, anisotropic, and tunnel magnetoresistive element

used to remedy faults such as unbalanced supply motors, squirrel cage motor broken bars of the electrical motors [3].

1.2 Data Acquisition and Preprocessing

A pertinent strategy for data acquisition, i.e. data collection and the storage, is more important than an effort to collect extra data. It is essential for industries to install a reliable acquisition system that ensures credibility, validity, and reliability of the collected data and consequently a more accurate and efficient monitoring. The monitoring process must be carefully set up considering external information such as system operating conditions, according to the expert advices.

After collecting and storing, the raw data is injected into processing procedure, one of the most important steps in condition monitoring. The first step of data processing serves to identify the errors, which significantly affect data quality, and consequently defines an appropriate processing plan for correcting them. The major error types for CM data are human, transmission, recording and sensors errors. There is no generalized approach to effectively manage these errors. Indeed, data processing is a meticulous process that requires different skills and experience from data scientists, and depends on the orientation of technical and business experts. However, they can generally be grouped as follows:

1. *Data inspection.* This first step aims to detect unexpected, inconsistent and incorrect data. It is performed by two techniques: summary statistic and data visualization [4].
    - *Summary statistics* provides an overview of data. It allows identifying the number of missing values, the characteristics of the data such as range, mean value, distribution, and relationship between variables.
    - *Data visualization* is the graphical representation of data. It establishes a systematic mapping between graphical marks and data values to visually and vividly demonstrate the properties of the data. Nowadays, thanks to numerous statistical and information graphic tools, complex data has become more accessible, understandable, and usable.
2. *Data cleaning.* The second step is crucial, especially for treating sensor noises. It is applied to remove the following errors:
    - *Irrelevant data* is the one that is not consistent with the context of the studied problem. For example, the name of repairman is unnecessary to predict the system remaining useful life time. In practice, it is not trivial to detect the irrelevant data. Hence, this task should be carefully performed according to expert advices and after meticulously considering the correlation between the variables.
    - *Duplicates* are the records repeated in a dataset, normally caused by a combination of different sources or due to data entry. The deduplication is often manually performed using filtering techniques. However, for large databases, it is preferable to use statistical or machine learning techniques [5, 6].
    - *Syntax errors* always occur in datasets due to a manual data entry. They can lead to unexpected consequences when mining the data, especially for categorical variables. Generally, syntax errors might be corrected using matching techniques.
    - *Outliers* are those values which are outstandingly different from all other records. They might be attributable to noises, negative effects of the environment or caused by anomalies in the monitored system. Therefore, handling outliers should be carefully performed to avoid the loss of crucial information [7].

- *Missing values* are unavoidable errors in datasets when some observations are not available for one or more variable. Neglecting them might significantly distort the conclusion drawn from the data. Hence, it is necessary to handle the missing values. The missing values can be filled by statistical values or interpolation using regression methods [4, 8, 9]. In some particular cases where missing data is informative in itself, it is necessary to develop appropriate algorithms that allow recognizing this omission.

3. *Data transformation.* The last step aims to transform the data into a format that is ready to use and most convenient for a specific purpose. It includes the following tasks:

   - Standardizing: it aims to ensure that the data are uniform for every variable.
   - Min-max scaling and Z-transform: it serves to transform the data within a specific scale to enhance their consistency.
   - Re-sampling: it is used to enhance the hidden trend of the variables and also balance their record numbers of for further analysis by machine learning tools. For example, the vibration signals, with high frequency of 25 kHz, could be segmented and re-sampled by the interval time when temperature measurement is recorded.
   - Dimension reduction: it is an essential task to reduce the number of variables by obtaining a set of necessary ones to facilitate further visualization and analysis. It can be performed by using methods based on statistics, information theory, dictionaries, and projections [10].

2 Signal Processing Techniques for Feature Extraction

Figure 2 presents an overview of signal processing techniques used to extract meaningful features from the recorded raw measurements. In general, these methods can be classified into three categories: time, frequency, and time-frequency domain [11].

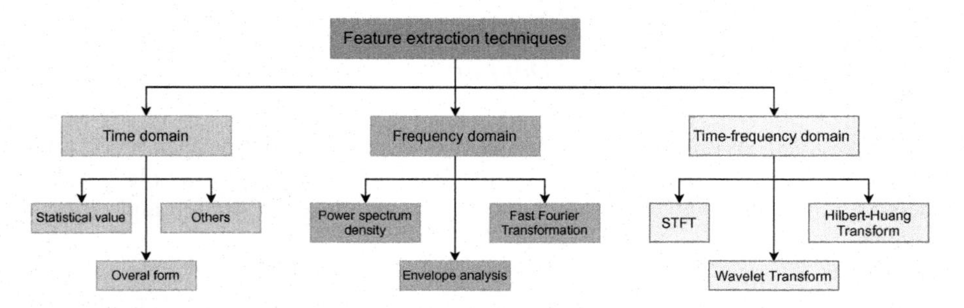

Fig. 2 Signal processing techniques for extraction of features.

2.1 Features in Time Domain

Signal processing techniques for raw data in time domain are classical, fast and simple. They play a critical role for fault detection and diagnostic in different components such as gears, bearings, machining tools, etc.

Among features in time domain, the statistical indicators are widely used, thanks to its capacity to reflect the incipient characteristic of signals [12]. For example, root mean square (RMS) value allows measuring the power content in the vibration signature, and consequently can be effective for detection of an imbalance in rotating machinery. The standard deviation (STD) is used to evaluate the dispersion of a signal while the kurtosis (KUR) aims to evaluate the peakedness of signals for fault detection and severity quantification. Furthermore, skewness (SKW) is used to measure the symmetry and asymmetry of signal distribution. Besides, the energy ratio (ER) is effective for the detection of heavy faults.

In addition to the statistical values, the factors that represent the overall shape of signals, such as the Crest factor (CF), Peak to Peak value (PP), Shape factor (SF), and Impulse factor (IF), are powerful to capture changes in the signal pattern [12, 13] when anomalies occur. The conventional scalar indicators, e.g. RMS, KUR, CF, and Peak are also combined to create new indicators, called TALAF or THIKAT, that aim to predict future failures and track defects from the first signs of degradation to the end of life [11]. In [14], the authors proposed to use entropy features extracted from vibration signals for bearing failure prognostics.

Although it does not require many computational time and resources to evaluate the temporal features, these indicators are sensible to noisy signals. It could require other techniques to enhance signals before the evaluation of indicators. For example, in [15], the deterministic and random parts of the vibration signal are separated by an autoregressive model (AR); and then the fault indicator is calculated by an energy ratio between these two parts. The papers [16] propose to use the Park's and Concordia transform for stator current signals and then use the current pattern for fault detection and diagnosis. The authors in [17] evaluate the RMS value of current signals after a noise cancellation step using Wiener filter.

2.2 Features in Frequency Domain

Frequency analysis is a common technique to convert time-series measurements into frequency values that are sensitive to the anomaly appearance. For example, bearing defects generally generate characteristic frequencies in the vibration and current signals.

Among frequency analysis, Fast Fourier Transform (FFT) is widely used to decompose physical signals into spectrum of continuous frequencies or number of discrete frequencies [18]. After performing FFT, the magnitude values at the characteristic frequencies are used as common indicators for bearing fault detection and diagnostic [19]. In [20], the authors propose a frequency feature called PMM, that is the ratio between the maximal value of FFT magnitudes at the characteristic frequencies and the mean of the entire magnitude frequency value, for defect and severity classification and for bearing performance assessment. The authors in [21] used the spectral kurtosis for mechanical unbalance detection in an induction machine.

In addition to FFT, power spectrum density (PSD) that describes the spectral energy distribution into the signal's frequency components is also used to evaluate the features for fault detection and diagnostics. In [12], the authors propose to use the maximal value of PS for bearing fault detection based on vibration signals. The PSD magnitude of the non-stationary current-demodulated signals at the characteristic frequency is proved as an efficient features for fault diagnosis in [22]. On the other hand, using Welch's periodogram of the stator current, the authors in [23] developed a new fault detection indicator, which is calculated as the sum of the centered reduced spectrum amplitude around the characteristic frequencies.

Envelope analysis allows reflecting the energy concentration in narrow bande while there are multiple repetitive vibration impulses generated due to a contact between a localized defect and another surface. For example, the maximal value of the envelope spectrum magnitude is used as an indicator for bearing fault detection in [12]. Numerous articles show that the envelope spectrum magnitudes at the characteristic frequencies are powerful to detect and classify the bearing failures [24, 25]. On the other hand, in [20], the ratio between the maximal value of envelope spectrum magnitudes at the characteristic frequencies and the mean value of the entire magnitudes is also used for diagnostic and degradation assessment.

Although frequency analysis techniques are widely used for fault detection and diagnostics but they require signal spectrum knowledge and are limited to the equipment having fault characteristic frequencies. Moreover, they are not suitable for non-stationary signals due to the loss of information when transforming the time-series signal into frequencies.

2.3 Features in Time-Frequency Domain

To capture the non-stationary characteristic of signals, it is necessary to present them into two-dimensional function of time and frequency. Therefore, numerous signal processing techniques in time-frequency domain have been developed in literature.

Short time Fourier transform (STFT) firstly decomposes the signals into a set of data within a fixed window length and secondly performs the FT on every data window. Then, the spectrum magnitude at FFT characteristic frequencies is proposed to use as an effective feature for bearing defect diagnostic [26]. However, the selection of the fixed window length before performing STFT can strongly affect the method performance.

Wavelet transform (WT) is recommended to overcome the above mentioned limit of STFT thanks of its capacity to flexibly adapting the resolutions of time and frequencies for signal analysis. This technique can be divided into three groups: continuous

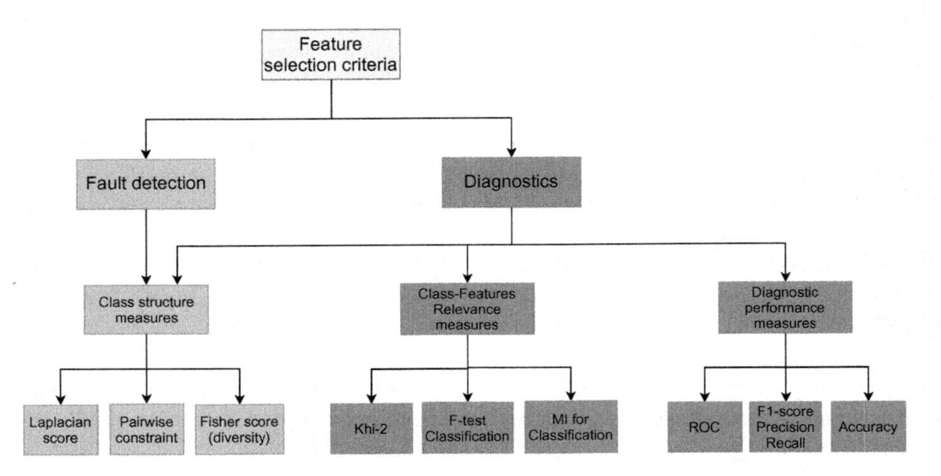

Fig. 3 Feature selection criteria for fault detection and diagnostics.

wavelet, discrete wavelet and wavelet packet transform. For bearing fault detection, the features are extracted based on statistical evaluation of the wavelet coefficients [27]. Besides, features based on the wavelet energy are also extracted from vibration or stator current signals [20].

As the performance of wavelet analysis method strictly depends on the choice of the mother wavelet, it is suitable to use Hilbert Huang transform (HHT) for analyzing non-stationary signals when we do not have a prior information about the signal shape. The HHT technique includes two phases. Firstly, the input signal is decomposed into a set of intrinsic mode functions (IMFs) using Empirical mode decomposition (EMD). Secondly, the instantaneous frequencies of the IMFs are extracted through Hilbert spectrum analysis (HSA). Therefore, HHT becomes a powerful tool for analyzing and characterizing the signal spectrum in time [28, 29].

3 Feature Selection

Feature selection (FS) aims to identify a subset of features which allows reducing effects from irrelevant ones, and therefore providing good prediction results [30]. The classification of FS techniques can be performed according to the characteristics of target models (supervised, semi-supervised, and unsupervised), the search strategies (forward increase, backward deletion random, and hybrid), the selection criteria, or the learning methods (filter, wrapper, and embedded) [31, 32].

3.1 Common Criteria in Literature for Feature Selection

Figure 3 presents an overview of the criteria widely used in literature for feature selection. According to the fault detection and diagnostics purposes, these criteria can be grouped into three following categories:

1. *Class structure measures*. The criteria in this group aim to capture local and global structure of the data space. Among them, Fisher score is the most widely used criterion. It aims to find a subset of features such that in the data space limited by the selected features, data points in the same class are close are possible while the distance between data points in different classes are large as possible. Besides, Laplacian score allows evaluating the features according to their locality preserving power while pairwise constraints provide information about link-constraints between observations to identify whether a pair of data samples belong to the same class or different classes.
2. *Class-Feature's relevance measures*. This second group aims to measure the relevance between features themselves and also their pertinence for the classification. For example, the Chi-square test allows us selecting the best features to build the model by testing the relationship between them. A feature having high Chi-square value is more dependent on the response and consequently can be selected for model training. Another test, F-Test, is also widely used for feature selection. It check whether a model created by a feature is significantly different to the one build just by a constant. Hence, the significance of each feature in improving the model is consequently evaluated. Besides, mutual information is used to quantify the "amount of information" obtained about diagnostic information by observing the feature. It is a non-negative value, and equal to zero if and only if two random variables are independent, while higher values mean higher dependency [33].
3. *Diagnostic performance measures*. The final group aims to evaluate the performance of a diagnostic model build by a subset of features) and based on it, the feature subset contributing to create the highest-performance model is selected. Among the performance metrics, the accuracy is widely used to evaluate the ratio of the number of labels predicted that exactly matches the true labels. Besides, the precision metric, which indicates how accurate the model is out of those predicted positive, is preferred when the cost of False Positive is high; while the recall metric is more suitable for the cases when there is a high cost associated with False Negative. When we consider a balance between false positive and false negative cases, F1-score might be a better measure. Finally, the AUC (Area Under The Curve) - ROC (Receiver Operating Characteristics) curve is one of the most important evaluation metrics for checking the model performance. ROC is a probability curve and AUC represents the degree or measure of separability.

3.2 A Proposed Metric for Evaluating Feature Performance

The paper [11] proposes a metric that allows directly evaluate the performance of the extracted features for fault detection and diagnostic. The proposed feature satisfies the following requirements: 1) convenient and simple for prompt evaluation and 2) independent from the feature units. Thus, it allows comparing heterogeneous features in different domains.

This feature, called the distance ratio (RD), is evaluated by the ratio between the Euclidean distance from a such feature value (point i), extracted from monitoring signal, to the median value of the set of nominal feature values (S_{FV}) characterized healthy state, and the standard deviation of nominal feature value set, $\sigma(S_{FV})$:

$$RD = \frac{||i - \text{median}(S_{FV})||_2}{\sigma(S_{FV})} \quad (1)$$

For fault detection and diagnostic, if the extracted feature allows clearly distinguishing the normal and abnormal state, then the feature performance is well highlighted. Considering Eq. 1, if the distance of a such feature value to the median value of the healthy set is significant while the standard deviation of distances between nominal feature values is small, then the correspondent RD is high. In other words, the greater RD value is, higher is the feature performance. Therefore, RD can be used as an effective measure for feature performance ranking when considering fault detection issues. Regarding diagnostic problems, it can be extended based on the evaluation of distances from a such group to others.

3.3 Feature Selection Techniques

Figure 4 presents an overview of the performance of the feature selection techniques grouped according to the learning methods. Among three groups, the filter models are the simplest and fastest ones but their performance is lower than wrapper and embedded models. Nevertheless, the wrapper methods provide better results compared to filter approaches but hey are very computationally expensive to implement, especially for data with a large number of features [34]. To benefit the advantages and overcome the shortcomings of the filter and the wrapper groups, embedded approaches were developed. Therefore, they achieve low computational cost than wrapper models and high performance than filter models.

In the first group, the filter methods rank features based on certain statistical criteria, such as Laplacian, Fisher score [35], or the mutual information [19, 36] to eliminate the inappropriate features. In [12], the authors introduced an ordering metric, that measures the separability between the normalized feature vectors characterizing healthy and faulty classes. An analysis and comparison of multiple features for fault detection and prognostic in ball bearings was presented in [11]. However,

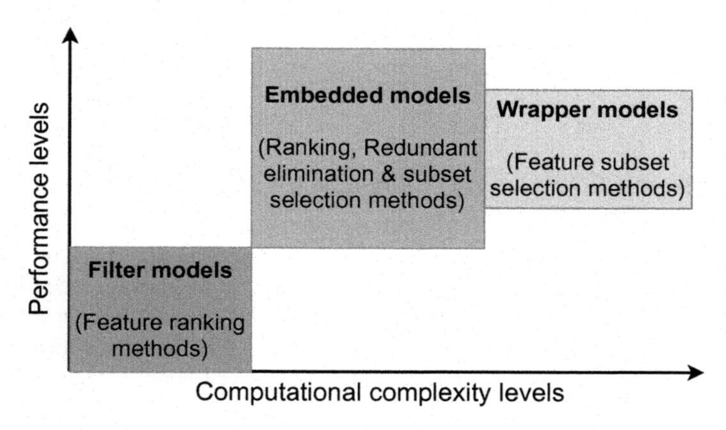

Fig. 4 Characteristics of feature selection techniques.

one of the important limitation considering these above methods, it is not trivial to identify the number of selected features. Furthermore, in practical applications, the important features could have the low ranking according to certain criteria evaluation but are more informative when combined with others for a specific learning purpose [30].

The wrapper methods propose to incorporate a specific learning algorithm in the process of feature subset selection [37, 38]. The feature search component will create a subset of features while the feature evaluation component will use the accuracy of the predetermined learning algorithm to assess the quality of this subset. For search component, a wide range of sequential and heuristic algorithms are developed in literature. For example, the study [39] developed Sequential Floating Forward Selection algorithm to decide whether a feature is added to the selected subset or not. In [40, 41], the authors proposed to use the Genetic Algorithm (GA) to find the optimum feature subset that maximizes the accuracy of the given classification algorithm.

The embedded models can be considered as hybrid models by combining the filter and the wrapper ones. It integrates the optimization of feature subsets during the learning process to avoid the training of the model each time when exploring a new feature subset. Therefore, they achieve lower computational cost than wrapper models and higher performance than filter ones. In [42], the authors present two feature selection approaches, concave minimization and support vector machine approach, for finding a separating plane that discriminates two classes in an n-dimensional space. Besides, the regularization models, that allow minimizing fitting errors and simultaneously penalizing the coefficients corresponding to features, receive increasing attention thanks to their good performance [34]. For example, Lasso [43, 44], bridge [45], and elastic net [46] are the popular and efficient regularization methods that are widely applied for feature selection. However, the embedded methods lack the generality as the method of the optimal feature subset selection is specific for a given classification algorithm [32].

3.4 A Proposed Algorithm for Feature Ranking

The paper [11] proposes a fast, simple, and effective ranking algorithm to access the performance of multiple features for fault detection and diagnostics, see Fig. 5.

This algorithm is based on the principle: *the higher mean value of RD (Eq. 1), of a feature is, the greater performance of this feature is*. In detail, for ranking N features, the RD measures are firstly evaluated for every feature. Then, N features are sorted according their RD mean values such as the RD mean value of i-th feature is greater than the one of $i+1$-th feature ($\mu_i > \mu_{i+1}$). Next, it is necessary

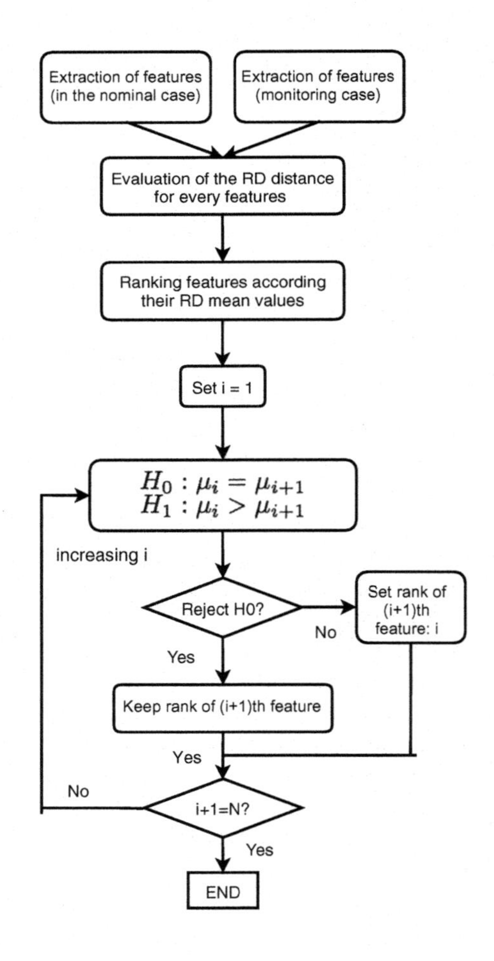

Fig. 5 Feature ranking algorithm [11].

to verify the confidence level of this feature ranking list. In other words, we consider whether the results ($\mu_i > \mu_{i+1}$, $i \in [1, N-1]$) are statistical significant or only random observations:

- If $H_0 : \mu_i = \mu_{i+1}$ is rejected, the rank of i and $(i+1)$-th feature are maintained.
- If not, we can not conclude that the i-th feature is better than the $(i+1)$-th one, then they have the same rank.

4 Health Indicator Construction

4.1 Taxonomy of Existing Methods

In general, the health indicator (HI) construction techniques can be classified into four categories: statistical projection, mathematical model based, deep learning based or optimization methods of feature combination.

In the first group, mathematical model based methods focus on developing mathematical expression that allow capturing the relations between CM measurements of the system and its health states. For example, the authors in [47] evaluate the distance between the vibration signals of degraded and nominal bearings, and then smooth it by an exponential model. In [48], the authors manually choose the relevant features, and then construct the HI using a weighted average combination of the chosen ones. The HI is also developed based on expertise knowledge about physical behavior of system [49, 50] and about the relevant feature used for creating effective HIs [51]. In a recent study [52], the authors propose to use the multivariate state estimation method, which is a non-parametric regression modeling technique, to generate useful HIs. However, the above studies are based on assumptions about degradation forms over time or the expertise knowledge about signal processing techniques, data analysis, and system behaviors. Then, they might not be suitable for complex systems where an automatic process is preferable.

Deep learning (DL) methods, provides alternative solutions for automatically extracting and constructing useful information without the expertise knowledge in the case of abundant data. For example, Long Short Term Memory (LSTM) Encoder-Decoder [53] or Recurrent Neural Network (RNN) Encoder-Decoder [54] can be used for automatic creating the HI. Besides, the Convolution Neural Network (CNN) is also applied to create HIs using raw vibration signals [55, 56] or using time-frequency features extracted from data [57, 58]. From these studies, it can be seen that the Deep Learning approaches can take advantage of abundant data to automatically generate health indicators without much expert knowledge about the system. Nevertheless, the deep features created by these works are difficult to understand and cannot be interpreted as physical characteristics of the system.

Besides, the statistical projection methods aim to represent the observations from the high-dimensional to a lower dimensional space. For early studies, principal component analysis (PCA) was proposed to find lower dimensional representation of features for condition and performance assessment [59]. To overcome the PCA limitations when facing nonlinearities and time-varying properties, numerous PCA variations, e.g. Kernel-PCA, PCA-based KNN and PCA-based Gaussian mixture models, were developed to handle data [60]. Besides, other non-linear combination techniques such as Isomap or Linear Locally Embedding (LLE) are developed for finding manifold embedding of lower dimensionality [61, 62]. However, the new features created by these mentioned statistical projection methodologies are not interpretable, which can lead to a deal-breaker in some settings.

The final group that is based on the optimization methods of feature combination aims to automatically find the best mathematical expression that combines low-level features to form more abstract high-level prognostic features [63–65]. These methods are flexible in formatting mathematical functions and allows easily integrating expertise knowledge about the HI formulations by defining an appropriate initial population. Furthermore, the created HI, which is an explicit mathematical function of low-level features, can be interpreted for further studies [33].

Figure 6 presents an overview of the HI construction methods in two aspects: interpretation ability and expert-knowledge requirements. Among four groups, the mathematical based methods allows creating the HIs that can be easily interpreted by physical characteristics of a specific system. However, they require thorough expertise knowledge about the system and its degradation trend. Contrastingly, the deep learning based methods require little expert knowledge, but high computational resources; and moreover it is difficult to interpret their created HI. Besides, the statistical projection methods and the optimization algorithms of feature combination can be considered as the alternatives solutions that have the acceptable interpretation level and do not require many expert knowledge about systems. Among them, the genetic programing techniques are promising candidates and will be detailed in next subsection.

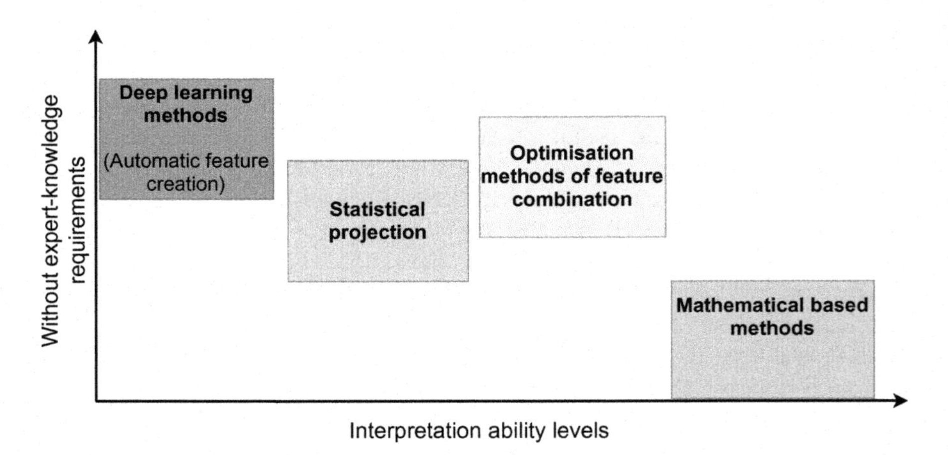

Fig. 6 Characteristics of HI construction methods.

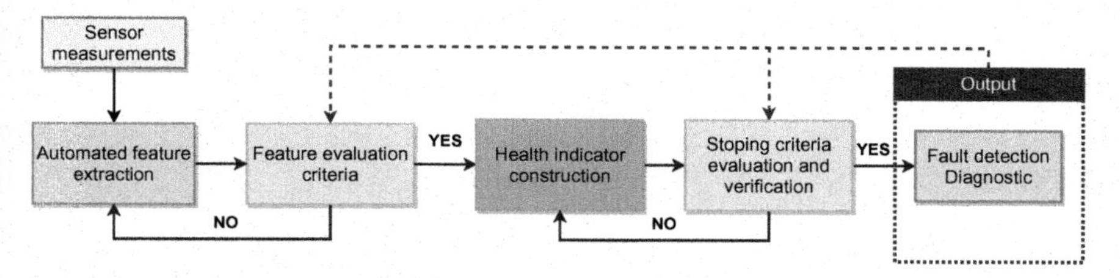

Fig. 7 Overview of the HI construction method.

4.2 Automatic Health Indicator Construction Method

In [33], the authors developed a new HI construction framework, including a complete automated process from extraction of low-level features to construction of useful HI. The proposed framework does not require the expertise knowledge but allows facilitating its integration if available. In addition, it takes into account multi criteria for a better HI performance evaluation. Moreover, it can be easily deployed for various systems.

An overview of the proposed framework is presented in Fig. 7. The automated HI construction framework is based on the two-stage Genetic Programming that are flexible in creating new mathematical functions. In the first stage, pertinent low-level features are automatically extracted from raw sensor measurements. It starts with a population of individuals that are tree-like representations of the feature extraction functions, their parameters and the sensor signals. Next, it evaluates them based on some evaluation criteria and generates a new population by using evolutionary operators on high scoring individuals and then eliminating the low scoring ones. In the second stage, GP is used to derive reasonable mathematical formulations of the features extracted in the first stage to create powerful health indicators. Each individual HI is defined using an expression tree constructed by combining the values of low-level features and a set of mathematical operators.

First Stage: Automated Feature Extraction

The first stage aims to identify the best combinations of the feature extraction functions and their relevant parameters for raw sensor signal. This process is based on genetic programming, as show in Fig. 8.

Firstly, an initial population including n_p individuals is randomly created. Every individual, which is a combination of 1) a feature extraction (FE) function, 2) a sensor signal output and 3) a value of window length parameter n, represents a way for extraction of features. Its performance is then evaluated through one (or a combination) of the evaluation criteria for fault detection and diagnostic purposes presented in Subsect. 3.1.

From the initial population, n_o offspring are generated in different ways through crossover, mutation of FE function, terminal mutation and reproduction. Note that for all operators (crossover, mutation and reproduction), the individual having the

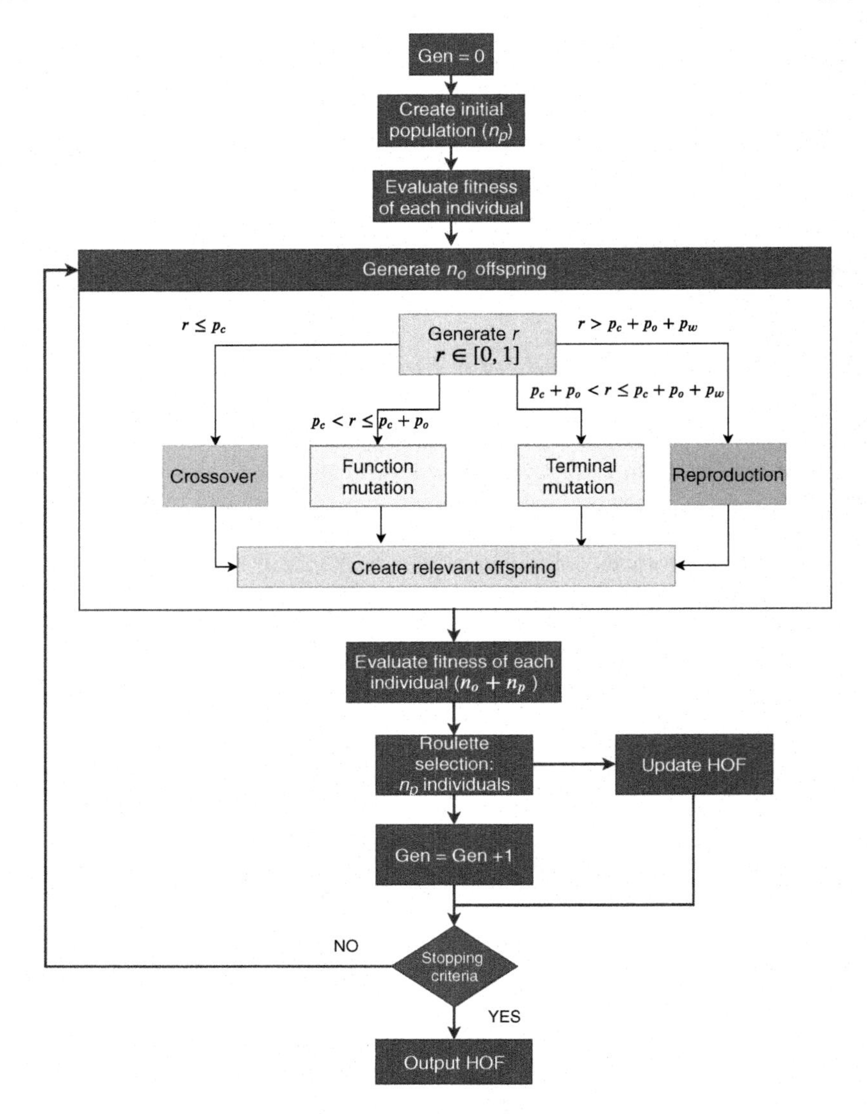

Fig. 8 Flow chart of the automated feature extraction algorithm [33].

better finest function is chosen with the higher probability. Beside, the operator is randomly chosen based on the following principles:

1. **Crossover.** If a random value r, $r \in [0, 1]$, is less than the crossover probability, p_c, then two individuals are chosen from the initial population to exchange the terminals that belongs to the same type, e.g. the window length parameter of

parent 1 will be replaced by the window length parameter of parent 2. After crossover operation, two offsprings are created.

2. **Mutation of FE function.** If r is superior to p_c but inferior to its cumulative sum with the probability of the FE-function mutation p_o, the FE function of the chosen parent will be replaced by another FE function.
3. **Mutation of terminal.** If r is superior to this sum $(p_c + p_o)$, but is inferior to the cumulative sum including the probability of the terminal mutation p_w, the terminal (i.e. sensor output or window length parameter n) will be replaced by another same type terminal.
4. **Reproduction.** If r is superior to the sum of crossover and mutation probability $(p_c + p_o + p_w)$, the chosen parent will be copied to create its offspring.

After creating n_o offsprings, all individuals including the ones in the parent population n_p, and the one in the offspring population n_o, will be evaluated to update the hall of frame (HOF) that is the best solutions through all generations. The HOF number, i.e., number of extracted features, is defined by users. Among $n_o + n_p$ individuals, n_p individuals will be randomly chosen as the parents of the next generation. Note that the individual having the better finest function will be kept with the higher probability. For a new generation, the above procedure will be repeated until the stopping criteria (the maximal number of generations) is attained.

Second Stage: Automated Health Indicator Construction

The second stage of the proposed methodology aims to find best mathematical functions that allow combining the low-level features extracted from the first stage to derive the powerful HI. To prevent not-a-number values that can be created by random combinations, several variants of basic mathematical operators are proposed in Table 2.

The HI combination stage is summarized in Fig. 9. It aims to find the best multi-level-combinations of mathematical operators defined in Table 2 and the low-level features extracted during the first sage. To do that, we implemented the following evolutionary operators:

Table 2 Summary of mathematical operators to create the HI. [33]

Operators	Formulation
Addition	$x + y$
Subtraction	$x - y$
Multiplication	$x \times y$
Protected division	x/y if $y > 10^{-9}$, otherwise 10^9
Protected exponential function	$\exp(x)$ if $x < 100$, otherwise 10^9
Protected logarithmic function	$\log(\lvert x \rvert)$ if $\lvert x \rvert > 10^{-9}$, otherwise -10^9
Power function	$x^a, a \leq 10$
Negative function	$-x$
Squared function	$\sqrt{\lvert x \rvert}$

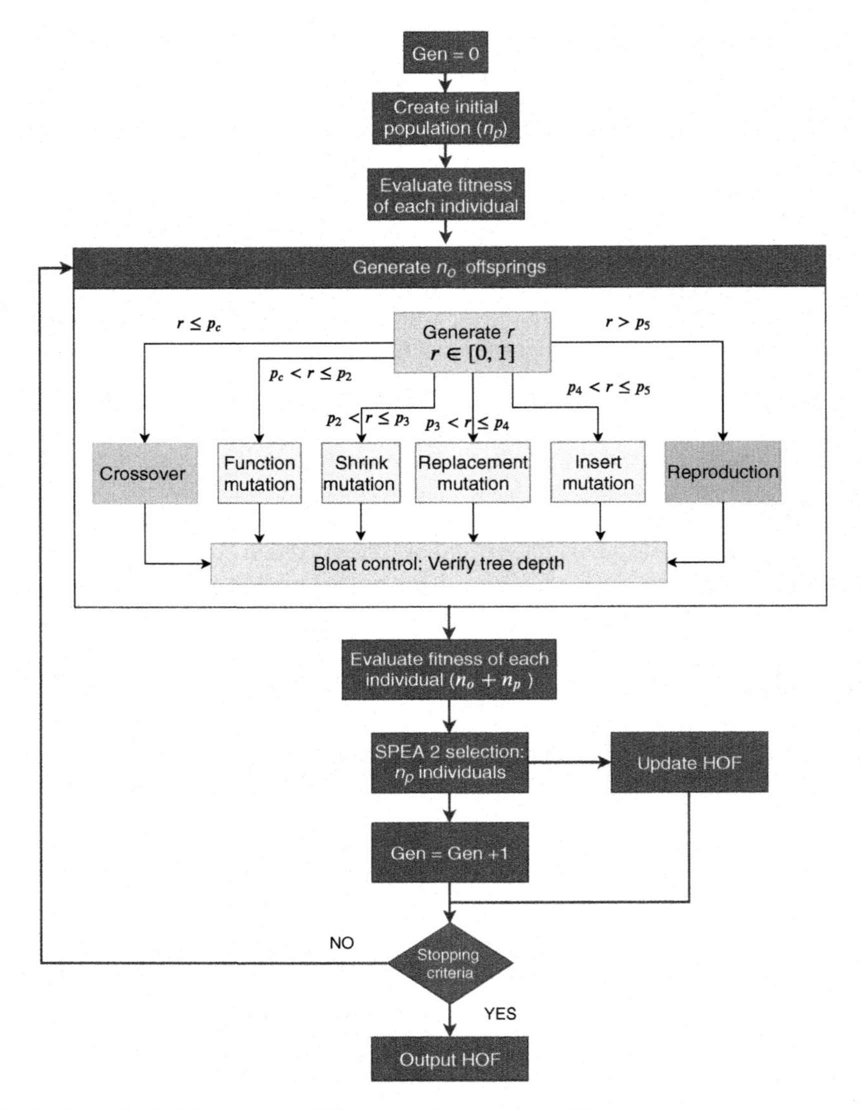

Fig. 9 Flow chart of the automated HI construction algorithm [33].

1. **Crossover:** If the random value r is inferior or equal to the probability of crossover ($r \leq p_c$), the crossover operator will be performed. It randomly selects one point in each parental individual and exchanges their relevant subtrees.
2. **Function mutation:** If the random value r is superior to the probability of crossover and inferior or equal to the sum of the cross over and function mutation probability, noted p_2 ($p_c < r \leq p_2$), the mutation of mathematical functions will

be performed. It replaces a randomly chosen mathematical function from parental individual by a randomly chosen function with the same number of arguments.
3. **Shrink mutation:** If the random value r is superior to p_2 and inferior or equal to the cumulative sum of the cross over, the function mutation and the shrink mutation probability, noted p_3, ($p_2 < r \leq p_3$), the shrink mutation will be performed. This operator shrinks the parental individual by choosing randomly its branch and replacing it with one of the branch arguments (also randomly chosen). It allows investigating if a simplified formulation provides a promising solution.
4. **Replacement mutation:** If the random value r is superior to p_3 and inferior or equal to p_4, the cumulative sum p_3 including replacement mutation probability, ($p_3 < r \leq p_4$), the replacement mutation will be performed. It randomly selects a point in the parental individual, then replaces the subtree at that point as a root by the expression generated by a random combination between mathematical functions and low-level extracted features.
5. **Insert mutation:** If the random value r is superior to p_4 and inferior or equal to p_5, the cumulative sum including the insert mutation probability, ($p_4 < r \leq p_5$), the insert mutation will be performed. It inserts a new branch at a random position in parental individual. The original subtree will become one of the children of the new mathematical function inserted, but not perforce the first (its position is randomly selected if the new function has more than one child).
6. **Reproduction:** If the random value r is superior to p_5, the offspring is created by a copy of the parental individual.

In addition, as the length of the individuals can be rapidly explored through generations, then too complicated expressions that can not be interpreted might be created. To prevent this issue, a bloat control of the expression depth is performed to eliminate the long offsprings. Besides, as one prefers simple expressions having acceptable HI performance, the individual length can be set as a one of the fitness functions. The proposed algorithm finds the best combinations that minimize the individual length and maximize the HI evaluation criteria presented in Subsect. 3.1. This multi-objective optimization problem is handled by using the Strength Pareto Evolutionary Algorithm (SPEA II) that allows locating and maintaining a set of non-dominated solutions.

4.3 Applications of the Automatic Health Indicator Construction Method

The two main challenges of the automated HI constructions are: 1) the ability to correctly chose the pertinent measurements among various sensor sources, and 2) the capability to handle raw data from high-frequency sensors. Hence, in [33] the authors investigated whether the proposed methodology can address these challenges through two benchmark case studies.

4.3.1 Case Study 1 - Turbofan Engine Degradation

This case study is widely used in PHM field [66]. It presents various degradation scenarios of the fleet of engines from a nominal state to a failure in the training sets and a time before the failure in the test sets. Both of training and test sets consist of 26 columns that describe the characteristics of the engine units. The first and second column respectively represent the ID and the degradation time steps for every engine. The next three columns characterize the operation modes of the engines while the final 21 columns correspond to the outputs of 21 sensors. Among these numerous measurements, it is necessary to correctly chose the pertinent features. Therefore, the subset FD001 of this dataset is used to verify whether the proposed method can automatically chose the informative features or not.

Figure 10 illustrates raw signals recorded from the four first sensors (among 21 sensors) of the turbofan engine dataset FD001. One can recognize that the first sensor provides useless information that should be automatically eliminated during the first stage of the proposed methodology. Indeed, Fig. 11 presents four examples of the results obtained after the feature extraction stage. They are respectively the results of the smoothing function (v_{SM}) for sensors 2, 3, 11 and 21 that are also the ones recommended by the previous works in literature [66]. These results highlight the performance of the FE stage that automatically chooses the appropriate sensors and based on these measures, extracts the pertinent low-level features. In fact, comparing Fig. 11 with Fig. 10, we find that the extracted features are more monotonic, smooth and robustness than the raw sensor measurements. In other words, these features can better represent the characteristics of the turbofan engine's degradation process.

After extracting low-level features, the second stage of the proposed methodology allows using these features to create the powerful HI according to the predefined criteria. Figure 12 presents one example of the created HI for the case study of turbofan engines. One can recognize that the created HI represents well the degradation

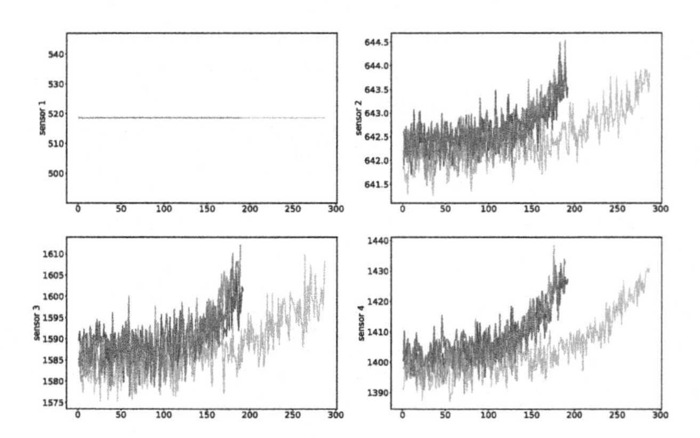

Fig. 10 Turbofan engine's raw data [33]

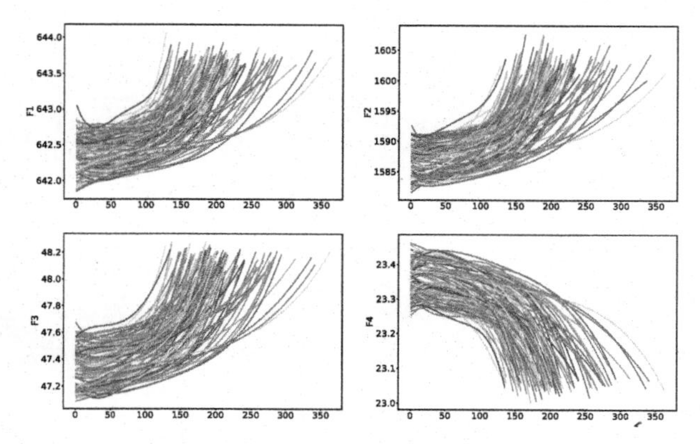

Fig. 11 Features extracted from Turbofan engine dataset [33]

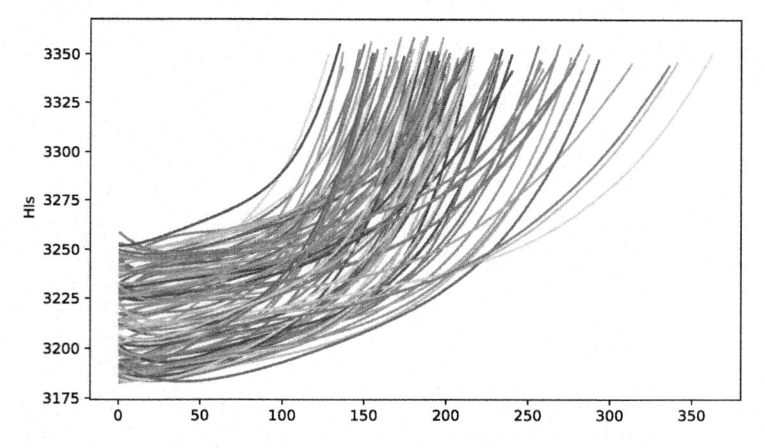

Fig. 12 Illustration of the created HI for turbofan engines [33]

processes in this case. It is monotonically increasing over time and almost end at the same failure threshold.

4.3.2 Case Study 2 - Bearing Degradation

This case study was generated by the NSF I/UCR Center for Intelligent Maintenance Systems (IMS) and used as a common benchmark data in PHM field [67]. It includes three sub datasets that describe three test-to-failure experiments of four bearings installed in an AC motor that operates at 2000 RPM of rotation speed with a radial load of 6000 lbs. In contrast to the previous case (Turbofan engine), in this case, there is only a high-frequency sensor type (i.e. vibration) to monitor the bearing state. Concretely, 1-second vibration signal snapshots is recorded at specific intervals by

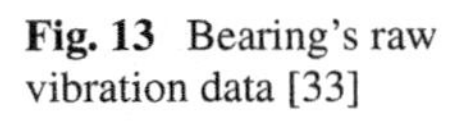

Fig. 13 Bearing's raw vibration data [33]

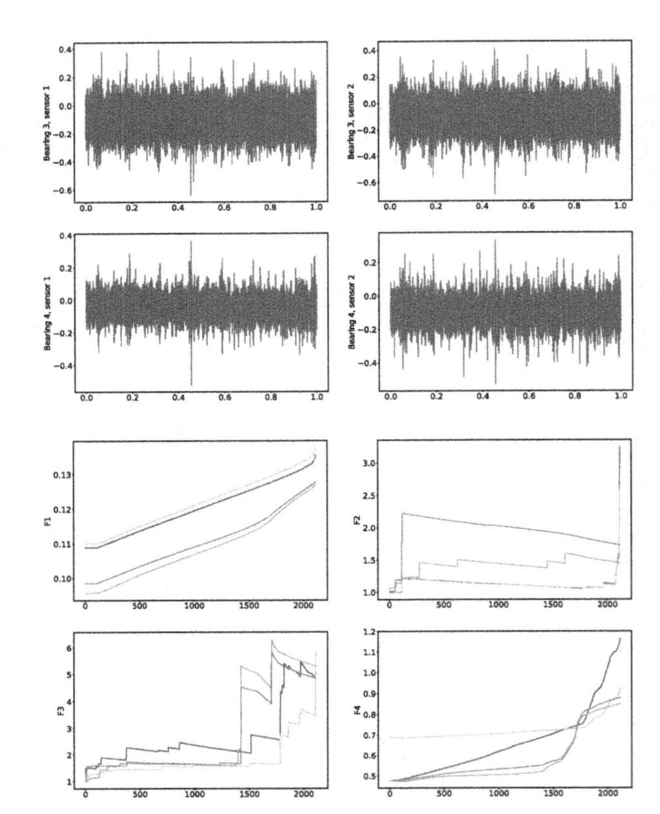

Fig. 14 Features extracted from IMS bearing dataset [33]

NI DAQ Card 6062E with the sampling rate set equal to 20 kHz. To verify if the proposed methodology can handle the high-frequency raw data, the sub dataset No.1, that is the longest test, is used in [33].

Figure 13 presents raw vibration signals recorded from two channels of bearings 3 and 4 during one second in the first test of IMS bearing dataset. Note that these bearings fail after more than one month of running test-to-failure experiment. Thus, for every sensor channel, more than 2100 samples of 1 s-signal are recorded during the experiment. It is impossible to directly visualize the degradation process of bearings 3 and 4 with these raw data (Fig. 13), while the extracted features after the first stage allow representing the evolution of bearing degradation over time (Fig. 14). The extracted features are respectively the results of the v_{MA}, v_{RMS}, v_{CF} and v_{IF} functions applied on 2100 samples of 1 second of vibration signals. Hence, the proposed method's performance is one more time emphasized. It allows automatically handle the high-frequency signals and deriving useful features after the first stage.

Finally, Fig. 15 shows the created HI that well captures the degradation trend of bearings. In addition, their trajectories almost end at the same failure threshold. Hence, this HI allows monitoring bearing conditions for fault detection and contributing to improve the precision of prognostic models.

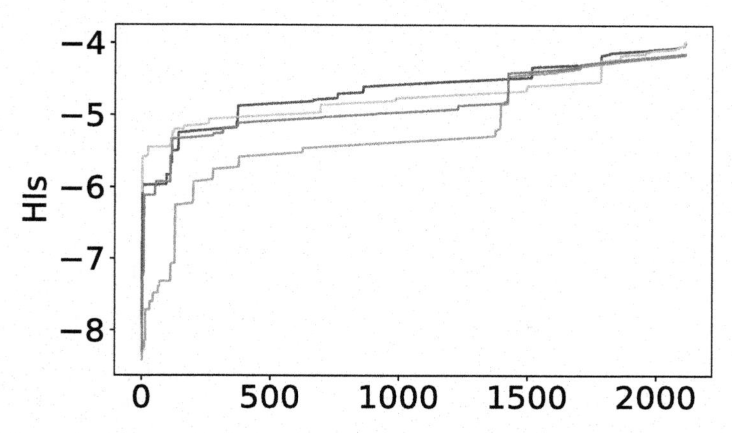

Fig. 15 Illustration of the created HI for bearings [33]

5 Conclusions

This chapter presents a tutorial on feature engineering and health indicator construction in condition monitoring. In addition, it also provides a systematic and comprehensive review of existing techniques for feature extraction, feature selection, and health indicator construction. This thorough review presents a general guideline for practitioners on how to properly implement condition monitoring systems for fault detection and diagnostic purposes.

Among feature extraction methods, time-domain techniques are classical, fast and simple. They can be applied for different fault types of various systems but are not viable for noisy measurement. On the other hand, the frequency-domain analysis is effective to detect system anomalies when knowing their fault characteristic frequencies. However, it is not suitable for non-stationary signals. On the contrary, time-frequency transformation techniques are powerful for analyzing and characterizing the signal spectrum in time. Nevertheless, those techniques often require a lot of computational power as well as experience to choose the appropriate set of hyperparameters.

For feature selection, filter approaches are used for unlabeled data or when there is no correlation between features and labeled data. Besides, embedded approaches are appropriate for scenarios in which high accuracy and inexpensive computation are required. They are however not suitable for high dimensional data which can be addressed by using wrapper methods based on heuristic search algorithms.

Finally, for health indicator construction, methods based on statistical projection are good candidates for high-dimensional but non-abundant data, especially when there is no expert knowledge about the system to construct mathematical health indicators. Otherwise, deep learning based methods are suitable for scenarios with high-dimensional and abundant data. Besides, methods based on optimization of feature combination, particularly genetic programming, are promising solutions for giving interpretation ability of the created health indicators. Furthermore, they can also be extended to integrate the possibly available knowledge.

References

1. Soualhi M, Nguyen KTP, Soualhi A, Medjaher K, Hemsas KE (2019) Health monitoring of bearing and gear faults by using a new health indicator extracted from current signals. Measurement 141:37–51
2. Cheng S, Azarian MH, Pecht MG (2010) Sensor systems for prognostics and health management. Sens. (Basel, Switz.) 10:5774–5797
3. Soualhi M, Nguyen KT, Medjaher K (2020) Pattern recognition method of fault diagnostics based on a new health indicator for smart manufacturing. Mech Syst Sig Process 142:106680
4. Nisbet R, Elder J, Miner G (2009) Data understanding and preparation. In: Nisbet R, Elder J, Miner G (eds) Handbook of statistical analysis and data mining applications. Academic Press, Boston, pp 49–75
5. Yakout M, Berti-Équille L, Elmagarmid AK (2013) Don't be SCAREd: use SCalable automatic REpairing with maximal likelihood and bounded changes. In: Proceedings of the 2013 international conference on management of data - SIGMOD 2013. ACM Press, New York, p 553
6. Haas D, Krishnan S, Wang J, Franklin MJ, Wu E (2015) Wisteria: nurturing scalable data cleaning infrastructure. Proc VLDB Endowment 8:2004–2007
7. Subramaniam S, Palpanas T, Papadopoulos D, Kalogeraki V, Gunopulos D (2006) Online outlier detection in sensor data using non-parametric models. In: Proceedings of the 32nd international conference on very large data bases, VLDB 2006, VLDB Endowment, Seoul, Korea, pp 187–198
8. Berti-Équille L, Dasu T, Srivastava D (2011) Discovery of complex glitch patterns: a novel approach to quantitative data cleaning. In: 2011 IEEE 27th international conference on data engineering, Hannover, Germany, pp 733–744
9. Ratolojanahary R, Houé Ngouna R, Medjaher K, Junca-Bourié J, Dauriac F, Sebilo M (2019) Model selection to improve multiple imputation for handling high rate missingness in a water quality dataset. Expert Syst Appl 131:299–307
10. Sorzano COS, Vargas J, Montano AP (2014) A survey of dimensionality reduction techniques. arXiv:1403.2877 [cs, q-bio, stat]
11. Nguyen KTP, et al (2018) Analysis and comparison of multiple features for fault detection and prognostic in ball bearings. In: PHM society European conference, Utrecht, The Netherlands, vol 4, pp 1–6
12. Shukla S, Yadav RN, Sharma J, Khare S (2015) Analysis of statistical features for fault detection in ball bearing. In: 2015 IEEE international conference on computational intelligence and computing research (ICCIC), Madurai, India, pp 1–7
13. Mahamad AK, Hiyama T (2008) Development of artificial neural network based fault diagnosis of induction motor bearing. In: 2008 IEEE 2nd international power and energy conference, Johor Bahru, Malaysia, pp 1387–1392
14. Qian Y, Yan R, Hu S (2014) Bearing degradation evaluation using recurrence quantification analysis and Kalman Filter. IEEE Trans Instrum Meas 63:2599–2610 (2014)
15. Li R, Sopon P, He D (2012) Fault features extraction for bearing prognostics. J Intell Manuf 23:313–321
16. Silva JLH, Cardoso AJM (2005) Bearing failures diagnosis in three-phase induction motors by extended Park's vector approach. In: 31st annual conference of IEEE industrial electronics society. IECON 2005, p 6
17. Zhou W, Habetler TG, Harley RG (2007) Bearing condition monitoring methods for electric machines: a general review, pp 3–6

18. Gebraeel N, Lawley M, Liu R, Parmeshwaran V (2004) Residual life predictions from vibration-based degradation signals: a neural network approach. IEEE Trans Ind Electron 51:694–700
19. Kappaganthu K, Nataraj C (2011) Feature selection for fault detection in rolling element bearings using mutual information. J Vibr Acoust 133:061001 (2011)
20. Yu J (2012) Local and nonlocal preserving projection for bearing defect classification and performance assessment. IEEE Trans Ind Electron 59:2363–2376
21. Fournier E, Picot A, Régnier J, Yamdeu MT, Andréjak JM, Maussion P (2015) Current-based detection of mechanical unbalance in an induction machine using spectral kurtosis with reference. IEEE Trans Ind Electron 62:1879–1887
22. Gong X, Qiao W (2013) Bearing fault diagnosis for direct-drive wind turbines via current-demodulated signals. IEEE Trans Ind Electron 60:3419–3428
23. Picot A, Obeid Z, Régnier J, Poignant S, Darnis O, Maussion P (2014) Statistic-based spectral indicator for bearing fault detection in permanent-magnet synchronous machines using the stator current. Mech Syst Signal Process 46:424–441
24. Cong F, Chen J, Dong G (2012) Spectral kurtosis based on AR model for fault diagnosis and condition monitoring of rolling bearing. J Mech Sci Technol 26:301–306
25. Leite VCMN, et al (2015) Detection of localized bearing faults in induction machines by spectral kurtosis and envelope analysis of stator current. IEEE Trans Ind Electron 62:1855–1865
26. Yazici B, Kliman G (1999) An adaptive statistical time-frequency method for detection of broken bars and bearing faults in motors using stator current. IEEE Trans Ind Appl 35:442–452
27. Deekshit Kompella KC, Venu Gopala Rao M, Srinivasa Rao R (2018) Bearing fault detection in a 3 phase induction motor using stator current frequency spectral subtraction with various wavelet decomposition techniques. Ain Shams Eng J 9:2427–2439
28. Refaat SS, Abu-Rub H, Saad MS, Aboul-Zahab EM, Iqbal A (2013) ANN-based for detection, diagnosis the bearing fault for three phase induction motors using current signal. In: 2013 IEEE international conference on industrial technology (ICIT), Cape Town, South Africa, pp 253–258
29. Elbouchikhi E, Choqueuse V, Amirat Y, Benbouzid MEH, Turri S (2017) An efficient Hilbert-Huang transform-based bearing faults detection in induction machines. IEEE Trans Energy Convers 32:401–413
30. Chandrashekar G, Sahin F (2014) A survey on feature selection methods. Comput Electr Eng 40:16–28
31. Venkatesh B, Anuradha J (2019) A review of feature selection and its methods. Cybern Inform Technol 19:3–26
32. Khaire UM, Dhanalakshmi R (2019) Stability of feature selection algorithm: a review. J King Saud Univer Comput Inform Sci 1–14
33. Nguyen KTP, Medjaher K (2020) An automated health indicator construction methodology for prognostics based on multi-criteria optimization. ISA Trans 1–16
34. Tang J, Alelyani S, Liu H (2014) Feature selection for classification: a review, data classification: algorithms and applications 37–64
35. Knöbel C, Marsil Z, Rekla M, Reuter J, Gühmann C (2015) Fault detection in linear electromagnetic actuators using time and time-frequency-domain features based on current and voltage measurements. In: 2015 20th international conference on methods and models in automation and robotics (MMAR), Miedzyzdroje, Poland, pp 547–552
36. Gao W, Hu L, Zhang P (2018) Class-specific mutual information variation for feature selection. Pattern Recognit 79:328–339
37. Kohavi R, John GH (1997) Wrappers for feature subset selection. Artif Intell 97:273–324
38. El Aboudi N, Benhlima L (2016) Review on wrapper feature selection approaches. In: 2016 international conference on engineering MIS (ICEMIS), Agadir, Morocco, pp 1–5
39. Zongker D, Jain A (1996) Algorithms for feature selection: an evaluation. In: Proceedings of 13th international conference on pattern recognition, Vienna, Austria, vol. 2, pp 18–22

40. Huang C-L, Wang C-J (2006) A GA-based feature selection and parameters optimization for support vector machines. Expert Syst Appl 31:231–240
41. Cerrada M, Sánchez RV, Cabrera D, Zurita G, Li C (2015) Multi-stage feature selection by using genetic algorithms for fault diagnosis in gearboxes based on vibration signal. Sensors 15:23903–23926
42. Bradley PS, Mangasarian OL (1998) Feature selection via concave minimization and support vector machines. In: Proceedings of the fifteenth international conference on machine learning, ICML 1998, San Francisco, CA, USA. Morgan Kaufmann Publishers Inc., pp 82–90
43. Zou H (2006) The adaptive lasso and its oracle properties. J Am Stat Assoc 101:1418–1429
44. Tibshirani R (2011) Regression shrinkage and selection via the lasso: a retrospective. J Roy Stat Soc Ser B (Stat Methodol) 73:273–282
45. Huang J, Horowitz JL, Ma S (2008) Asymptotic properties of bridge estimators in sparse high-dimensional regression models. Ann Stat 36:587–613. http://arxiv.org/abs/0804.0693
46. Xiao H, Biggio B, Brown G, Fumera G, Eckert C, Roli F (2015) Is feature selection secure against training data poisoning? In: Proceedings of the 32nd international conference on international conference on machine learning (ICML), ICML 2015, JMLR.org, Lille, France, vol 37, pp 1689–1698
47. Medjaher K, Zerhouni N, Baklouti J (2013) Data-driven prognostics based on health indicator construction: application to PRONOSTIA's data. In: 2013 European control conference (ECC), Zurich, Switzerland, pp 1451–1456
48. Liu K, Gebraeel NZ, Shi J (2013) A data-level fusion model for developing composite health indices for degradation modeling and prognostic analysis. IEEE Trans Autom Sci Eng 10:652–664
49. Pan H, Lü Z, Wang H, Wei H, Chen L (2018) Novel battery state-of-health online estimation method using multiple health indicators and an extreme learning machine. Energy 160:466–477
50. Niu G, Jiang J, Youn BD, Pecht M (2018) Autonomous health management for PMSM rail vehicles through demagnetization monitoring and prognosis control. ISA Trans 72:245–255
51. Atamuradov V, Medjaher K, Camci F, Dersin P, Zerhouni N (2018) Degradation-level assessment and online prognostics for sliding chair failure on point machines. IFAC-PapersOnLine 51:208–213
52. Sun J, Li C, Liu C, Gong Z, Wang R (2019) A data-driven health indicator extraction method for aircraft air conditioning system health monitoring. Chin J Aeronaut 32:409–416
53. Malhotra P, et al (2016) Multi-sensor prognostics using an unsupervised health index based on LSTM encoder-decoder. arXiv:1608.06154 [cs]. http://arxiv.org/abs/1608.06154
54. Gugulothu N, TV V, Malhotra P, Vig L, Agarwal P, Shroff G (2017) Predicting remaining useful life using time series embeddings based on recurrent neural networks. arXiv:1709.01073 [cs]. http://arxiv.org/abs/1709.01073
55. Guo L, Lei Y, Li N, Yan T, Li N (2018) Machinery health indicator construction based on convolutional neural networks considering trend burr. Neurocomputing 292:142–150
56. Han T, Liu C, Yang W, Jiang D (2019) Learning transferable features in deep convolutional neural networks for diagnosing unseen machine conditions. ISA Trans 93:341–353
57. Mao W, He J, Tang J, Li Y (2018) Predicting remaining useful life of rolling bearings based on deep feature representation and long short-term memory neural network. Adv Mech Eng 10 https://doi.org/10.1177/1687814018817184
58. Li X, Zhang W, Ding Q (2019) Deep learning-based remaining useful life estimation of bearings using multi-scale feature extraction. Reliab Eng Syst Saf 182:208–218
59. Li W, Yue HH, Valle-Cervantes S, Qin SJ (2000) Recursive PCA for adaptive process monitoring. J Process Control 10:471–486
60. Thieullen A, Ouladsine M, Pinaton J (2012) A survey of health indicators and data-driven prognosis in semiconductor manufacturing process. IFAC Proc Vol 45:19–24
61. Benkedjouh T, Medjaher K, Zerhouni N, Rechak S (2013) Remaining useful life estimation based on nonlinear feature reduction and support vector regression. Eng Appl Artif Intell 26:1751–1760

62. Zhang Y, Ye D, Liu Y (2018) Robust locally linear embedding algorithm for machinery fault diagnosis. Neurocomputing 273:323–332
63. Smart O, Firpi H, Vachtsevanos G (2007) Genetic programming of conventional features to detect seizure precursors. Eng Appl Artif Intell 20:1070–1085
64. Firpi H, Vachtsevanos G (2008) Genetically programmed-based artificial features extraction applied to fault detection. Eng Appl Artif Intell 21:558–568
65. Liao L (2014) Discovering prognostic features using genetic programming in remaining useful life prediction. IEEE Trans Ind Electron 61:2464–2472
66. Ramasso E, Saxena A (2014) Review and analysis of algorithmic approaches developed for prognostics on CMAPSS dataset, Technical Report, SGT Inc Moffett Field United States, SGT Inc Moffett Field United States
67. Cong F, Chen J, Dong G (2012) Spectral kurtosis based on AR model for fault diagnosis and condition monitoring of rolling bearing. J Mech Sci Technol 26:301–306